Springer Series in Synergetics

Editor: Hermann Haken

Synergetics, an interdisciplinary field of research, is concerned with the cooperation of individual parts of a system that produces macroscopic spatial, temporal or functional structures. It deals with deterministic as well as stochastic processes.

R. K. Mishra D. Maaß E. Zwierlein (Eds.)

On
Self-Organization

An Interdisciplinary Search
for a Unifying Principle

With 71 Figures

Springer-Verlag
Berlin Heidelberg New York
London Paris Tokyo
Hong Kong Barcelona
Budapest

Professor Dr. R. K. Mishra

All-India-Institute of Medical Sciences, New Delhi 110029, India

Professor Dr. D. Maaß

University of Kaiserslautern, Department of Computer Science,
Pfaffenbergstrasse, Gebäude 32, D-67618 Kaiserslautern, Germany

Professor Dr. E. Zwierlein

University of Kaiserslautern, Department of Social Sciences, Economics and Engineering,
Pfaffenbergstrasse, Gebäude 3, D-67618 Kaiserslautern, Germany

Series Editor:

Professor Dr. Dr. h. c. Hermann Haken

Institut für Theoretische Physik und Synergetik der Universität Stuttgart,
D-70550 Stuttgart, Germany and
Center for Complex Systems, Florida Atlantic University,
Boca Raton, FL 33431, USA

ISBN 978-3-642-45728-9   ISBN 978-3-642-45726-5 (eBook)
DOI 10.1007/978-3-642-45726-5

Library of Congress Cataloging-in-Publication Data. On self-organization: an interdisciplinary search for a unifying principle/R. K. Mishra, D. Maass, E. Zwierlein, eds. p. cm. — (Springer series in synergetics; v. 61) Includes bibliographical references. ISBN-13: 978-3-642-45728-9 1. Self-organizing systems — Congresses. I. Mishra, R.K., Professor. II. Maass, D. (Dieter), 1930- . III. Zwierlein, Eduard. IV. Series. Q325.05 1994 003'.85—dc20 93-38258

© Springer-Verlag Berlin Heidelberg 1994

Softcover reprint of the hardcover 1st edition 1994

Typesetting: Camera ready copy from the authors/editors using a Springer T$_E$X macro package
57/3140 - 5 4 3 2 1 0 - Printed on acid-free paper

Preface

The thread of self-organization which is now recognized as permeating many dynamical transformations in diverse systems around us seems set to unleash a revolution as influential as that of Darwin in the last century. Darwin removed the 'originator' of a species; self-organization now seeks to remove the 'organizer' from an organism.

Methods of nonlinear dynamics have played a crucial role in opening up this field and if these methods have a progenitor it is Henri Poincaré (1854 - 1912) whose first substantial compilation amongst his prolific production was Les Méthodes Nouvelles de la Mécanique Celeste, Vol 1. This work appeared in Paris in 1892, a century ago, thus offering us the opportunity to celebrate a centenary of this extraordinary revolutionary of science. Many of the extensive contributions of Poincaré helped to lay the foundation of this field of nonlinear dynamics, for example the mathematical theory of dimensions, (qualitatively) global aspects of phase space dynamics, topological analysis, fixed point theorems, bifurcation concepts (to be used later for example for Andronov-Poincaré bifurcation), difference equation mappings in phase space, surface of section, rotation number of maps, indexing closed curve in a vector field, etc. These are all of structural significance in the methodology of a large part of the field, and homage is due even if one were to forget his introduction of the concept of 'relativity' later made famous by Einstein.

Centered around the concept of dynamical systems, the conference on which this book is based raised and discussed many issues. To give an overview: **Bremermann** examines self-organization in neural nets, perceptrons, the immune system (a different network), dynamical chaos, evolution and even market economics. The last of these is treated in detail by **Hinterberger** in the framework of evolutionary economics, now a burgeoning field, in a manner that underlines its self-organization. **Haken** reviews succinctly the synergetic route to self-organization and explains stability, instability, the slaving principle and order parameters enabling self-organization to serve as a bridge between natural and social sciences. **Weidlich** addresses himself to the very difficult task of implementing the ideas of self-organization in the socio-political field. He examines trends, utilities, motivation and equations of motion for social dynamics in the context of transitions from a totalitarian to a liberal regime.

Whether in the social sciences or in biology, and particularly in some problems in the former, computation cannot be avoided. Indeed many solutions in deterministic chaos and fractals would be intractable and unimaginable without a computer. **Richter** takes up an analysis of the role of the computer in symbolic and subsymbolic reasoning. Artificial intelligence, connectionistic paradigms and neural nets are discussed. Special reference is made to the PATDEX system of the University of Kaiserslautern in relation to two types of case bases; strategy and diagnostics. Strategy guides the diagnostic process for completing information. PATDEX organizes complete information from incomplete information, so that a sufficiently correct diagnosis can be made, an objective of the MOLTKE system of which PATDEX is a part.

Turning to biology, **Conrad** examines the need for quick and efficient information processing and control, and, in the context of molecular orbitals in submolecular biology, shows that the parallelism which is inherent in the electronic wave function can speed up the self-organization of a complex between reactants such as an enzyme and a substrate, a crucial issue for the functioning of living systems. He suggests that all modes of self-organization – evolutionary, dynamical, algorithmic – are facilitated by this quantum speed-up effect. The role of infrared, heatbath, low frequency radiation, colour groups and deuterium substitutions is predicted. **Mishra** stresses the organization of living systems formed on the basis of a 'process structure' subservient to a hierarchical tree of functional parameters. Considering the energies involved, the swiftness necessary and the need for an 'instantaneous' whole-body coordination, the living state is proposed to be rooted in a field of quantum excitations pumped by metabolic and dissipative processes. **Pohlmann and Tributsch** use elements of approaches by Prigogine, Haken and Eigen to present a model of cooperativity in electron transfer in photosynthesis, particularly in the last step of oxygen evolution. Their approach could be generalized for application to other situations. **Saunders and Hao** discuss successive bifurcations especially in Drosophila and show that they lead to segmentation which is not explicable by a reaction-diffusion mechanism (Turing). As they point out, there may be more to the origin of an organism's properties, including regulation, than the natural selection of Darwin.

In the discussion of methodologies some special features are highlighted. **Landsberg** refers to a wide variety of mathematical concepts used in the field, but in particular discusses a central point in chaos and self-organization: The question of entropies-statistical, conditional, information, 'q', Rényi, algorithmic information gain or relative entropy and variance. **Ebeling** stresses the subtle but important point that self-organization is concerned with the value of the energy inside a system. This broadens the field of application by referring to value in favour of information, or selection (in biology), or exchange (in economy) in various fields. He then discusses optimisation strategies of the thermodynamic description and regards these strategies as twin approaches essential for self-organization. **Hao** considers symbolic dynamics in which phase space is coarse grained and then each region given a label from an alphabet. This field is important as a method of studying complexity, the quantification of which may be crucial. A bridge is needed between dynamics and the underlying grammar. This is provided by a transfer matrix (the Stefan matrix). **Bullough** presents numerous aspects of non-linear dynamics of immediate relevance to the field. Non-integrable Hamiltonian dynamics, solitary waves and infinite dimensional KAM theory, KdV solitons, topological breathers, quantum breathers, dressed solitons and dissipation into phonons are all analysed. Any or all these may turn out to be of relevance to self-organization especially in biology.

Li presents an interesting case for order-to-order transitions using a quantum uncertainty formalism. **Antoniou and Tasaki** from Prigogine's group present an important treatment of unavoidable intrinsic irreversibility in unstable systems which is of great significance for the fundamental question concerning the Arrow of Time which dominates organisational dynamics and transformations.

No scientific discussion of the late 20th century can ignore the properties of the observer. **Nalimov** discusses the role of consciousness in self-organisation, along with filters of space, time, numbers and probability which regulate consciousness. These ideas are discussed in relation to the biosphere and the universe in the framework of self-organization. **Josephson and Carpenter** in the context of music and meaning speak of man's ability to listen, play and create music. They invoke the self-organization of a subsystem for such aesthetic experience in man. **Zwierlein**, after an analysis of various relevant terms, introduces 'Lebenswelt' (Husserl, Heidegger) comprising history, language, culture, prescientific theory and practice of life and approach to world and mind, actually a 'hall of living' opening into various rooms, a unity where one executes one's living 'In-der-Welt-Sein' (Heidegger). It is from here that we understand and act and whence two streams extend our existence: one of 'being' and another of 'consciousness'. Ethics and epistemology should therefore remain dominant textures of self-organization and our understanding of it.

It is a pleasure for the organizers to acknowledge with thanks the generosity of the University of Kaiserslautern in sponsering this conference. For the local organization we are indebted to Dipl.-Wirtsch.-Ing. Werner Weiss and other colleagues for their excellent cooperation. In preparation of the papers we were efficiently supported by a team from the Regionales Hochschulrechenzentrum Kaiserslautern (RHRK), in particular by Dipl.-Phys. Klaus Uttler and Cand. Ing. Inf-Masch. Rüdiger Folz.

This conference could not have been held without the financial support of

Minister für Wissenschaft und Weiterbildung Rheinland-Pfalz
Daimler-Benz AG Stuttgart
Stiftungsfonds IBM Deutschland

for which we thank them. We are also grateful for the cooperation and understanding of our publishers Springer-Verlag, especially Dr. A. M. Lahee whose impressive, expert advice, guidance and collaboration are in a large measure responsible for the final form of this book.

One of us (R. K. M.) is grateful to Dr. F. A. Popp of the International Institute of Biophysics, Siegelbach for a long period of hospitality for the purpose of this conference.

New Delhi   *R.K.Mishra (Chairman)*
Kaiserslautern   *D.Maaß*
November 1993   *E.Zwierlein*

Contents

List of Contributors

Prof. Dr. I. E. Antoniou
Theoretische Naturkunde and Faculté
des Science
Free University of Brussels
C.P. 231
Campus Plaine
Boulevard du Triomphe
1050 Bruxelles
BELGIUM

Prof. Dr. H. J. Bremermann
Department of Molecular and
Cell Biology
101 Donnor
University of California
Berkeley CA 94720
USA

Prof. Dr. R. Bullough
Department of Mathematics
University of Manchester (UMIST)
Institute of Science and Technology
PO Box 88
Manchester M60 1QD
UNITED KINGDOM

Prof. Dr. T. Carpenter
Department of Music
Royal Holloway and
Bedford New College Egham
Surrey TW2 0EX
UNITED KINGDOM

Prof. Dr. M. Conrad
Department of Computer Science
Wayne State University
College of Science
Detroit, Michigan 48202
USA

Prof. Dr. W. Ebeling
Institute for Theoretical Physics
Humboldt-University
Invalidenstraße 42
10115 Berlin
GERMANY

Prof. Dr. H. Haken
Institute for Theoretical Physics and
Synergetics
University of Stuttgart
Pfaffenwaldring 57
70550 Stuttgart
GERMANY

Prof. Dr. B. Hao
International Centre
for Theoretical Physics
PO Box 586
34100 Trieste
ITALY

Dr. F. Hinterberger
Department of Economics
Justus Liebig University
Licher Str. 74
35394 Gießen
GERMANY

Prof. Dr. B. D. Josephson
Cavendish Laboratory
Madingley Road
Cambridge CB3 0HE
UNITED KINGDOM

Prof. Dr. P. T. Landsberg
Faculty of Mathematical Studies
University of Southampton
Southampton S09 5NH
UNITED KINGDOM

Prof. Dr. K. Li
Technology Centre
International Institute of Biophysics
Opelstraße 10
67661 Kaiserslautern
GERMANY

Prof. Dr. D. Maaß
Department of Computer Science
University of Kaiserslautern
67618 Kaiserslautern
GERMANY

Prof. Dr. R. K. Mishra
All-India-Institute of Medical Sciences
New Delhi 110029
INDIA

Prof. Dr. V. Nalimov
Laboratory of Mathematical Theory of
Experiment
Moscow State University
Moscow 119 899
RUSSIA

Dr. L. Pohlmann
Hahn-Meitner-Institut
Abt. Solare Energetik
Glienicker Str. 100
14109 Berlin
GERMANY

Prof. Dr. M. Richter
Department of Computer Science
University of Kaiserslautern
67618 Kaiserslautern
GERMANY

Prof. Dr. P. T. Saunders
Department of Mathematics
Kings College
Strand
London WC2R 2LS
UNITED KINGDOM

Prof. Dr. S. Tasaki
International Solvay Institute for Physics
and Chemistry
C.P. 231
Campus Plaine
Boulevard du Triomphe
1050 Bruxelles
BELGIUM

Prof. Dr. H. Tributsch
Hahn-Meitner-Institut
Abt. Solare Energetik
Glienicker Str. 100
14109 Berlin
GERMANY

Prof. Dr. W. Weidlich
Institute for Theoretical Physics
University of Stuttgart
Pfaffenwaldring 57 III
70550 Stuttgart
GERMANY

Prof. Dr. E. Zwierlein
Department of Social and
Economic Sciences
University of Kaiserslautern
67618 Kaiserslautern
GERMANY

Introduction

Eduard Zwierlein

> For after all, what is man in nature? A mere nothing in comparison with the
> infinite, an all in comparison with the infinitely small, a midpoint between
> nothing and everything. Infinitely far from grasping the extremes, the end of
> things and their beginning are for him invincibly hidden in an impenetrable
> secret; and he is equally incapable of seeing the nothingness out of which he
> is drawn, and the infinite in which he is swallowed up.
>
> *"Blaise Pascal"*

The idea for an interdisciplinary and international conference on self-organization was born about two years ago. It seemed to us that during recent decades
there has begun to spread a new idea of how to view things in almost all sciences:
emerging paradigms, synergetics, neuronal networks, quantum computing, chaos,
non-linear dynamics etc., many aspects that might refer to self-organization
shedding light on many old riddles, e.g. evolution, living systems, consciousness,
artificial intelligence, social and technical systems. And so we wanted to callex-
perts on different subjects under the coordinating idea of self-organization, pre-
senting a conference in Kaiserslautern, Germany, with the title: Self-Organization
As A Paradigm For Sciences.

But when the organizers of the conference first met in Kaiserslautern they
did not expect the adventure which awaited them, an adventure of coordinating
intellectual power from all over the world, of finding agreements, and gathering
the financial support necessary. But we finally succeeded in creating a unique
opportunity, not only because of the outstanding scholars sharing their ideas
with one another at the conference, but also because of the unique interdisciplinary composition of the conference itself. It would have been arrogant and
imprudent to ignore the interdisciplinary or even transdisciplinary character of
self-organization. Self-organization compels us to broaden our horizons beyond
a limited consideration of the powers of nature and to listen to an open and
investigative discourse without the bounds of prejudice.

Regretably, at least two disciplines, architecture and psychology, were missing from the conference although their contribution could have been important
in identifying some of the practical and theoretical aspects of self-organization.
On the other hand, we probably made a big step towards overcoming that nasty

Springer Series in Synergetics, Vol. 61   **On Self-Organization**
Eds.: R.K. Mishra, D. Maaß, E. Zwierlein   © Springer-Verlag Berlin Heidelberg 1994

gap between the so-called "two cultures". C.P. Snow's idea of opposing natural and technical sciences to social sciences and humanities has no original or systematic truth in itself, but only reflects a certain and finally misleading development of some Western philosophy of the last five hundred years. Setting up an interdisciplinary dialogue must bear in mind the roots of all sciences. The honest search for truth always originates in the interest to serve a good life, and is at least in no opposition to both: survival and good life of mankind in the context of the surrounding nature.

But nevertheless, it is true that some of the different sciences have no "common language" at all. Bringing viewpoints from different faculties with specific background knowledge, a specific access to reality, and specific "languages", there is a great challenge to avoid the confusion of the Tower of Bable Although only one language was used at the conference (English), it took a lot of time for many of the participants to begin to understand one another. Terminology and different scientific background indeed proved to be a major problem and numerous propositions and definitions were developed, proposed and rejected. The discussion was very much colored by semantic problems and also by methodological difficulties. The variety of approaches and also the different context of different countries contributed to the general diversity of the conference, being not only international and interdisciplinary, but also intercultural.

Most probably we are no longer in the state immediately after building the Tower of Bable when all languages became confused, but unfortunately we are still far from reaching the new state of Pentecosts. And though many efforts have been made and many "translations" in order to create a mutual understanding and a discussion full of openness and clarity, we feel that there is still a long way to go. Three days was insufficient to resolve the considerable differences between the disciplines represented and sometimes within the disciplines themselves. But as you observe some progress in this quest, you will see that the real and decisive gap to overcome is that we are, so to speak, scientific, technical, and economic giants, on the one hand, but mere ethical dwarfs, on the other. I will touch upon this issue later on.

I shall not attempt here to give a thorough and authoritative exposition of the conference. Impossible briefly reviewing the relevant theoretical and practical ideas outlined in lectures, statements, and discussions in the conference, but mainly collected in the book you have at hand, I only want to mention two points we had in mind while planning the conference on self-organization and which might capture its broader implications: the idea of a common paradigm for the sciences and the idea that self-organization could be a key to the survival of mankind.

To overcome the confusion of different methods and terminology does not mean to create a universal proto-language for everyone. Between a postbabylonian and a prepentecostical state of exchanging ideas and facts self-organization was hotly debated. And indeed we could not find, and actually did not expect to find, agreement about how to define self-organization precisely. There was no simple definition that all could accept and at no point did a consensus emerge whether this could be found at all. But throughout the conference the dominant questions were related to a deep and thorough understanding of self-organization comparable to an approach starting from different points, coming from different directions, making various efforts, and using different ways. In this way the forty or so participants represented a perfect microcosmos of the global scientific community. Behind this game of various perspectives it became obvious that we do not presently need a universal protolanguage (which might be the result in the end), but a common goal, a commonly shared problem that in itself possesses the power to overcome the traditional borders between the sciences and that therefore could eventually serve as a paradigm for all sciences.

In the search for a common paradigm, self-organization should be investigated as a fruitful candidate. Discovering and developing such a common paradigm would be also a contribution to the unification of the sciences. And there are many indications that self-organization can be designed as a general theory and explanational framework for the sciences. A unifying paradigm is not a uniform paradigm and therefore does not imply a homogeneous modelling for all sciences (self-organization is no matrimonial agent for lonely sciences!), but shaping a focus or framework where eventually all our scientific problems and all the ideas that are dancing around might converge towards something like a crystallization point. And of course, a unifying paradigm does not imply the prejudice that reality in itself is a unity without any contradictions. That we do not know. But looking for a unifying paradigm, a single window or looking glass to understand reality, might characterize the scientific search for truth in the way we characterize an optical prism (i.e. the paradigm) splitting a beam of ordinary white light (the ultimate reality) into light of different wavelengths. Each colored light of the colored band or spectrum of light (representing the different sciences) being the result of the dispersive power of the medium (i.e. the paradigm) must try to go back its own way through a commonly shared medium or paradigm towards their ultimate and common source.

A similar point of general importance emerged concerning ecological aspects and also social problems. As Kant says: our practice is a bad one only because we are lacking a good theory. This touches the question of ethics and the responsibility of sciences. Can we consider self-organization as a key to survival? As we all have abused our planet and recklessly squandered what we have inherited from our forefathers and actually borrowed from our children, it is necessary to strive for responsible behavior. To be a good human being in an integer environment – this is becoming more and more difficult to achieve with the garbage, air and water pollution, deforestation, with the loss every hour of another species to extinction, with disorder, poverty, and starvation – the list goes on and on.

Time is running out, but it's stil not too late. We can and must turn things around. Environmental protection, future responsibility, long-term responsibility, protection of the forthcoming generations, sustainable development, and the precaution principle – these are some of the key words of our global responsibility. But being responsible for the environment and our beautiful planet means in the first place responding to its inherent laws. The investigation of self-organization and how nature organizes its complexity and ordering processes by a wise management may lead us to two conclusions. First of all, we have once more to learn respect and awe for the highly sophisticated and astonishing processes of self-organization, and to avoid their disturbance or even destruction while taking them carefully into consideration. On the other hand it is possible not only to take care of nature's laws watching and obeying them, but also to learn from its wise management. We do not have the answers to everything and we can learn so much from the wonderfully evolved laws and rules of nature, considering nature simultaneously as a prototype and as the heritage of millions of years experience in survival. It is surely worth initating and following the ideas which nature has developed.

Our discussion of self-organization may be beneficial in helping prevent further destruction of our planet as we must learn to accept and to value all forms of life. The remedy proposed is already at hand. There is a strong and important message here that needs to be understood: "At this time living Nature beckons us, as it has done always, to its methods of silent organization and tools of self-reflection, self-renewal, self-excitation, self-control and self-organization, within any constraints that may be encountered to ensure its stability, survival, enfoldment of options and dynamic processes and manifestations. Nature does this without being irksome to any subsystems and optimizes their communication and brings about their growth with rational abundance for all. The voice of Nature is now strident; there is no time to lose, tomorrow may be to late for us to get up by the bootstraps." (R.K. Mishra)

Although social and ecological challenges demand a quick reaction, the discussion about self-organization will be an ongoing process perhaps in the long run aiming at an ultimate belief of truth. At present, the comprehensive debate must go on with all the scientific investigators freely sharing with an open and tolerant mind that what they are able to see. There is always blindness in our knowledge, incompleteness, concealed, dark areas that we try to shed light upon, difficult puzzles, that we want to unravel. We are very glad that the Kaiserslautern Conference made a small step in this ongoing process of revealing and that it joined with its small contribution the adventurous journey of a sailing ship in virtually unknown waters.

Self-Organization in Evolution, Immune Systems, Economics, Neural Nets, and Brains.

Hans J. Bremermann

1 Introduction

Self-organization is *creation without a creator attending to details.* More precisely: the formation of patterns and structures that form from the initial state, without intervention, in a dynamical system, or through the interaction of finite state automata. The term self-organization is also applied to *supervised learning by neural nets,* and to *genetic algorithms* [1, 2, 3].

In neural net training, the "creator" arranges the overall net configuration (number of layers, number of hidden units, etc.). The individual connection weights are set iteratively by a training process. During training samples are presented to the input units of the net. Net responses are evaluated by a separate "evaluator" or evaluating circuit, which informs the net whether the output is correct. The net then modifies the weights according to an internal learning algorithm. The "creator" does not communicate to the neural net what the values of the individual connection weights should be. In fact *the "creator", except in special cases, would have no way of knowing what the values of the connection weights should be without actually running the training process.*

Evolution is another example. No outside agency "tells" an organism directly what the values of the individual nucleotides in its genome should be. These values are determined indirectly: DNA sequences are transcribed into RNA, which is translated into amino acid sequences which determine the conformation of proteins. The functional properties of proteins depend upon conformation. Myriad interactions of proteins create a cell, and the interactions of cells during development create macroscopic organisms, which then interact with each other, with predators and prey and with the physical environment. In the end, some organisms produce more offspring than others. In this way some nucleotide sequences emerge to be more numerous than others, and some drop out altogether.

Evolution can be viewed either as "supervised" or as "unsupervised" self-organization. Evolution is "unsupervised" if we consider the biosphere as a whole. It is "supervised" if we look at individual species. Here the habitat of the species acts as an "evaluator" that decides which genotypes are "selected".

Unsupervised self-organization can be seen in *dissipative structures* [4, 5] in *synergetics* [6], in *embryogenesis,* where the anatomical structure of an organism unfolds from the interaction of individual cells, and in *artificial life* [7], the emergence of patterns and images on a computer screen from the interaction of self-reproducing automata. Supervised self-organization is the governing principle in the *immune system* that combats bacteria, viruses, and fungi. Immunoglobulins are not preprogrammed in the inherited genes but are generated by random assortment of gene segments and by hypermutation. The "evaluator" is the strength of antigen binding. A molecular mechanism then causes those B-cells to divide whose antibodies most effectively bind the invading antigen.

In *market economies* self-organization can adjust supply, demand, and prices. Economists have shown that the "invisible hand" of a free market can lead to prices where, supply and demand are in equilibrium ("equilibrium prices"). There is a cost in human terms: competition can lead to inequality, the losers often pay a high price, and winners may acquire unreasonable privileges. *Wealth is "autocatalytic"*, and the ideal of an "equal chance for all" requires special *rules for the game of competition*, such as access to education regardless of economic status.

Undesirable social and environmental side effects of capitalism can be mitigated by legislation, redistribution of income through taxes and subsidies, and services by public agencies (health, education, welfare). Unfortunately, taxes, regulations, and bureaucracies seem to take on a life of their own: *they too are autocatalytic*. More about this problem later.

Though the side effects of competition in a free market can be severe, the alternative of a centrally planned economy has proven to be worse. Self-organization is important. Centralized micro-management of modern industrialized economies has failed. Prescribed prices and centrally planned production and distribution have proven to be disastrous, as history has shown. With time, the relationships between prices and cost became severely distorted. Future historians may sort out the relative importance of the factors that contributed to this distortion: human selfishness at all levels, corruption, the terror of labor camps, the stifling of initiative in authoritarian structures, perhaps the sheer complexity of economic processes, lack of feedback, slowness of communications, and the inability to make wise decisions by even the most idealistic of bureaucrats.

To understand why self-organization in economic affairs produces an abundance of goods and services while centralized micro-management does not, a comparison with other instances of self-organization can be illuminating. Complexity is the common denominator. Equilibrium prices are the result of a myriad of interacting factors: human behavior, culture, traditions, laws and regulations, technology, resources, and the interaction with economies around the world.

The complexity of the behavior of dynamical systems was generally ignored prior to about 1960. The world seemed predictable if one only knew the correct laws of physics, chemistry, biology, economics and sociology. Theoretical physicists thought in terms of solutions of differential equations. When the dynamics became complex, as in thermodynamics, there was a conceptual jump to randomness and a few macroscopic variables such as entropy, volume and pressure. There was virtually no mathematical and computational complexity theory before the 1960s.

In recent years the study of dynamical "chaos" has become popular. Chaos in nonlinear dynamical systems is common and can occur in very simple systems. In most cases it can be explored only with the help of computers. The exploration of chaos is a paradigm shift: from logical-mathematical deductive inference to exploration by computer simulation. Similarly, the computer is indispensable for studying self-organization even in relatively simple and well defined systems, such as trainable, feed-forward neural nets.

In nature the principle of self-organization is ubiquitous. Neither the details of immune defenses nor the myriad details of neuronal organization are preprogrammed in inherited genes. Self-organization plays a role from fine tuning of muscle coordination in motor skills to organizing knowledge representations at higher levels in the human brain.

The idea of self-organization, in abstracto, can hardly be objectionable, but it has encountered resistance in all the fields mentioned in the title of this paper. Evolution is controversial, even today. Thirty years ago, many immunologists subscribed to the "instructive theory" of the immune response, which assumed that immunoglobulins were preprogrammed in the genes. Communist doctrine ignored the self-organizing capabilities of markets. In computer science, self-organizing neural nets ("Perceptrons") became accepted only after considerable debate. In psychology, behaviorists, who have been rather oblivious to questions of computational complexity, resisted "connectionism". Nowhere, however, has the idea of self-organization encountered as much resistance as in the question of human evolution.

About half of the population of the USA does not accept evolution "as fact". It is not only the Christian fundamentalists who reject evolution. That humans should have arisen through a process of self-organization of DNA is indeed difficult to comprehend. It is psychologically less satisfying than an act of creation, some 200 to 300 generations ago, by a *creator* who shaped the first human being, our common ancestor, *in his own image.*

Evolution is the most complex of all self-organizing processes. For a better understanding, as in the case of economics, it is useful to compare biological evolution with other, simpler instances of self-organization, such as simulated evolution in a specified environment ("genetic computer algorithms") and the well studied example of trainable neural nets (which were among the first self-organizing systems to be studied extensively with computer simulations) [8].

Widespread experimentation with self-organization has become possible only recently. It requires plentiful affordable computing power. For example, training of neural nets on non-trivial tasks was not successful until the nineteen-eighties, when inexpensive but powerful personal work stations became available. Pioneering attempts in the fifties and sixties by Frank Rosenblatt [9] and many others suffered from expensive and cumbersome computing. In spite of heroic efforts, the pioneers never discovered the remarkable performance of multi-layer neural nets with "hidden units" which Rumelhart, Sejnowski, and others demonstrated in the eighties [10].

Self-organization in the brain was Rosenblatt's primary motivation to explore his "Perceptron" model. His sponsors also had other objectives in mind: the development of useful trainable pattern recognition devices. The latter became a reality in the eighties. Conceptually, Rosenblatts's model of "learning through synaptic modification" had a renaissance in the eighties and together with other neural net models by Hopfield and Tank [11] (with roots in thermodynamics) had a profound impact on the understanding of the brain, as well as commercial and military applications.

From Aristotle [12] to Hilbert to Goedel and Turing, logical reasoning has been considered the highest level of intelligence, if not the basis of all human intelligence. In the forties and fifties the *Turing machine* appeared to be the appropriate model of the brain in view of Turing's remarkable discovery that there exist Turing machines (fairly simple ones for that matter) that are *universal [13]*. Such a machine can emulate any other Turing machine and hence *compute* (in principle) *anything that is computable [14]*.

Followers of the artificial intelligence "paradigm" of the fifties, sixties, and seventies therefore saw it as their mission to explore the art of programming. To search for *universal principles* in developing *algorithms*. No need to build special purpose machines [15]. Von Neumann-type computers (like Turing machines) were universal. After some unexpected and remarkable early success stories, the going was slow. All the early AI problems generated trees of nodes that must be searched. These trees grow exponentially as one probes deeper or solves harder problems. The exponential growth of trees quickly overwhelms advances in the speed of computation.

The present author was a witness to the dispute of brain model paradigms, working one summer with Frank, and enjoying on many occasions the hospitality of Marvin Minsky. What struck him was the realization that both algorithmic artificial intelligence as well as trainable nerve nets have *extraordinary computing requirements* which literally are "out of this world".

We live in a finite universe of a limited age, with a finite speed of light. Quantum theory imposes minimum requirements of time and energy on the measurement of physical markers with sufficient precision. Putting all this together one can derive a "fundamental quantum limit" for the transmission and recovery of stored information, which is rather restrictive in the face of some of the computationally expensive artificial intelligence and net training programs. The author first announced the limit thirty years ago at a conference "Self-organizing Systems 1962" [16]. The limit touches upon some very fundamental questions in quantum mechanics and quantum gravity. Bekenstein has connected it with the entropy of black holes and has gone to great length in investigating the limit. For a recent review see: Bekenstein and Schiffer 1990 [17]. For a discussion of the entropy of the universe see: Penrose, 1989 [18].

What does an absolute upper limit on computing imply for mathematics and for the theory of knowledge? In a book on applied mathematics, Lin and Segel, pondering the "unreasonable effectiveness of mathematics in the physical sciences", state that through the ages, many philosophers have given an answer which can be summarized as "God is a mathematician" [19]. Once the author asked Marvin Minsky's: "If God is a mathematician, where does He do His computing? It is easy to propose tasks which require an amount of computations that would exceed the power of any computer the size of the universe [20]. Is evolution His algorithm, are we His computation?" Marvin said: "Maybe He has another universe to do His computations in".

Classically, mathematicians have not worried about physical processes ("computation") being required for deductive inference, and many do not do so today. This "faith" in mathematics per se is illustrated by a recent review of a book by King [21] Martin Gardner, writes: "King shares physicist Eugene Wigner's puzzlement over the 'unreasonable effectiveness' of mathematics. This puzzlement puzzles me. Not only is the Universe mathematically structured, it is made entirely of mathematics. Matter consists of fields and their particles which are not made of anything except equations. What is so mysterious about the application of these equations to a Universe from which our minds, in turn made of mathematics (!), originally extracted them? If mathematical patterns are not discovered, but are entirely human creations like poetry and music, then their powerful effectiveness in science and technology is indeed what Wigner called an undeserved miracle."

Let me comment only this: there is an observer problem, both in physics, and in the Platonic universe of mathematics. And there are the problems of consciousness and ethics. We refer the reader to a book by Margenau and Varghese [22] in which the authors have collected the responses of over 60 scientists (including 24 Nobel Laureates) on questions about the origin of life, evolution, the universe, consciousness, morality, and God.

In the final sections of this paper we deal with higher levels of self-organization in the *brain*. If knowledge is limited by computational resources, then an interesting question is: why would the human brain waste precious computing time during sleep? *Or does it?* If time is not wasted, then one must ask: What purpose, if any, is there to the fantasies of dreams, to the monsters that arise during the night, to the bizarre surrealism of situations and images, that are so typical of dreams?

We attempt to give an answer which we believe to be original and which is published here for the first time [26]. The main points are:

1) Dreaming has a physiological basis (REM sleep) that is common to mammals [27]. Mammals, generally fine-tune their inborn drives and behavior patterns in accordance with the circumstances of their local experiences. This requires a considerable computational effort which is done in an off-line mode.

During REM sleep the outgoing nerve fibers to the muscles are totally blocked and *motor movements are paralyzed*. This blocks the acting out of the *virtual reality-like dream simulations*. Such blockage is necessary if the simulations explore hypothetical behavior that is being "tried out" in fantasy and which has to be separated from reality.

2) The fantastic, *surrealistic nature* of dreams is *necessary* in order to *simulate hypothetical situations* and to *keep them separate from factual knowledge*, which is not only verbal, but also consists of real experiences, real images, and prototypical examples.

3) The power of dreaming rests on the brain's ability to express itself through metaphors. The objects, images, persons that appear in dreams usually *do not*

represent themselves, but are *metaphors.* When metaphors are taken literally, or reality is interpreted metaphorically, intelligence suffers and mental disturbances can result.

The human brain has a dual nature, one dominated by metaphors and analogy, the other by language and logic. The duality of the mind's representations makes possible self-references without the danger of falling into the well-known paradoxes that occur in mathematical logic when self-reference is permitted [28]. A rational censor, a filter sorting out the creative figments of the imagination, thus, has a role in the self-organization of knowledge. This function is as important as standing watch over the illicit stirrings of a libidinous ego. Freud saw dream symbolism as a disguise to fool the sensor. We believe that beyond that, it is a basic ingredient of the self-organization of the SELF.

2 Biological and Artificial Evolution

2.1 Natural evolution

Life descends from prior life. Existing genomes are the result of a chain of replication reaching back to the origin of life. Mutants, that is changes in the DNA sequences, are "selected" or "rejected" through their effect upon the comparative reproductive success of the organism that carries the genome. The reproductive success, (technically known as "fitness"), is the result of the complex interactions of organisms with each other and with their habitat, the organism is the product of a long chain of molecular, intermediate steps between "genotype" and "phenotype". In the models of classical mathematical genetics the "fitness function" is replaced by a linearized, almost trivial approximation, of the relation between genotype and phenotype and its effect on survival and reproductive success. Consequently these models have a limited range of validity.

Evolution is inevitable since DNA replication has an intrinsic error rate which is not zero (though very small). The error rate is *necessarily non-zero* because thermodynamic fluctuations affect the chemical reactions that are involved in replication. DNA is a better memory molecule than RNA, since it allows proofreading and error correcting enzymes to reduce the remaining mutation rate to very low levels. RNA has an intrinsic replication error rate that is several orders of magnitude larger than that of DNA. The intrinsic mutation rate of the genetic memory molecule determines the maximum size of the genome. Hence there are no modern prokaryotes or eukaryotes with RNA genomes, only viruses.

DNA probably was preceded by an "RNA-world", when RNA was the genetic memory molecule. Messenger RNA, self-splicing of RNA introns, the ability of RNA to synthesize proteins, and the mixture of proteins and RNA in ribosomes, may be surviving mechanisms from the ancient RNA world. Manfred Eigen and his associates at the Max Planck Institute for Biophysical Chemistry at Goettingen have investigated RNA evolution theoretically and in the test tube under cell free conditions. They have obtained ab initio replication and evolution of RNA, provided nucleotide monomers and a single modern protein enzyme are supplied [26, 27, 28].

At the origin of life, this enzyme could not have arisen spontaneously. Hence the path to the RNA world is believed to lie in a prebiotic evolution of *autocatalytic systems* in the primordial organic soup. Their synthesis in the test tube remains a challenge.

2.2 Evolution takes time, is non-predictable in the long run

The combinatorially possible genomes (DNA sequences) vastly outnumber the actual genomes that have ever come into existence. Evolution on earth, since the beginning of life, has explored only a minute portion of an enormous phase space. On another planet, under similar conditions, the result of evolution, therefore could be different.

The stochastic search of the phase space (by DNA base-pair substitutions, recombination, frame shift mutations, and some transduction of DNA by viruses) takes time. The fossil record shows that breakthroughs can occur after long periods of stagnation (something that can also be observed in the training of neural nets, simulated evolution, and computer games of "Artificial Life") [29].

The fossil record documents that there was microbial life 3.5 billion years ago, and maybe earlier. (In other words, life appeared not long after the newly formed planet had cooled). However, at least 2.5 billion years passed before the great breakthrough towards multicellular organisms populated the earth with a plethora of creatures, large and small.

Primates have been around for about 60 million years, hominids for 3 or 4 millions, but anatomically modern man perhaps only 50 to 150 thousand years. The breakthrough from non-human primates to modern man involves *communications*. Actually, animals communicate, but through limited and semantically fixed signals (chemical, acoustic, behavioral displays, etc.). Human communications are *semantically open*. The signals of human communications (not only words, but images, gestures, metaphors) also lend themselves to the *self-organizing representation of knowledge in the brain*. (More about this in section 7).

The evolutionary breakthrough towards anatomically modern man is remarkable not only because of the far-reaching magnitude of its effects, but also for the very small portion of the genome that is involved: humans and chimpanzees differ in only 1.8% of the base pairs in their DNA.

2.3 Artificial evolution

One can gain insights into the global course of evolution by *simulating* it in model systems. Such simulations should be part of every biology curriculum. Genetic algorithms can be more than instructive: "Nature's own" evolution method has been applied, for example, to the optimization of strategic war games (Richardson model) [30]. For decades the world tethered on the brink of nuclear annihilation. Nuclear war was unthinkable, yet the missiles were real and ready to be fired at an instant. This was the doctrine of "mutual assured destruction", or

"MAD" for short. Strategists formulated mathematical models of missile warfare. Different strategies and parameters could be chosen. How could such models
be tested and strategies optimized? Only simulations could shed some light on
it, and different strategies have been explored through genetic algorithms.

John Holland also pioneered genetic algorithms applied to the creation and
optimization of artificial intelligence programs [31]. Little is known about optimality of AI (artificial intelligence) programs. Therefore it is often difficult to
judge whether genetic algorithms give optimal results or not.

This author also proposed genetic algorithms, as early as 1958 [32], which anticipated some of the contemporary developments of "Artificial Life" and neural
net training. It was proposed to apply genetic optimization to Rosenblatt's mul-
tilayer perceptrons. At about the same time, a paper by Friedberg [33], reported
the actual implementation of applying a genetic algorithm to the evolution of
a very simple computer program. *Amazingly, Friedberg's evolution had taken a
1000 times longer than exhaustive search.*

Discouraged by Friedberg's results this author decided to back off and to
explore the simulation of genetic algorithms in more transparent situations first,
and to replace the non-analytical, non-transparent perceptron performance function by something much simpler: The squared sum of the residues of a system of
linear equations. Amazingly, even for this simple quadratic, unimodal function
(with a single global minimum), the evolution process tended to stagnate rather
than converge [34, 35, 36].

This made a strong impression on the author and alerted him to the phenomenon of computational complexity which was then very poorly understood
and far from people's minds, at a time when computer speed seemed to increase
by a factor of two or so, each year. The simulations also explored the effect of
sexual recombination upon the speed of convergence. This had a dividend. They
led the author to a new theory of sex.

2.4 Evolution and sex

In the seventies a lively debate occurred about the evolutionary significance
of sex. It was then commonly accepted that "sexual recombination speeds up
evolution so that man could evolve before the universe burns out". However,
in his simulations this author had seen that speed-up is a rare effect, without
immediate, evolutionarily stable selective value. This led him to look for an
alternative theory of the function of sexual recombination in biology: [37, 38, 39].

Most natural populations have a large amount of allelic polymorphism. Classical theory had been at a loss of an explanation for some time. The difficulties
can be traced to the oversimplified, linearized fitness function that was generally
employed in mathematical genetics.

It was proposed that polymorphism is a deliberate defense of multicellular
organisms against micro-parasites, such as bacteria, viruses and fungi. *The latter
have much smaller genomes, shorter generation times and higher mutation rates.*
As a consequence they can evolve and adapt much faster than their hosts which
as multicellular organism have much larger information requirements and hence

larger genomes. The maximum allowable mutation rate is inversely proportional to the size of the genome. *Hence co-evolution of host and micro-parasite is an unequal game.* The older theory had not taken this into account.

Hosts hold off parasites by numerous molecular defenses. Parasites can breach defenses. When the host evolves a new molecular conformation that throws off a virus or fungus, it is usually not long before the parasite has breached it by its own adaptive evolution. Solution: *Polymorphism* in molecular conformations of molecules that are targets for the intruder. In this way a "molecular key", that allow a parasite to enter one individual does not fit the next. Polymorphism also is characteristic of the key molecules of active immune defenses (the class I and class II MHC (major histocompatibilty complex) molecules of the mammalian immune system), which is maintained by sexual recombination. This theory of sex was independently proposed by W.D. Hamilton [40]. It seems to be standing up well to experimental scrutiny. The moral: In discussions of co-evolution it is important to pay attention to different speeds of adaptability (which depend upon genome size, mutation rates and generation times).

2.5 The adaptive speed of the AIDS virus

Another example of the importance of adaptive speed is the human immuno deficiency virus (HIV). This virus is an RNA (retro)-virus which mutates *a million times faster* than the genomes of mammals. In terms of nucleotide evolution it covers as much ground during its persistence within a single host (about 10 years), as primates have covered since the separation of the ancestral lines of chimpanzees and humans. This has led to an extraordinary variety of different strains of HIV. In-host evolution is also responsible for the rebound of the virus several months after AZT treatment begins, and for the ultimate inefficacy of AZT and similar drugs in clearing the virus from the host and in their inability to arrest disease progression not only temporarily, but permanently.

3 The Immune System

3.1 Vaccines and theory

The practice of vaccination predates understanding of the immune system by a long time. 200 years ago Jenner vaccinated a boy with cow pox virus, challenged him with the human virus, and found him protected against small pox (thanks to cross reactivity). Jenner succeeded decades before Schleiden and Schwann announced their cell theory, and almost a century before the germ theory of disease was established. More than a century has passed since Pasteur developed his rabies vaccine, which prevents the otherwise fatal disease, even when vaccinations take place after a person has been bitten by a rabid animal. More recently many infectious diseases: polio, tetanus, cholera, measles, hepatitis, and more, have been conquered by vaccines, and small pox had been wiped out completely, world-wide.

In spite of these triumphs and in spite of the advances of molecular biology, protection against HIV, the human immuno deficiency virus, remains elusive. It constitutes a new kind of challenge, a retrovirus that manages to persist in active infection for several years before it kills its host. In the author's view, the challenge requires a deeper understanding not only of antibodies, but of the activation and regulation of the cytotoxic T-cell response. Antibodies bind to antigens *externally.* T-cells detect viral infection *inside cells.* and kill cells that have been found to contain suspicious, non-self peptides.

In a forthcoming paper [41] this author proposes that the dynamics of HIV replication, neutralization by antibodies, and activation and decay of the cytotoxic T-cell response, has a stable, non-zero steady state, provided a certain threshold condition for the rate constants is satisfied. During the early, flu-like phase of the infection, HIV apparently weakens the immune system irreversibly, such that the threshold condition becomes satisfied. This allows the virus to persist in a low level, chronic infection. The weakened immune system remains partially functional, fending off opportunistic infections and keeping the host alive for several years.

During this period (known as "latency period"), one component of the immune system, the CD4+ cell population, declines in number, for reasons that are not well understood. When only about 20% of the CD4+ cells remain, opportunistic infections become stronger and eventually kill the host as the CD4+ count approaches 0%.

To conquer AIDS, a better understanding of the full range of interactions in the immune system may be required. In this respect immunology is a very young science. The molecular mechanism of antigen presentation by class I and class II MHC molecules, and the functioning of the T-cell receptor (TCR), have been known for less than five years.

3.2 Self-organization through clonal selection

In the nineteen-fifties, about a century after Darwin's "Origin of Species", Mac-Farlane Burnett proposed the "clonal selection theory" of the B-cell response. B-cell clones expand through a proliferation of those cells whose surface immunoglobulins bind to the invading antigen. *The immune system thus is self-organizing.*

The analogy with Darwinian evolution, however, is limited. The response is no "free-for-all" evolution. Instead, selection, expansion, maturation, limitation, and finally preservation of memory cell populations, is carefully orchestrated. The organism imposes a network of controls upon the response. In this respect the *immune response is more like a modern market economy*, where competition is constrained by laws and regulations. Yet the immune response would *not be possible without self-organization*. The information content of the genome is not large enough to be able to mount an "instructive", genetically pre-programmed *immune response*. The number of possible antigens is simply too large and unpredictable (recall from section 2 that pathogens evolve much faster than their host species and therefore can generate ever novel "molecular surprises" for their hosts).

The immune system has been studied extensively in the mouse. Transferability of knowledge *from mice to men* rests on the *common ancestry* of both species and upon the fact that evolutionary *change has been slow.* Here we encounter again the phenomenon of complexity and the speed of evolution, in this case with desirable consequences.

3.3 Immunological identity and self-recognition

The human immune system has about as many cells as the human brain. It consists of B-cells and T-cells, natural killer cells, macrophages, and a collection of interleukins, that is hormone-like substances that play a role in regulation. Each individual (except identical twins) has a unique set of class I and class II MHC (major histocompatibility complex) alleles which define the "self" identity of cells and which distinguishes them from the cells of all other individuals.

The MHC "diplotype" arises from the combination of the haplotypes of egg and sperm. Variety is generated through the random assortment of parental genes during meiosis. This author [42] has postulated that in addition some mechanism analogous to the self-incompatibility locus that is present in flowering plants, may contribute to the preservation of variety in small populations where inbreeding is a problem. The necessity for this variety derives from the ability of pathogens to attack genetically homogeneous populations through superior speed of co-evolution (as discussed in section 2).

The mechanism of generating self-defining molecular identity implies that the immune system has to "learn" to distinguish *"self"* from *"non-self"*. This is accomplished by a process of *clonal deletion*, during neonatal development, of all those B-cells and T- cells that are *self-reactive*. In addition there may be mechanisms that induce *"anergy"*, that is a state where the cells are not deleted but "turned off".

Unfortunately, in many auto-immune diseases the mechanism of self-recognition becomes defective and the immune system attacks the organism's own cells. This happens in many forms of arthritis, lupus erytomatosus, some forms of diabetes, and more. It is presently not clear what role, if any, viruses may play in auto-immune diseases in turning the immune system against its own self.

3.4 Clonal selection

Clonal selection acts on a reservoir of some 10^8 clones of B-cells and T-cells of different specificities. Such a large number of clones is necessary to cover the great variety of possible antigens. In addition, in the maturation of an immune response, selection of somatic mutants among B-cells further improves the binding affinity of antibodies.

The *selection* of B-cells is *local,* and *self-organizing:* Those B-cells that bind antigen are "activated", they are stimulated to divide. The immune response is a race of proliferation between the replicating intruding antigen and the production of antibodies which inhibits the antigen. Early during the B-cell response both secretion of antibodies and proliferation takes place, while in the later stages antibody secretion takes precedence. This switch between proliferation and production can be modeled as an operational control problem and shown to be optimal.

The *regulatory processes* that accompany a B-cell response are not yet fully understood. Besides the stimulus of antigen binding, interleukins and interactions with T-helper cells (the CD4+ T-cells) play a role. Interleukin 2 is an activating signal that is secreted by activated T-cells, and which in turn stimulates other T-cells, putting them on alert.

The immune system as a whole is sensitive to light, mental states, circadian rhythms. The latter can be influenced by very weak, low frequency electromagnetic fields [45]. Here may be a challenge to establish whether there are measurable activation effects of the kind of fields that are associated with the living state [46].

When an antigen has been eliminated, not all the cells of the expanded B-cell clone survive. On the other hand the clone does not return to its "pre-challenge" population level but remains at some intermediate, expanded level. This constitutes "immunological memory".

3.5 Limitation of clonal expansion must be self-organizing

B-cell specificity is determined by their receptors, which have variable parts that arise from random assortment of gene segments and somatic hypermutations. Clonal expansion is straight forward: When enough of the receptors bind antigen, especially when multi-valent antigen caps and immobilizes several receptors on the B-cell membrane, then B-cells are activated and divide.

Since B-cell identity is generated by a random process, *the organism has no genetically programmed means of distinguishing between B-cell clones.* The clonal selection theory of McFarlane Burnett therefore has a counter part: *The idiotypic network theory* of Jerne (who, like Burnett, won a Nobel prize). According to this theory, *antibodies* themselves *become antigens* by carrying *epitopes that are antigenic,* and which therefore stimulate other B-cell clones, which in turn carry epitopes, which stimulate further clones, and so on. The result is an "idiotypic network" of B-cells and antibodies that bind to each other and which stimulate and inhibit each other.

How can such a network remain stable, maintain immunological memory, yet be able to respond quickly to any novel or known antigen challenge? Mathematical models have been formulated and explored by simulations. There remain some difficulties. For reviews see [47, 48]. The empirical evidence for a "deep" network (rather than the existence of the first layer of anti-antibodies) remains somewhat shaky.

Analogously to B-cells, T-cells have receptors whose variable parts are generated by *random assortment of gene segments.* The control processes that reign in an expanded clone after it has served its purpose remain to be clarified, but it is clear that the process must be self-organizing. Without controls one would expect an imbalance between clones of different specificities over time, yet all specificities, especially those that have never been stimulated, must remain at sufficient strength to counter the unpredictable and ever inventive antigens that can strike at any moment.

4 Economics

4.1 The modern mix

Modern economies resemble not so much *"Social Darwinism"*, as the regulated, yet self-organizing *immune system*. Growth and decline of enterprises operate under an umbrella of regulatory constraints early capitalism was less inhibited, when many people had to work long hours for low wages, without the benefits of social security, health, unemployment, and disability insurance. Since then most industrialized countries have passed social legislation.

These laws and benefits have improved the quality of life, but they are expensive. Through the years taxes and public debt have risen. High taxes inhibit economic growth. Growth stimuli, in the form of tax cuts, cause deficits. Slow growth, besides unemployment, generates lower tax revenues and extra costs for welfare. At present, all industrialized economies are running deficits. Meanwhile voters press for revitalization of the economy, limitation of taxes, and more services. Does anybody have a mathematical model that shows a stable future?

Modern economies are of recent origin, they remain tentative and experimental. A hundred years ago 9 out of 10 people lived on the farm or were born on the farm. The long term stability of the contemporary mix of markets, taxes, government services, health care, and social security remains to be proven.

4.2 Autocatalytic growth of bureaucracy

Market economies can cope with constraints as long as everybody plays by the same rules. There are three problems with rules and regulations:

1) Their effect may be quite different from what lawmakers had in mind when they were written.

2) They keep on changing, and are getting ever more complicated.

3) Laws may require complex administrative procedures, both for government to enforce, and for those who have to comply.

Many new construction projects now require an "environmental impact report." Analogously "economic impact reports" for new legislation should be required. Unfortunately, the legislative process usually is not based upon a rational analysis, but upon deals between representatives of special interest groups.

As regulations and tax laws are being "improved", they become ever more complex, requiring more bureaucrats. As administrators are being overwhelmed by work they require assistants (usually not one but two, for reasons of status in the hierarchy). So bureaucracies grow like a pyramid, with a broadening base at the bottom. The system is *"autocatalytic"* the more administrative jobs there are, the more new ones are being created.

Eventually administration becomes an end into itself. The purpose of regulations that gave rise to the rules and procedures are forgotten. Procedures and paperwork become rituals, to be performed without questioning. And the rituals provide jobs, and benefits for those who perform them. Efficiency is measured by "output" in terms of forms and memos and paperwork done.

What happens in public administration also happens in private business, especially in large companies. In a truly competitive environment, however, companies, may be forced into reorganization, go bankrupt or merge.

4.3 Growth, chaos, and feedback delays

Wealth, bureaucracies, real estate prices, and overall economic growth are *autocatalytic*, like biological population growth. When resources are limited, population growth, can be described by the logistic equation:

$$x'(t) = rx(1 - x/K),$$

where r is the intrinsic growth rate, K the carrying capacity, and x is the magnitude of the population. This equation gives a *qualitative fit* to many population growth curves, whatever the myriad details of the interactions that bring them about. It is a bit like the macro-variables of thermodynamics, which also obey simple laws.

The solution trajectories of the logistics equation *asymptotically approach the carrying capacity*, and they do so *smoothly*, for every starting point and for every growth rate. This situation changes dramatically when the *differential equation* is converted into a *difference equation* (which would describe population growth or decay from one season to another). The same is true for *delay differential equations*, where growth retardation or stimulation is *not instantaneous* but takes *effect after some time has passed* [49]. Depending upon the length of the delay and upon the growth rate, solutions can become *oscillatory* and at certain values of r, *chaotic*. May, and Yorke, Smale, and Roessler, in the seventies, called attention to the phenomenon, that chaos is ubiquitous in nonlinear systems, and occurs already in the very simple and basic discretized logistics equation. Since then, dynamical chaos has been studied intensively. Chaos, limit cycles, and strange attractors are important in many fields: physics, meteorology, biology, and economics. For a historical account of the idea, see Gleick [50].

4.4 Controlling autocatalysis

The business cycle of booms and busts has been running for as long as anyone can remember. How can the cycle be dampened? Monetarists rely on a very simple tool: The "money supply" is autocatalytic. Central banks, like the Federal Reserve and the Bundesbank, can manipulate the growth rate of the money supply by raising and lowering interest rates. In the discretized logistics equations when, the intrinsic growth rate is small, growth is steady, when it is larger, periodic oscillations appear, at even larger values the trajectory becomes chaotic. To avoid large oscillations and chaos, both growth and decline must be kept within bounds.

The Federal Reserve monitors the money supply and attempts to control its rate of growth by manipulating the Federal discount rate. The "Fed" controls only a *single currency, a single interest rate*. Real estate prices are autocatalytic, too, and have their own cyclic dynamics. These interacting dynamical systems fit a notion of *synergetics*, the interaction and "enslavement" of different auto-catalytic dynamical system by some dominant "organizing parameters" [51].

The European monetary system is currently struggling to keep fixed ex-change rates between currencies of different countries, that have different rates of inflation, different interest rates, and different levels of unemployment. The Maastricht treat envisions "convergence" of all these parameters in different countries, and a single currency, and a strong Euro-bureaucracy in Brussels. Events in ethnically diverse countries, notably Russia and Yugoslavia, and to a lesser degree in Canada, Belgium, and Spain show that people are not willing to sacrifice their local cultural identities, which go with territory and language, to a modern super-state. It would be interesting to see a rational analysis, that is mathematical models simulating various scenarios of retaining multiple national currencies. Can the common market live indefinitely with *different currencies, with different interest rates, margin requirements, different rates of inflation?*

Carrying this idea a bit further: Within the same country, could there be different kinds of currency: one kind of currency for real estate, (whose price cycles are very unsettling), others for savings, for pensions, for food? Conversion rates could change. The savings and pension currency should retain its value. Profits could be taxed at different rates.

Such a monetary system would be more akin to the *interwoven autocatalytic hypercycles* that have been studied, both experimentally and theoretically, in the context of prebiotic evolution, by Manfred Eigen [52], Hofbauer, Schuster [53], Sigmund [54], and others. These models could be translated into models of economies, markets and firms with the objective of understanding economic cycles in the framework of a general theory of autocatalytic phenomena.

Perhaps a "hypercycle economy" with multiple currencies would give the central bank more control than the "grass roots" want to surrender. The question is: has anyone thought about (simulated) such a model? Or is Europe, with the demise of fixed exchange rates, with different rates of inflation in its member countries, with the right of citizens to have bank accounts and loans in different currencies, already on the way towards such a system? How long will it be till governments discover that currency conversions could be a splendid source of new tax revenue, as well as controls? Shouldn't these questions be thoroughly analyzed before events run their course, driven by the same kind of "logic" that billions of years ago shaped autocatalysis at the origin of life?

4.5 Markets and information

Granted that supply and demand should be mediated by markets. Some markets, like stock markets and commodity markets are well organized with instant quotations, transaction records and clearing mechanisms for the bids of buyers and sellers. What would the world be like if, for example, cars were sold the same way? What a waste to look through newspaper ads, haggle with salesmen, trying to determine which dealer offers the best buy! Why not have a computerized search for what is available at what price, and where. If we can search in seconds through hundreds of stock quotations and transactions, why not for cars, and appliances, and furniture, computers, TVs, and anything else for that matter? Are people ready for this? Or is the hullabaloo, the hype, the color, the whole advertising culture, a part of life that fills emotional needs, too important to give up?

4.6 The information society

Economics, once upon a time, was barter, like chemical kinetics in a prebiotic system. Then came money and credit. Now we are in a transition to an information society. Commodity options are bought and sold without taking actual delivery of pork bellies, or soy beans, etc. In many stores sales are immediately entered into the computer, which keeps track of supply and demand, and which may reorder automatically. Why should we go to a store? Perhaps for entertainment, but soon coaxial cable and individual video should make actual shopping trips superfluous. Combine this with computerized comparison shopping. An information society, where all material transactions are overlaid by informational transactions, seems to be on the horizon. We may not be ready for it, and may not see the consequences how it will affect the quality and forms of life. *Technologically the information society is feasible today.*

Prebiotic, autocatalytic systems have not survived into the present. Information molecules, RNA and DNA have taken over. Cells are miniature computers [55]. DNA stores the "systems software" and the instructions for making enzymes, structural proteins, receptors, regulatory proteins, and more. Information dominates. However, as an old saying goes, what God created during the day, the devil spoiled at night. There must have been parasites right from the origin of life [56]. Viruses are "information criminals", they are composed of RNA and DNA and a minimum of additional equipment that is necessary to get into a target cell. Once inside, they insert their own programs and subvert and exploit the host cell's metabolism towards their own replication.

Will there be similar phenomena in the information society? Computer viruses, which invade computers and destroy files already exist, like the infamous 1992 Michelangelo virus. This seems to have been the work of mischievous pranksters. The possibility, however will not be lost on creative criminals, who will try to invade computers for their personal gain, to divert funds and the shipments of goods. "Antiviral software" sold briskly at the time when the Michelangelo virus was scheduled to strike. Forecast: this will be a growth industry.

5 Neural Nets

5.1 Prolog

This author, some thirty years after his initial suggestion of applying genetic algorithms to perceptron training, in collaboration with R.W. Anderson, found a possible answer to the question how the *human brain adjusts* its synapses [57]. The following is a summary, and a general discussion of neural nets as self-organizing systems.

5.2 Death and resurrection of a paradigm

In the early sixties many groups, perhaps as many as a hundred world-wide, experimented with perceptrons, applied to anything from recognition of hand printed characters, speech, weather prediction, to the recognition of patterns in geological data indicative of oil bearing strata, and more. Results with simple perceptrons proved disappointing, and practical or analytical results for multi-layer perceptrons never emerged.

Then came Minsky and Papert's analysis "Perceptrons" [58], a book that could have been entitled: "What Simple Perceptrons Can't Do". It helped lay to rest research on *simple* perceptrons.

More than a decade later, Rumelhart, Hinton, McClelland, Sejnowski [59] and others had surprising successes with neural nets with "hidden units" (which correspond to multi-layer perceptrons). They *replaced the discontinuous step function of the linear threshold elements* (McCulloch and Pitts neurons), that Rosenblatt had employed, *by a continuous activation function* and continuous rather the binary response. In this way they were able to compute gradients. With optimization of net performance by steepest descent along gradients, an algorithm which has come to be known as "back-propagation of error", they obtained startling results [60]. Since then neural nets have taken a place in psychology ("connectionism") and computer science, and have found many commercial and scientific applications.

5.3 Back-propagation is unbiological

Unfortunately, computation of the gradient and adjustment of synaptic weights, through "back propagation of error", would require a neural network as complex as the "feed forward net", and the method becomes inefficient and incredibly complex in network architectures that allow lateral and feedback connections. Hinton [61], one of the pioneers of neural networks, acknowledges that "as a biological model, back-propagation is implausible". Mel [62] asks, "is it, then, a fundamental law that neural associative learning algorithms must be either representationally impoverished or mechanistically overcomplex?" In an article "The recent excitement about neural networks", Francis Crick [63] writes:

"It is hardly surprising that such achievements have produced a heady sense of euphoria. But is this what the brain actually does? Alas, the back-prop nets are unrealistic in almost every respect ... *Obviously what is really required is a brain-like algorithm which produces results of the same general character as back propagation*" [emphasis added].

5.4 How the brain adjusts synapses – maybe

In a recent paper Bremermann and Anderson [64] have described such an algorithm. We have experimented with a method that is *so simple and straight forward* that it seems *ideal for implementation in a biological structure*. DNA can generate neural circuits through morphogenesis (though evolution would have taken many millions of years). Thus, in principle, DNA could have organized back-propagation networks in the brain. But, there is a trade-off between simplicity and whatever increased performance might be obtained from complex neural circuitry. The algorithm that we proposed is of utmost simplicity. This is its merit. It does not require much "genetic information" (as would back-propagation).

The algorithm that we have proposed and investigated, is a *random walk through the space of synaptic weights*. It relies on random *trial fluctuations*. They have a Gaussian distribution and can be *generated locally at the synapses*. Trial fluctuations that are "successful" (as measured by an independent observer circuit), are then *fixed by a global signal:* "To all synapses: *Freeze the last trial fluctuation and make it the starting point for new trials*".

How can such a simple rule possibly be substitutable for back-propagation? We were able to show through simulations, that this algorithm, augmented by a few additional features, *performs almost as well as, and sometimes better, than back-propagation*.

5.5 Results

We have successfully used Gaussian random walks and a modification, called "chemotaxis algorithm", for the training of nets for a variety of tasks: discrimination of earthquakes from underground explosions [65, 66] generation of time-varying signals from "recurrent networks", which allow lateral and feedback connections, and the generation of control signals for open-loop, dynamical systems [67].

For the latter tasks no comparative performance data for the back-propagation algorithm are available in the literature. To obtain a comparison we have applied our method to two benchmark problems: "n-bit parity" and the "encoder problem". The former is of maximum difficulty and statistics of the performance of back-propagation (with fine tuned "momentum term" and "learning rate") [68] are available in the literature. For "n-bit parity" performance of our algorithm is slower by a factor of 5 to 10 (independently of increasing n). The "encoder" is an identity input-output map through several intermediate layers whith fewer units and connections). For the encoder the "back- propagation" algorithm gets terribly complex, and our algorithm performs better, especially if there are several intermediate layers. For detailed data see Bremermann and Anderson [69].

This results are very encouraging since the essential point of optimization by Gaussian random walk is its utter simplicity of biological implementation.

5.6 Self-adjustment of topographic cortical projections

The encoder can also be seen as a model of how neural connections are established that project from the retina, through the optic nerve, the lateral geniculate, and several cortical projection areas. We propose that a Gaussian random walk may help explain how these maps can be "self-adjusting" such as to maintain and refine topographic projections.

5.7 Net expansion and contraction during training

We conducted experiments for n-bit parity (interactively, with simple random walks as well as chemotaxis) where *during the training of a net additional units are added or removed.* This process does not unduly disturb training progress. In fact, when a task proved too difficult for a net (not enough hidden units), adding units would help achieve ultimate convergence. Also, when training reached a prolonged plateau, the algorithm could be made to converge by reducing the net (and the problem) in the course of training, then expanding the net and training it on the full problem again.

These findings are consistent with observations by Merzenich [70] who showed that loss of sensory input leads to reallocation of the vacated space in the cortex. Our algorithm allows dynamic re-allocartion of space in the crowded neocortex, without disruption of basic capabilities.

Is this, then, how the brain trains its nets? The answer can only come experimentally. In the anatomy of the brain there is no evidence of complicated back-propagation circuits. The local synaptic modifying mechanisms are being studied intensively, as are global signals for synaptic modification in long term potentiation. However, due to the difficulties of neurophysiological observations, confirmation will take time.

6 Representation of Knowledge

6.1 Communications and representation of messages

When messages are received or sent, there must be high level neuronal representations in the brain that are the internal equivalents of the externalized images or words. They are supported by lower level structures, all the way down to individual connections between dendrites and axons and synapses of varying strengths. Before and after messages are received or sent, underneath the output level, operations on stored representations of knowledge occur that are not necessarily conscious.

Whatever these deep structure representations of knowledge are, at the communication end they must be matched to the *media of communications,* which are thus a *window* on high level neuronal processing. In the following we make a case, that not only *spoken language* is such a window, but that other media, which are based more on *"imagination"*, (icons, pantomime, gestures) are important too.

6.2 Hominid communications: analog to digital

Hominids are the "missing links" between ape-like ancestors, several million years ago, and Homo Sapiens sapiens, anatomically modern man, who replaced Neanderthals some 50,000 to 100,000 years ago.

Whether Neanderthals could speak like Homo Sapiens sapiens, is a subject of hot debate. Neanderthal brains were as large or larger than modern brains, but their is some doubt about the anatomy of the larynx and the resonant cavities that produce speech sounds in modern humans [71]. It is clear, however, that speech could not have been the first medium of communications in the evolution of hominids, imitative gestures and imagery must have evolved first [72, 73].

Why has this early medium been partially replaced? The reasons for speech getting established are technical, they are based on the possibility of *communication in the presence of noise* through error correction, when *transmission is digital*. The reasons for speech being superior are essentially the same why *digital sound is superior to analog sound,* why digital recordings are better than analog recordings, why compact disks are better than analog cassettes, or vinyl records.

Shannon's communication theory [74] has shown the way how digital messages can be communicated over noisy channels, and restored, through error correcting codes, to original, pristine purity. Such error correction is made possible by elaborate encoding [75] and decoding, which requires computation, and hence computer chips [76]. Analog signals do not require such coding and decoding chips, and the analog technology developed when computer chips were not yet available. The disadvantage of analog sound is that it becomes contaminated by noise when transmitted over noisy channels, and the contamination cannot be removed.

Speech is digital. Words are transmitted as strings of phonemes (rather than strings of bits or bytes). All of the several thousand languages on earth employ only about 30 distinct phonemes. Shannon's theory works the same for binary digits as for any number of symbols, as long as they are distinct and finite in number. The principles for error correction are the same: they require redundancy, like "check bits" that allow the receiver to restore the original strings when there are errors or dropouts. Speech can tolerate as much as 50% errors and dropouts and still remain intelligible [77]. *Speech can be understood in this way over considerable distances, against ambient noise, during day and night, in any kind of weather.*

On the down side: *the speech code is not genetic, it must be learned, it varies from tribe to tribe, changes in time.* For example, New Guinea, which was settled only a few thousand years ago, has developed hundreds of distinct languages.

Communication through imitative gestures and icons requires little or no learning. When people meet, who have no spoken language in common, they can fall back on the older media. Icons are also very popular in communicating with personal computers and for the same reason: they are much easier to remember than the often arcane lists of typed computer commands.

6.3 Reason and intuition

The human mind can reason abstractly and comprehend intuitively. The laws of reasoning are well known, the mechanisms of intuitive insights are not. They are based on cognitive capabilities that are not well understood and cannot be simulated well by machines. Computers are excellent at logical calculations, poor at cognition, pattern recognition, speech recognition, reading hand-written characters, etc.

Reasoning power is a novelty in the evolution of the mammalian brain. Aristotle [78] rated it as the highest form of intelligence, possessed only by humans. In contrast vision, touch, and taste are capabilities that have been refined through tens, maybe hundreds of millions of years of evolution. Humans have not lost these capabilities, on the contrary, they form the basis of intuitive intelligence.

6.4 Representation of knowledge

Before words are spoken, they must have some kind of representation in the brain of the speaker, and likewise in the brain of the listener, so he can relate the received phoneme strings to tokens in his network of neurons. Such tokens may be thought of as groups of neurons, connection or activation pattern Words do not stand alone. They form conceptual networks (also known as Quillian nets) [79]. Logical reasoning draws upon such networks, and logical inference can derive novel connections in the network. The laws of logical reasoning were first formalized by Aristotle.

How is *intuitive knowledge* represented? Little is known about this. It is clear however, that the sender of a descriptive gesture has some sort of physical representation of an image inside him (or her), which is in turn connected to other internal representations of remembrances of images of things and events, prototypes, and condensed prototype examples. Both media of communication, speech and imagery, must have corresponding physical structures in the neural nets that are at the transmitting and receiving end of communications. In other words, paraphrasing an expression by Marshall McLuhan: *The medium is the representation of knowledge.*

6.5 Reality versus potentiality

Many languages make a distinction between *indicative and subjunctive.* The former mode is used when subject matter is asserted as *fact,* the latter when the statement may or may not be true, when one is talking about *potentiality rather than actuality.* In English the distinctions between indicative and subjunctive are not as sharp as in Latin, or Spanish, where the subjunctive is very strong.

We propose that the intuitive side of the brain, like the logical side, has a mode that makes a distinction between that what is real and that which is not, between actual behavior and possible behavior, whether it is acceptable in accordance with internalized norms, or whether it is wishful and possibly illicit behavior.

Distinctions between possible and sensible or permissible actions are important, both in dealing with the physical and the social environment: Even under Stone Age conditions people could not always have acted impulsively, but had to foresee the consequences of their actions.

6.6 The Self

The innermost agency that is in charge of behavior and which must mediate between conflicting drives, instincts, taboos, goals and self-images of one's destiny is THE SELF. The self processes experiences, especially traumatic ones, and integrates them into its general body of policies. It may be caught in contradictions between goals, convictions, and instincts, or may have been traumatized by unfortunate experiences.

A person who cannot resolve these conflicts seek psychiatric treatment that is designed to reveal and to resolve these contradictions. However, even in the well adjusted person the self constantly processes recent experiences and integrates them with its earlier store of policies and with basic, internalized goals and drives.

In order to be able to separate that what is *real and acceptable* from *potential scenarios,* the self employs something *analogous to the subjunctive mode.* However, the self is an evolutionarily ancient institution, hence it employs imagery rather than language. *How could imagery distinguish what is real from what is not?* The answer, we suggest, can be found when we examine the "imagination" of dreams.

7 The Self

7.1 Virtual reality

There are two kinds of sleep: REM ("Rapid Eye Movement") sleep and ordinary sleep. REM sleep is accompanied by rolling of the eyeballs, a distinctive pattern of brain waves, and immobilization of exterior motor activity: the body remains, as if paralyzed, rigidly in the same position. REM sleep alternates with ordinary sleep 4 to 5 times a night and lasts 10 to 40 minutes each period. (REM stages get progressively longer) [80].

When a sleeper is aroused by a highly emotional dream, or by an external stimulus such as a loud noise, a touch, etc., the current content of ongoing activity may become aware and may be remembered.

If a sleeper is awakened during successive periods of REM sleep, a thematic progression through the night may be observed. Past events are processed with some delay. The "post traumatic stress syndrome" is a familiar example. The unconscious mind may come to terms with traumatic events days, weeks, months or even years after they happened.

Data processing, especially of images, takes time. Off-line processing makes sense. "Dream work," as Freud called this process, may be incompatible with the computational requirements of the "quick response," alert state of the wakeful mind, when a person (or animal) is actively interacting with the environment.

REM sleep is not limited to humans, but has been observed in most mammals, though the percentage and timing vary between species. According to Winson [81], REM sleep first arose about 140 million years ago in a common ancestor of the marsurpial and placental mammals. It may be interesting to note [82], that REM sleep is affected by magnetic fields which are able to induce symptoms mimicking REM sleep deprivation [83].

During ordinary sleep, humans frequently change position, which relieves strain on muscles. When awakened, sleepers report only "mild ideation" versus the sometimes dramatic images of dreams. People can train themselves to remember the content of their dreams routinely. This is known as "lucid dreaming." Not all the dreams are violently emotional, but a common denominator is an eerie quality of unreality.

Dreams have something in common with "virtual reality," a new creation of the computer age. To experience the latter, a viewer must wear goggles that project stereo images on small TV screens inside the goggles. The viewer also wears a glove filled with sensors that let him manipulate the images that are projected in his goggles. In this way the viewer can interact with the images of objects in his field of vision, as if they were real. He can also manipulate an image of himself, can fly through space, escape from gravity, do all sorts of things, interact with "objects of the imagination." In other words, virtual reality has much in common with dreams. No harm is done by the imaginary actions, they are scenarios, simulations of possible, and perhaps impossible events and actions.

7.2 Two kinds of sleep, two modes of self-organization of knowledge

The two kinds of sleep constitute different modes of off-line processing. They correspond to the indicative and subjunctive modes of spoken language. One deals with actuality, the other with potentiality. One mode, during ordinary sleep digests fact and experiences as additions to the store of factual knowledge that is already in place. The other, during REM sleep deals with simulated behavior, with virtual reality.

While potentially dangerous actions are explored through hypothetical simulations, inhibition of all exterior motor activity is a must. In cats the inhibition can be inactivated by a lesion in the lower brain stem. During REM sleep these animals at times rise, attack and appear startled by invisible objects, while remaining unresponsive to real visible stimuli [84]. A video showing a cat "acting out" its dreams is commercially available [85].

The results of dream simulations must be evaluated and if necessary "censored" before they can become policy. Analogously, during the dark days of the cold war the Pentagon played computerized version of simulated strategic warfare. The answers could not be obtained by logical analysis of mathematical game models (such as the "Richardson model") alone. The games had to be as realistic as possible. At the same time any accidental launching of missiles had to be categorically inhibited. The inhibition of all exterior motor activation during dreaming serves an analogous purpose.

The hubris of the human species versus animal intelligence has generally belittled the accomplishments of animal brains. Many of the capabilities of image and sensory signal processing and motor control in animals are as sophisticated as those of their human counterparts, and sometimes more so. The human species, however, is heir to the *evolutionary breakthrough in communications:* from the 30-odd innate vocal and postural signals with fixed meanings, that many species have at their disposal, to the *semantically open media of iconic communications and speech.* Once this step had been taken, brains could process vast new expanses of knowledge and transcend into realms beyond the confines of the "software" that fits into the limited memory of the genes.

7.3 Surrealism and self-organization

Communication through imitative gestures requires the ability to guess meaning from sketchy and abstracted renditions. When speech developed, perhaps no earlier than 100,000 years ago, this interpretative capacity would have been "unemployed" or "under employed," had it not been adapted to a new function, the interpretation of metaphors. Metaphors in ordinary language, represent pieces of reality by analogy.

The images in dreams, are not only metaphors, they do not represent reality, or actual policies of the self, but fantasies, simulations of "what would happen if." They also play out battles between conflicting drives, instincts, fragments of the personality. The images do not represent themselves, they are never to be taken literally. They are tainted with a mode of surrealism that, like the subjunctive in speech, should prevent confusion between the real self and simulations that are part of the self-improvement, the self-organization of the self.

Why is the self in need of development? Genes can program instincts and drives. However, not enough information fits into the genetic memory for flexible, intelligent behavior. Mammals improve their behavior through learning. In humans programmability of personality in response to experience and socialization has reached new highs, but at a price.

The metaphorical images of the self are open to *confusion between literal and figurative meaning.* A breakdown of the barrier allows *metaphorical substitutions* for the targets of drives ("sublimation"). Real hunger could become hunger for possessions or knowledge. Love for one's family could be retargeted towards the tribe, a nation, or all mankind. Many great scientists, artists, musicians, overachievers of many kinds, must have retargeted their drives. Freud speaks of the *sublimation of drives,* especially the sex drive. This can produce *neurosis,* but *sublimation is functional. It provides society with all kinds of talents that are useful for the family, the tribe, or society at large.*

The mind also has become receptive to symbolic, dreamlike images from the outside. Initiation ceremonies that mark the end of childhood, are laden with ritual symbolism, in tribal ceremonies around the world. We suggest, that at the same time when our ancestors became receptive to symbolism, they also began to *externalize their fantasies* through paintings, sculptures, carvings, adornment, body painting, etc. All this appears with anatomically modern man. A "creative explosion" occurred some 50,000 years ago [86].

Not all artistic expression has been benign: It is hard to rationalize human sacrifice as having any redeeming social significance. It is difficult to see the humanity of ritual mutilation.

The rational mind, as evaluator and censor, can get into battles with the unconscious mind. At their best, the two capabilities work in tandem, each contributing what each side does best. At worst, the bizarre fantasies are acted out, or the conflict leads to suppression of the intuitive mind which then may take revenge by inducing neuroses and symptoms of physical disease through somatic conversion.

We suggest that the self needs a model of its functioning in order to avoid unnecessary internal conflicts. Freud and Jung contributed much towards this end. However, some 90 years have passed since Freud's "Interpretation of Dreams" [87]. Since the fifties, computer technology, artificial intelligence, and self-organizing systems have made giant strides. There should be a synthesis [88, 89].

7.4 The origin of symbols

"Träume sind Schäume" – "dreams are nothing but foam and bubbles" goes a saying, which is a denial, not a statement of fact. Nearly a century ago Freud and Jung discovered the power of dream interpretation for psychoanalysis, and the similarity of dream symbolism with other instances of symbolism, especially in mythology, religion, and art. Together they developed the beginnings of psychoanalysis and the practice of interpreting dream symbols and other forms of symbolic expression of their patients.

They disagreed about the origin of the symbols they encountered. Their disagreement was so fundamental that it broke their friendship. Freud proposed the "sexual theory": Dreams are hallucinatory wish fulfillment, especially of illicit sexual desires. A punishing "censor" watches over the goings-on in the unconscious mind, which, however, can fool the censor by means of disguises (the symbols) which the censor has difficulty to penetrate. When it does, it tries to block the illicit fantasies, and the unsatisfied drives search for substitutes that are more acceptable to the censor. Substitutes, however, are not the real thing, and the "ego" remains in a state of more or less mild dissatisfaction. The continued struggle can be creative, energizing great accomplishments, or it can lead to neurosis and somatic symptoms.

Jung de-emphasized the sexual theory. He saw dreams as containing messages to the conscious mind to restore mental equilibrium. The same symbols that occur in dreams are found in myths around the world, and there is a striking similarity even when the tribes and peoples involved have had no contact with each other, due to geographic or historical separation. Where, then do these myths and symbols come from? Are they inherited, but how?

Jung opted for a theory of the origin of symbols in past experiences of the species. He did not rule out something like extrasensory perception to explain how a contemporary mind would capture past events. To Freud this was anathema: "My dear Jung, promise me never to abandon the sexual theory. That is the most essential thing of all. You see, we must make a dogma of it, an unshakable bulwark ...against the black tide of the mud of occultism" [90, 91].

The lack of agreement has not prevented psychotherapy from flourishing. Both Jungians and Freudians interpret symbolic messages from then unconscious mind. The rational mind does not have the metaphorical capabilities of the unconscious mind (which therefore can "fool the censor" by metaphorical disguises). The unconscious mind, on the other hand, may be unable to extricate itself under its own power from the contradictions that prevail between drives, experiences, values, and conflicting demands.

The therapist, like a mediator in a conflict, can bring the various sides together, he can bring to "light" (make conscious) traumatic events that have been suppressed because they are too terrible to be faced, which, however, form part of a person's store of experiences and knowledge and cannot simply be "forgotten". Or, the therapist may, by role playing (assuming the role of a father or mother figure), or by reaching the mind with his own symbolic messages ("You don't need a penis to drive a bus"), help to straighten out the unconscious mind.

The process takes time: Like the systems software in a computer, the self, the structures that define a personality, can be modified only at the danger of disastrous consequences. Systems software is usually well protected against careless modification. Hence it is not surprising that analyses take time. Sometimes, however, a bit of "debugging", an interpretation that helps understanding the meaning of some cryptic message from the unconscious, can be helpful to quickly restore, as Jung proposes, "mental equilibrium". This point is emphasized by Erich Fromm [92], who points to dream interpretation in the Talmud, which says: "A dream which is not understood is like a letter which is not opened" [93]. Also, according to Fromm, "the Talmudic interpretation is apparently based on the idea that a symbol always stands for something else" (p 129).

For the practice of psychotherapy, answers to philosophical questions about the origin of symbols have few operational consequences [94]. However, for understanding our minds, our culture, our myths, our religions, our rituals, and folklore, a model of the SELF is indispensable.

References

1. W. Ebeling and R. Feistel, *Physik der Selbstorganisation und Evolution*, Akademie-Verlag, Berlin, 1982

2. I. Rechenberg, Evolutionsstrategie - *Optimierung technischer Systeme nach Prinzipien der biologischen Information.* Friedrich Frommann Verlag, Stuttgart-Bad Cannstadt, 1973.
3. M. Conrad, *Adaptability: the significance of variability from molecule to ecosystem,* Plenum Press, New York, 1983.
4. G. Nicholis and I. Prigogine, *Self-organization in non-equilibrium systems: from dissipative structures to order through fluctuations,* Wiley, New York, 1977.
5. I. Prigogine and I. Stengers, *Order out of chaos: man's new dialogue with nature,* New Science Library, Boulder, CO: Distributed by Random House, 1984.
6. H.Haken, these Proceedings.
7. S. Levy, *Artificial Life,* Pantheon, New York, 1992.
8. M. C. Yovits and S Cameron, editors, *Self-organizing Systems,* Pergamon Press, New York, 1960.
9. F. Rosenblatt, *Principles of Neurodynamics,* Spartan Books, Washington, D.C., 1962.
10. D. E. Rumelhart and J. L. McClelland, eds., *Parallel Distributed Processing, Vol 1 and 2,* MIT Press, Cambridge, MA, 1986.
11. J. J. Hopfield and D. W. Tank, Computing with Neural Circuits: A Model, *Science,* 8 August 1986: 625-633.
12. Aristotle, *Psychology, De Anima (On the Soul),* Book II.
13. M. L. Minsky, *Computation: Finite and Infinite Machines,* Prentice Hall, Eglewood Cliffs, NJ, 1967.
14. M. A. Arbib, *Brains, Machines, and Mathematics,* McGraw-Hill, New York, 1964.
15. M. L. Minsky, Steps towards Artificial Intelligence, *Proceedings of the IRE, Special Computer Issue,* p 8-30, 1961.
16. H. J. Bremermann, Optimization through Evolution and Recombination, Part I. Limitations on Data Processing Arising from Quantum Theory, in: M. C. Yovits, G. T. Jacobi, eds., *Self-Organizing Systems 1962,* Spartan Books, Washington, DC, 1962.
17. J. D. Bekenstein and M. Schiffer, Quantum limitations on the storage and transmission of information, *International J. of Modern Physics, Part C,* 1: 355-422, 1990.
18. R. Penrose, *The Emperor's New Mind,* Oxford University Press and Penguin Books, New York, 1989.
19. C. C. Lin and L. A. Segel, *Mathematics Applied to Deterministic Problems in the Natural Sciences,* Macmillan, New York, 1974, p 5.
20. H. J. Bremermann, Complexity and Transcomputability, in R. Duncan and M. Weston-Smith, *The Encyclopaedia of Ignorance, vol.1, Physical Sciences,* p 167-174, Pergamon Press, Oxford, 1977.
21. Book review by M. Gardner: J. P. King, *The Art of Mathematics,* Nature, 2 July 1992.
22. H. Margenau and R. A. Varghese, *Cosmos, Bios, Theos,* Open Court, LaSalle, IL,1992.
23. H. J. Bremermann. Book manuscript in preparation.
24. J. Winson, *Brain and Psyche,* Random House, New York, 1985.
25. D. L. Chester, *Formal logic and the representation of linguistic deep structure,* Ph. D. Thesis, Dept. Mathematics, Berkeley, CA 1973.
26. M. Eigen, J. McCaskill, P. Schuster, Dynamics of Darwinian Molecular Systems, *J. Phys. Chemistry.*

27. M. Eigen and P. Schuster, *The Hypercycle: A Principle of Natural Self-Organization*, Springer-Verlag, Berlin 1979

28. P. F. Stadler and P. Schuster, Mutation in autocatalytic reaction networks–an analysis based on perturbation theory. *J. Math. Biol.* 30: 597-632.

29. S. Levy, *Artificial Life*, loc.cit.

30. S. Levy, *Artificial Life*, loc.cit.,p 184-185.

31. J. H. Holland, Concerning Efficient Adaptive Systems, in M. C. Yovits, G. T. Jacobi, G. D. Goldstein, eds., loc. cit.

32. H. J. Bremermann, *The Evolution of Intelligence. The Nervous System as a Model of its Environment.* ONR Report no.1, contract Nonr 477(17), University of Washington, Seattle, WA, 1958.

33. R. M. Friedberg, A learning machine, part I and part II, *IBM Journal of Research and Development*, 2:2-13, and 3: 282-287.

34. H. J. Bremermann, Optimization through Evolution and Recombination, loc. cit.

35. H. J. Bremermann, M. Rogson, and S. Salaff, Search by Evolution , in M. Maxfield, A. Callahan, and L. J. Fogel, eds. *Biophysics and Cybernetic Systems*, Spartan Books, Washington, DC, 1965.

36. H. J. Bremermann, M. Rogson, and S. Salaff, Global Properties of Evolution Processes, in H. H. Pattee, E. A. Edelsack, L. Fein, and A. B. Callahan, eds. *Natural Automata and Useful Simulations*, Spartan Books, Washington, DC, 1966.

37. H. J. Bremermann, Sex and Polymorphism as Strategies of host-pathogen interactions. *J. Theor. Biol.* 87: 641-702.

38. H. J. Bremermann and J. Pickering, A Game Theoretical Model of Parasite Virulence, *J. Theor. Biol.* 100: 411-426.

39. H. J. Bremermann, The Adaptive Significance of Sexuality, in S. C. Stearns, ed., *The Evolution of Sex and its Consequences*, Birkhäuser Verlag, Basel, 1987.

40. D. Hamilton, Sex versus non-sex versus parasite. *Oikos* 35: 282-290.

41. H. J. Bremermann, paper in preparation.

42. H. J. Bremermann, The adaptive significance of sexuality, loc. cit..

43. H. J. Bremermann, paper in preparation, loc. cit.

44. F. Wong Staal, The AIDS Virus-What we know and what we can do about it. *Western J. Medicine,* 155: 481-487, 1991.

45. R. A. Wever, Light effects on human circadian rhythms a review of recent Andechs experiments. *J. Biol. Rhythms*, 4: 161-185, 1989.

46. R. K. Mishra, these proceedings.

47. A. S. Perelson, Immune network theory. *Immunol. Rev.*, 110: 5-36, 1989.

48. F. J. Varela and A. Coutinho, Second generation immune networks. *Immunology Today*, 12: 159-166, 1991.

49. R. M. May, *Stability and complexity in model ecosystems*, Princeton University Press, Princeton, 1973.

50. J. Gleick, *Chaos: making a new science*, Viking, New York, 1987.

51. H. Haken, these Proceedings.

52. M. Eigen, *The hypercycle, a principle of natural self-organization.* Springer-Verlag, Berlin, 1979.

53. J. Hofbauer, P. Schuster, and K. Sigmund, Competition and cooperation in catalytic self-replication. *J. Math. Biol.* 11: 155-69, 1981.

54. K. Sigmund, A survey of replicator equations, in J. L. Casti and A. Karlqvist, eds. *Complexity, Language and Life: Mathematical Approaches. Biomathematics Vol. 16,* Springer Verlag, Berlin, 1986.

55. M. Conrad, these Proceedings.

56. H. J. Bremermann, Parasites at the origin of life. *J. Math. Biol. 16:* 165-180, 1983.
57. H. J. Bremermann and R. W. Anderson, How the brain adjusts synapses–maybe. In Boyer, edtr. Automated Reasoning, Kluwer, 1991.
58. M. Minsky and S. Papert, *Perceptrons,* loc. cit.
59. Rumelhart and McClelland, *Parallel Distributed Systems,* loc. cit.
60. H. J. Bremermann and R. W. Anderson, loc. cit.
61. G.E. Hinton, Connectionist Learning Procedures, *Artifical Intelligence* 40(1): 143-150, 1989.
62. B. Mel, *Connectionist Robot Motion Planning,* Akademic Press. Boston, New York, 1990.
63. F. Crick, Therecent excitement about neural networks, *Nature,* 1989.
64. H. J. Bremermanmn and R. W. Anderson, loc. cit.
65. F.U. Dowla, S.R. Taylor and R.W. Anderson, Seismic discrimination with artifical neurtal networks: Preliminary results with region spectra data. *Bulletin of the Seismological Society of Anmerica* 80(5): 1346-1373, 1990.
66. R. W. Anderson, Ph. D. Thesis, University of California, Berkeley, 1991.
67. R. W. Anderson, loc. cit.
68. Tesauro and B. Janssens, Scaling Relationships in Backpropagation Learning. *Complex Systems* 2: 39-44, 1988.
69. H. J. Bremermann and R. W. Anderson, loc.cit.
70. M. M. Merzenich, G. Recansone, W. M. Jenkins, T. T. Allard, and R. J. Nudo, Cortical Representational Plasticity. In P. Rakic and W. Singer, *Neurobiology of Neocortex,* Dahlem Conference Report No. 42. Wiley - Interscience, New York, 1988.
71. A. Gibbons, Neandertal Language Debate: Tongues Wag Anew. *Science,* 256: 33-34, April 3, 1992.
72. P. Lieberman, *Uniquely human: the evolution of speech, thought, and selfless behavior,* Harvard University Press, Cambridge, MA, 1991.
73. H. J. Bremermann, book manuscript in preparation.
74. C. E. Shannon, *The mathematical theory of communication,* University of Illinois Press, Urbana, IL, 1949.
75. E. R. Berlekamp, *Algebraic coding theory,* McGraw-Hill, New York, 1968.
76. J. Adamek, *Foundations of coding: theory and applications of error-correcting codes, with an introduction to cryptography and information theory.* Wiley, Chichester, NY, 1991.
77. L. Brillouin, *Science and Information Theory,* Academic Press, New York, 1962.
78. Aristotle, De Anima, Book II, loc. cit.
79. J. F. Sowa, *Conceptual Structures: Information Processing in Mind and Machine,* Addison-Wesley, Reading, MA, 1984.
80. J. Winson, *Brain and Psyche,* loc. cit.
81. J. Winson, loc. cit.
82. R. K. Mishra, these Proceedings.
83. R. Sandyk, N. Tsagas. P. A. Anninos, K. Derpapas, Magnetic fields mimic the behavioral effects of REM sleep deprivation in humans. *Internat J. Neuroscience,* 65: 61-68, 1992.
84. J. Winson, loc. cit.
85. *Dream voyage,* 28 minute video, Films for the Humanities and Sciences, P.O. Box 2053, Princeton, NJ 08543, 1992.
86. J. E. Pfeiffer, *The Creative Explosion. An Inquiry into the Origins of Art and Religion.* Harper and Row, New York, 1982.

87. S. Freud, *Die Traumdeutung*, Wien 1900, Transl.: *The Interpretation of Dreams*, Standard Edition 4-5, London 1954, Science Editions, Wiley, New York, 1961.
88. R. Haskell, *Cognition and Dream Research*, reviewed by J. Montangero in *New Ideas in Psychology* 9(1)¿ 117-125, 1991.
89. D. Riemann, Traum Interpretation and experimentelle Traumforschung (A survey). *Zeitschrift für Psychosomatische Medizin und Psychoanalyse*, 36:21-38, 1990.
90. G. Wehr, *Carl Gustav Jung, Leben, Werk, Wirkung*, Koesel-Verlag, Munich, 1985, Translation: *Jung, a Biography*, Shambala Publications, Boston, MA, 1987.
91. C. G. Jung, recorded and edited by A. Jaffe, Random House, New York, 1961.
92. E. Fromm, *The forgotten language*, Grove Press, New York, 1951.
93. M. Niehoff, A dream which is not interpreted is like a letter which is not read (Biblical dream interpretation and rabbinic exegesis in the Pesher-Habakuk (Phab at Qumran), *J. of Jewish Studies*, 43: 58-84, 1992.
94. C. Rycroft, *Psychoanalysis and Beyond*, University of Chicago Press, Chicago, IL, 1985.

On the Evolution of
Open Socio-economic Systems

Friedrich Hinterberger

Abstract

Some applications of evolutionary and self-organization models to economic problems are outlined along with their major limitaitions, which require a broad concept of social science modelling. Among the concepts discussed are neural nets, systems dynamics, synergetics and Darwinian evolution.

The purpose of this paper is to explore the range of possible uses of the self-organization approach for social science applications, and especially for economics. After some general remarks (1), I will present two special cases. Neural nets could improve our understanding of individual economic decisions (2), while an analogy to biological evolution can help to explore certain aspects of economic processes on a societal level (3).

1 Self-organization in economics

Friedrich von Hayek, a Nobel Laureate in economics, claimed that social and economic order, such as law, language, and markets, are the "results of human action but not of human design" [1]. This means that he places the development of economic order between natural and artificial phenomena. He relates this view to Adam Smith, the founder of economics as a science, who used the famous metaphor of an "invisible hand", which drives society to an end which was no part of the individuals' intention. One famous passage reads as follows: The individual agent is "as in many other cases, led by an invisible hand to promote an end which was no part of his intention. Nor is it always the worse for the society that it was not part of it. By pursuing his own interest he frequently promotes that of the society more effectually than when he really intends to promote it" [2].

Although such a process is based on the private intentions and actions of the participating individuals under specified circumstances, the outcome is shown to be the product of a "process that aggregates the separate and 'innocent' actions of numerous and dispersed individuals into an overall pattern which is the very phenomenon we set out to account for" (Ullmann- Margalit [3], p. 265). Hayek's main point is that neither individual participants in the market process nor external spectators (for example, a scientist) can have the knowledge necessary to predict the outcome of uncoordinated individual behavior. As a scientific method for describing such a world Hayek repeatedly refers to biological theories of complex phenomena. In terms of contemporary science we may talk about "self-organizing systems".

Unfortunately, throughout the history of economic thought this view has been narrowed down to what we understand today as neoclassical economics, which has become the economic mainstream. This stream of thought views the individual as well as certain aggregates as functioning according to well specified mechanisms. The claim is that the functioning of an economy can be sufficiently described deductively rather than inductively. Deductively means that economic processes and their results can be derived from general principles. The axiomatic concept of rational individual decision-making provides such a general principle. In an extreme representation, there is no micro-macro dichotomy. Macroeconomics on the level of a national economy is nothing but the mathematical aggregation of the microeconomic results on an individual level. Such a process is often said to take place in an equilibriating economic system that works infinitely fast so that situations of disequilibrium remain unconsidered. Other methods explicitly miss the point by postulating, for example, an exogenous auctioneer who allows for trading only if every participant feels satisfied. The most rigorous member of that school is maybe Gary S. Becker [4], the 1992 Nobel Laureate in economics. Philip Mirowski [5] and others have argued convicingly that the methodology used in this kind of economics is a direct analogy to classical mechanics in physics.

According to the up-to-date version, individuals are seen as optimizing certain prespecified target functions (profits, utility) under various constraints (budget, information, transaction costs, institutional restrictions). The discussions between various schools of economic thought predominantly tackle the proper formulation of these target and constraint functions, but there is a wide consensus that thinking about these issues in terms of individual rational decision-making in the sense of optimizing individual evaluations, be they subjective (as utility) or objective (as profits), does make sense. Individuals may have relations, but the relations are pre-determined and determine the results of the individuals' rational decisions. A contemporary version of this kind of economic methodology, the so-called neo-institutionalist approach, stresses the impacts of institutional settings on economic decisions. Transaction costs, property rights and other constraints, which are partially unknown, make rational decisons impossible. Though we still lack a proper axiomatic foundation of so-called bounded rationality, which would be necessary for a proper formulation of a deductive description/explanation of an economy in which transaction costs and property rights play a non-negligible role. The macroeconomic outcomes are just the mathematical aggregates of the behavior of individuals. But also most other macroeconomic theories, which describe and explain the behavior of certain aggregates (for example, national income, unemployment etc.), do that in an equally mechanistic fashion.

Such a description would play an important role within economic theorizing if it could be embeddied into a broader framework. Economic theory has solved almost all relevant problems which are solvable within the mechanistic realm; in doing so mainstream economists often went too far by neglecting qualitiative differences in dealing with economic processes. We have to differentiate between idealizations describing a frictionless market mechanism in analogy to bodies moving without friction and descriptions of real world phenomena, where friction is important. In the following I hope to give some good reasons why many economic processes cannot be described by static or dynamic theories in a Newtonian tradition. In the model presented in the third section of this chapter the standard framework has its place within a more comprehensive one, although the neoclassical research program extends much further than the variety of phenomena to be really explained on the basis of its restictive assumptions. These restrictions are so fundamental that it would not be a big loss to drop some of the claimed generality: if mechanics is restricted within natural science, why should it provide a general principle in economics. Biological evolutionary theory explains many more concrete phenomena on the basis of fundamental principles than any general mechanistic economic theory.

In the following I will explore whether modern, post-Newtonian concepts of natural science allow for a new round in some old discussions between orthodox and various schools of non-neoclassical economists on a higher methodological level. In "Order out of Chaos. Man's New Dialogue with Nature", Ilya Prigogine and Isabelle Stengers [6] state:

> "Our vision of nature is undergoing a radical change toward the multiple, the temporal, and the complex. For a long time a mechanistic world view dominated Western science. In this view the world appeared as a vast automaton. We now understand that we live in a pluralistic world" (p. xxvii).

In his foreword to the same book Alvin Toffler explicitly relates this work to social problems:

> "What makes the Prigoginian paradigm especially interesting is that it shifts attention to those aspects of reality that characterize today's accelerated social change: disorder, instability, diversity, disequilibrium, nonlinear relationships (in which small inputs can trigger massive consequences), and temporality - a heightened sensitivity to the flows of time." (p. xiv-xv)

This can be directly translated to economics.

According to Ulrich Witt ([7], p. 8) an evolutionary theory is dynamic, based on an irreversible, historic conception of time that explains how innovations arise from the developments examined and what implications they have. And Kurt Dopfer [8] states: a

> "social science, such as economics, receives its meaning by offering the central question of why individuals relate to each other in a certain way" (p.59).

For Günter Schiepek [9] an aim of social synergetics is to explain the formation of stable psychological, social and external structures (see p. 126). This corresponds to the fundamental economic research program: the formation and development of the order of the social market economy which is based on uncoordinated actions. Social self-organization in this sense implies that a society (the "system" society) by itself and for itself creates structures which change the system and develop it further. These structures such as the state, enterprises, associations and lobbies, create regulations for their own organization as well as for their co-operation (corporate rules, laws, but also markets) and institutions for their enforcement (see Ekkehart Schlicht [10], p.220). That these processes are self-organization processes means that they are not based on human plans, as contractarian approaches, for example, suggest.

Contemporary self-organization theory also seems to be too narrow a framework for successfully discussing economic problems. Prigogine/Stengers [6] recommend care with simple applications of the thermodynamic paradigm to social sciences. Ernst Mayr [11] states that biology has its autonomy within the sciences because of certain characteristics of the world of life, which he sums up under the headings 'uniqueness and variability', 'systems and their hierarchical organization' and 'possession of a genetic program' (see pp. 55-60). Within a united science, biology would be the broader concept to which all the statements of physics apply. We can easily extend this to social science: all physical and biological laws and processes apply to social processes but special aspects of human abilities, human culture and social life require a broader theory that also allows us to explain the specialties based on the specifically human ability of active, conscious and intelligent decision-making. Therefore a third branch of explanations should be added to physical and biological analogies, which enables us to go beyond and develop explanations that are purely based on social science arguments.

"Science is rediscovering time" (Prigogine/Stengers, 1984, p.xxviii); "entropy increases in the direction of the future not of the past" (Prigogine/Stengers, p. 125). This is the crucial statement of thermodynamics with regard to the concept of time from which the self-organization paradigm is derived. In social science, theories of learning as well as the assumptions about the endogeny or exogeny of preferences, technologies and institutions depend upon the proper concept of time. Traditional economic and physical theories describe a timeless world where processes do not play any substantial role – even if intertemporal aspects are involved. The usual "trick" to evade the problems of historical time is to discount the evaluations of future events so that decisions can be made in a single initial point of time for the whole relevant future time span. Technically speaking, we use the concept of depreciation and argue that individuals are able to evaluate the possible future outcomes of their actions. Hence these approaches do not really allow for novelty; everything is in principle given at the beginning of the process – be it latent or activated.

Reversibility of time is a crucial property of mechanistic models. The deterministic and reversible "economic laws" of standard economic theories describe theoretical constructions that are similar to the motion of a frictionless pendulum in physics. They are based on a logical concept of time, which makes it possible to reduce the problems of an uncertain future to technical ones. As a consequence of reversibility, men and women – creatures of time themselves – are strangers in this world and empirical results based on these theories can only be approximations without any profound meaning (Prigogine/Stengers in their foreword to the German edition of "Order out of Chaos" ["Dialog mit der Natur"], p. IV). Stable and periodical objects of mechanistic physics are regarded as special cases in an unstable and developing world, as border line cases of an extended physical theory (see ibid, p. I). Irreversibility is granted a central role by major concepts in modern science (see the contribution by Antoniou and Prigogine in this volume) and should be included in a serious economic analysis. A similar question has been amply discussed in economic theory in the context of historical versus logical concept of time and I shall try to stress that modern physical, chemical and biological theories take a position favoring the point of view of historical time. New species can appear and vanish after some time, new knowledge can be invented and learned as well as forgotten; but the disappearance of a species is quite different from its appearance (see Konrad Lorenz [11]) and the process of learning is totally different from the process of forgetting – they are not simply inverse to each other. The difference is negligible if we are only interested in comparing two situations descriptively. But if we want to discover the reason or the probability of a certain transition from one situation to the other, we have to understand the underlying processes. From this point of view economics has to deal with a co-evolution of individual, economic, social, political and natural processes [12]. Comparative static analysis and dynamics in a Newtonian sense shows the effect of changes of certain variables on certain other variables given certain functional relationships between these variables. Self-organization economics has to show how variables (all of them considered as endogenous) and the connections between them are changed by some kind of interaction, e.g. in analogy to the evolutionary principle of "natural selection". For example, the process described by Schumpeter using the concept of a creative entrepreneur can be understood in this context – and is by no means to be described within a neoclassical model. A neoclassical model seems to be methodologically adequate for a description of economic results under certain given circumstances and to compare expected results coming from different given circumstances but it can never describe the appearance of novelty.

2 Neural nets as a basis for an alternative decision theory

Standard microeconomic theory is based on the construct of optimizing individuals (firms and households) that are more or less secure about the data they have to compute, so that optimal decisions on whether to participate in a certain market can be made. They at least have in mind probabilities for all relevant future events. Whenever an agent has to decide on whether to supply or demand certain commodities, bonds or currencies, he or she needs data about prices and quantities on that market and on certain related markets (because the markets for complementary and substitutive goods will influence the decision) as well as information on political developments and other factors external to the economy. The decision is to be made on the basis of that data about how much to sell and/or buy and at what prices. This way every agent is assumed to optimize expected profits and/or expected utility subject to certain contraints, costly resources, incomplete information and other restrictions. Within standard neoclassical economics[1] , the "rational behavior" of economic agents plays a crucial role for economic modelling, although there seems to be a broad consensus that rational behavior marks an idealization described by a set of basic axioms. These axioms are only apparently very unrestrictive; on the contrary, they actually reduce human behavior to simple response automatons, the only task of which is to calculate an optimum given certain preferences and restrictions.

The behavior of "rational" agents – according to this theory – can be made plausible by the following example: if an agent takes part in a lottery, the probability of winning a large amount is usually quite small. Now the rational expectations approach requires that the agent will not be frustrated by a loss and will not revise his or her decision due to the loss. But in reality many people will quit gambling after some "bad" experiences (as long as they are not addicted to gambling, which is not rational either). They realize that they are wasting their money; what was a rational decision in the first place is now evaluated differently.

The most important problem with the expected utility approach is that, as the process takes place in historical time, it is impossible to gain adequate information about the future which can only be aquired in a trial and error process. In such a world it is insufficient to describe economic agents simply as response automatons, as standard economic approaches usually assume.

I will not go into the technical details of concrete neural network approaches. It seems obvious that we can compare our problem with the problem of pattern recognition. Eric Baum [13] presents an algorithm of "back propagation" that is designed to yield an expert system to simulate the performance of medical diagnosis. Given input vectors of symptoms and output vectors of illnesses, the system is able to "learn" to make decisions. Thus the neural network approach could provide a possibility to describe a process in which an individual gathers necessary information step by step in order to achieve an optimal pattern of decision. What an economic agent does in such a process is to somehow perceive reality, then try to filter out necessary information and make his or her decision. The environment will re-act somehow on the agent so that some perceived feedback allows him or her to evaluate the new state of the system (and his or

---

[1] By "neoclassical theory" I mean every theory including the assumption of economic rationality according to the basic axioms of the expected utility approach apply. I do not want to blame more sophisticated "neoclassical" models, which avoid certain shortcomings (also because many non-neoclassical models share certain shortcomings). The aim of this paper is not to criticize certain theories such as the neoclassical theory in general, but to argue in favor of alternative models which may deliver additional insights. So what we should consider is a range of various theories rather than a clear-cut dichotomy between neoclassical and non-neoclassical economics

her own position within the system) and re-act in a certain manner. In this way economic agents re-evaluate their own decisions. The problem is that they are unable to know exactly whether their decison was optimal or whether a better decision is possible. They do not even know the extent to which the perceived change of the environment was due to their decision or to other changes within the system or its environment. Not only does perceived reality change: perception may also change if other information is recognized to be more relevant. There are many uncertainties connected with such a process. Without a feedback mechanism that tells us how to evaluate our own decisions, rational decison making would be impossible. But we still lack a proper theory of consciousness from which behavior of that kind could be derived.

This kind of "rationality" is totally different from the one described by neoclassical theory. Here optimizing behavior is a process, which takes place in historical time. Maybe it is better to avoid the term and use the more general term "conscious, intelligent decision-behavior" to make the difference clear. If, and only if, the environment remains stable, will experience push agents into the direction of an optimal decision, so that eventually the individual would reach a pattern of decision which can be described "as if" it were an optimizing agent as described by standard economic theory. In a dynamic, ever-changing world, however, this is simply not possible. The main problem for modelling such a process is that even the response of whether the recognition is "right" or "wrong" may fail.

In addition to that, we observe that individuals relate their behavior to each other's. How does a society's culture emerge and change as a result of a multitude of individual acts? We also face a bi-directional feedback relationship in this context: On the one hand, the culture of a society influences preferences of its individuals while the preferences on the other hand account for the entirety of culture of a society. Neoclassical economics cannot really comment on these basic facts of society, as it does not include endogeneous preferences (and technologies), because their acceptance could not be embedded into a maximizing model in which utility (and production) functions have to be given [14]. But endogeny is a crucial element in evolutionary economic models (see below), where individual preferences are not moved into the set of non-economic factors.

Michael Radzicki [15] has developed a system dynamics computer simulation model for examining such a setting. Actions or decisions are made as a result of a gap between the desired state of the system and the perceived state. If actions, which are supposed to alter the state of the system, result in a change of the perceived state of the system, further adjustments of behavior will be made if there is still a gap between the desired and the actual state. Hence, human behavior is modelled as dynamic and not as deduced from general principles. The subjective wants and perceptions directly create and change the socio-cconomic results. The individual is allowed to re-evaluate his or her own behavior step by step and thus learns by means of a trial and error process. If the computer simulation exhibits results similar to empirical data, it is plausible to argue that the structure of the models exhibits important features of the underlying processes. This model can be enlarged to a dynamic micro foundation of socio-economic processes. Instead of equilibria such models use the concept of attractors, which are defined as temporal situations of steady state, which can be related to the Marshallian concept of partial equilibria (see the next section of this chapter).

It is this point in particular that shows that the concept of self-organization could exceed the methodical narrowness of traditional economics because the relationship between changes of the economy and the society can be focussed: the complex systems of Western societies have created and continually changed their specific markets, governments, enterprises and institutions in the course of their self-organization. Radzicki [15] points out that "a synthesis of chaos theory and institutional economics can potentially produce ... an improved understanding of social conflict" (pp. 68-9). This agrees with a point often mentioned by non-neoclassical economists and other social scientists, while neoclassical economics according to the Newtonian research program provides a model of full harmony. All this does not inform us about the results we can expect from economic activity, but it tells us in which social structures this activity is embedded. It seems to be one of the most striking points of self-organization economics that it devotes such an importance to the description and explanation of economic structures (see, for example the methods developed by Hermann Haken, [16]). After these general theoretical considerations I will now come to a much more specific aspect regarding concrete economic processes.

3 A proposition for an evolutionary economic way of modelling

The idea of modelling cultural processes in analogy to biological evolution has a long tradition. We may refer to the fact that the evolutionary approach in a wide sense was first used for describing social developments back in the 17th century. Thomas Malthus introduced the idea that there is a struggle for food among populations which tend to grow geometrically. This principle was generalized by Charles Darwin in his "On the Origin of Species", on which the up-to-date Neo-Darwinian theory of biological evolution is based. So the step to transfer the concept back to economics is not a big one (see Michael Rothschild [19]). I want to start the following considerations with the warning that we should be careful with the conceptions used in this context. What follows is a suggestion to be discussed among biologists, systems theoreticians and economists in order to encourage the necessary transdisciplinary dialogue. It shall provide the basis for a common way of thinking about these issues. In that sense, a formal economic approach will be developed in analogy to biological evolution. I do not deny that there are substantial differences between natural and socio-economic processes. And I do not claim that every socio-economic phenomenon can be properly described in evolutionary terms. My only claim is that there are certain isomorphisms between biological, cultural and economic processes. These isomorphisms justify the application of an evolutionary view – especially when we discuss the relationships among these processes. As we observe similar phenomena (such as selection and adaptation) in nature and in the economy, it is reasonable to look for theoretically analogous descriptions. The major restrictions of such an undertaking shall also be made explicit. In particular, I use Richard Dawkins' [20] thesis that ideas, which he calls memes, are responsible for cultural evolution. Rothschild [19] and Kenneth Boulding [21] developed similar ideas for economics.[2]

This idea is not restricted to Dawkins' view of biological evolution, which is said to be a reductionistic one (see David Depew and Bruce Weber, [22]). Rather my co-evolutionary concept can be related, for example, to Lorenz's far more holistic view of these issues [11]. We may start from the fact that humans (besides other species for which what I say in the following is much less relevant) exhibit not only physical traits but also cultural traits. Cultural traits may be: social organizations and institutions, technical achievements, property rights etc. These traits are not inherited biologically. They are passed on via communication and via teaching/learning from parents to children, from teachers to pupils, but also via mass media and so forth. Hence, such a process can be interpreted as a Darwinian analogy and not as a Lamarckian type of evolution. Cultural traits heavily influence the relative success of humans against other species. A system-dynamic description of co-evolution between biological and cultural traits can describe this in a comprehensive way. (The master equation approach, for example, can be used to calculate how [the changes of] ideas of an individual are influenced by those of other individuals and thus how a certain distribution of opinions comes about; see [16] and the remarks in the preceding section of this chapter.)

How can a more concrete description of economic evolution be achieved? The following isomorphism makes it reasonable to apply the memetic concept of evolution to economic processes. Darwin tought us that the driving force behind biological evolution is the general tendency of nature to provide an incentive for every species to produce a number of offspring which is greater than the number that is able to survive on the whole. In economics we frequently assume that firms and households have an intrinsic "want" to produce more and/or to consume more than can be produced/consumed at that time on the whole. This can be interpreted as the driving force behind economic evolution. The memetic concept provides us a solid basis for making that more concrete. There have been some discussions among economists as to what the "genes" of economic processes might be. My answer is: we do not really need a replication process analogous to biological replication. Humans have brains to remember ideas and minds to communicate them. Production processes have routines to produce copies of certain blue-prints over and over again. This is technically very different from genetic replication but it provides the necessary function of the genes: storage

---

[2] The model presented here has been discussed on several workshops in Gießen, Heidelberg, Rome and Stockholm. One can see this section as the second in a sequence of three papers on the application of Dawkins' idea to economics. Ref. [17] presents a much more detailed discussion of Dawkins' own view and deals with some important broadenings. Ref. [18] deals explicitly with the interrelations between biological, cultural and economic evolution and applies the approach to problems of the economy-ecology relationship.

and transmission of information. What is much more interesting in an economic context is how selection and adaptation takes place.

We can formalize such a process as follows. Economists usually talk about preferences and technologies as ideas behind economic processes. These may serve as the relevant memes of economic evolution. On a market a certain type of commodities is traded; the market analyzed should be big enough to allow for substantial substitution on the demand side so that workable competition is ensured. To start with the supply side, (1) represents an adaptation and selection process with respect to a given initial distribution of commodities P_n^s, which are potentially supplied on the market. M_n^s is the set of technologies (i.e. the production memes) connected with these commodities (concerning the production processes and the commodities themselves, including the entrepreneurs' or firms' ideas on marketing etc.). We can say that the arrow between M_n^s and P_n^s represents a generalized production function. Note that this is not an aggregate view; M_n^s and P_n^s are vectors so that technologies and plans are modelled as distributed among the decision making units (firms and/or entrepreneurs) in a certain way. On the market the distribution of commodities actually traded (denoted by P^*) will in general differ from P_n^s.

$$M_n^s \rightarrow P_n^s \rightarrow P^* \rightarrow M_{n+1}^s \tag{1}$$

While ideas are unable to interact directly and, hence, are not directly subject to selection and adaptation, the commodities play the role of economic interactors, such as physical organisms are the interactors of biological evolution based on genetic replication (see [17] for a more detailed discussion of this differentiation). We can distinguish two fundamental mechanisms of selection. The first mechanism results in a direct elimination of products from the market due to bankruptcy of the respective firm in the worst case. Here the analogy to natural selection is quite striking. This alteration may be due to changes on related markets – as long as we do not consider the whole economy in (1). M_{n+1}^s, the technologies incorporated in the products remaining in the market, will therefore in general differ from M_n^s. In other words, technologies are selected out of the market and do not play any role in the next period. (Of course, prices play a substantial role within this process. They transfer information. But for the sake of simplicity, prices are not explicitly considered in our present descriptions.) But there is a second mechanism, working more indirectly and bringing genuine human abilities into the picture, which are usually not part of biological models. Entrepreneurs will have certain experiences in the market, which they attribute to the qualitative features of their products and/or production processes, and they may retrospectively change their plans. In other words, the experiences will influence their behavior in the next period, so that an additional change in the meme pool M_{n+1}^s will be induced.

Such a description implies a concept of rationality that is different from the assumption of maximizing "response automatons", in which the entrepreneurs have to know in advance what is profitable and what is not (or at least the stochastic distribution). Only in a special case will all entrepreneurs within a

specific market will be totally satisfied so that they do not change their behavior
(i.e., $M_{n+1}^s = M_n^s$ and $P_{n+1}^s = P_n^s$). Such a situation can be interpreted as
partial equilibrium: agents do not have incentives to change their behavior. (It
is not necessarily implied that P_n^s equals P^*. Furthermore, it is possible that
changes in the mental sphere do not change actual behavior – or vice versa. Entrepreneurs will be satisfied with limited non-realization of certain plans before
they change their behavior – especially when they face uncertainty.) Other phenomena can be described within this framework but will not be developed further
in this paper: Imitation plays an important role in such a process. Imitation can
be interpreted as "migration" of firms into the market of the respective good,
which directly affects P_n^*. Moreover, it would be possible to discuss analogies to
"mutation" or "memetic drift", which affect M_{n+1}^s independently from P_n^*; this
could be the effect of "pure inventions" based on creativity and pure research
(see [17] for a more detailed discussion of this point). There is another reason
why economic selection is somewhat weaker than natural selection. Technologies
are not only "remembered" as they are incorporated into commodities. They are
also descibed abstractly, verbally, visually etc. This way they are communicated
directly so that a technology may survive even if the respective commodity has
disappeared from the market.

Now we are able to describe the demand side of the market analogously to the
supply side (see the process described by (2)). Regarding household consumption the evolutionary economic process is maybe less evident and the mechanisms may be weaker than on the production side. Nevertheless, it is my claim
that similar processes can be described as the outcome of intelligent individual
decision-making. Here the memes are preferences of the individual participants
on the respective markets. Generally speaking, preferences are the mental description of what individuals would like to consume. These ideas, denoted by
M_n^d, determine demand plans P_n^d. The connection between M_n^d and P_n^d is a
general description of the individuals' psychology concerning demand behavior.
The demand plans, or ex-ante demand, are subject to a selection and adaptation
process on the market. Some individuals will be frustrated by the market results
(or effective demand) P^*, which will in general differ from P_n^d. Certain preferences will simply turn out to be unsatisfied. But to speak of direct elimination of
preferences is not very reasonable, whereas the indirect way of changes in meme
pools is important. Experiences on the market may force agents to revise their
plans, as we assume that preferences of the frustrated individuals can change.
From this follows that the meme pool M_{n+1}^d in the next period will in general
differ from M_n^d, if the market is not in partial equilibrium.

$$M_n^d \rightarrow P_n^d \rightarrow P^* \rightarrow M_{n+1}^d. \tag{2}$$

Such a feedback-loop refers explicitly to the decision-making behavior of conscious individuals as described above -assuming that they are able to revise their
decision on the occasion of a further purchase. Agents retrospectively evaluate
the former decision, and further decisions will depend on that evaluation. This
is the point where a purely biological analogy cannot hold: agents deliberately
and purposefully change their behavior, because they are at least partially able

to perceive the selection process as such and thus actively intervene in this process. Nevertheless, as the device to intervene comes from market signals, such a behavior should not be seen as unpredictable at all.

Supply and demand are, of course, closely interrelated; we can consolidate (1) and (2) to figure 1, in which demand of products does not only depend on preferences but also on actual supply, which in turn depends on the "production memes". In this sense, preferences can be described not only as depending on the demand conditions but also on the supply behavior, while technologies depend to some extent on demand conditions.

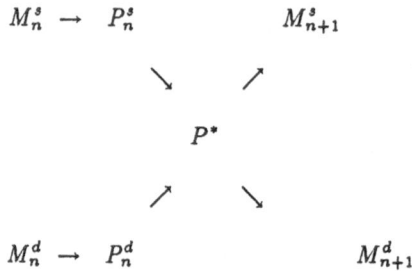

Fig. 1. Memetic feed-backs of supply and demand

This description is neither a substitute for traditional microeconomic analysis, nor is it necessary to adhere to the traditional view. The relationships (or mechanics) between P_n^s, P_n^d and P^* can be described by a traditional partial equilibrium description but can also be represented by a different partial analytic model of price formation or a model without prices. The value added by the evolutionary approach compared to comparative static partial market approaches results from the possibility of incorporating certain endogenies.

In terms of traditional economic theories, we can characterize such a theory as a Marshallian partial equilibrium approach see the interpretation of Marco Dardi [23] and John Foster [24] or disequilibrium in the sense of Richard Day [25], who draws a Schumpeterian analogy; see also [17]. Equilibrium in this sense shouldbe defined as a situation in which individuals do not have incentives to change their behavior. Hence, equilibria indicate evolutionarily stable situations. In other words, evolution, in the usual sense of the word, does not take place in situations of equilibrium. In every real economy we observe that equilibria are only local phenomena; individuals and organizations are in fact often forced to revise their plans and change their behavior. As Day puts it:

> In disequilibrium ... expectations are not fulfilled. Individuals and groups do not and cannot know what everyone is doing or what they will do in the future. They must observe, construct a record or image of the past which is of necessity simplified and imperfect. They must base

plans on data that may be misleading. They must construct controls that enable economic activity to continue when even the most carefully constructed and elaborated plans must be modified in the light of accumulating information or because they simply can't be carried out. If one wants to explain how and under what conditions an equilibrium circular flow emerges or fails to emerge one must take account of these fundamental facts. ([25], p. 62)

It is obvious that equilibria are not only necessarily local but also temporal. Because of the complex interrelationships within an economic system, other sectors will be affected by changes in a disequilibrium sector and this will trigger further adaptations. Partial equilibria allow us to examine certain stable situations which are necessary for the stability of the whole system. Examining the processes in the absence of equilibrium is even more interesting because it allows us to make statements on the developments of evolutionary systems. Partly, such a behavior can be described within the standard framework of any microeconomic (neoclassical) modelling. These models start from given elasticities of demand (with respect to prices and incomes) and describe adjustment of behavior to the effects of certain exogenous shocks. In our model, memes would not change in this case; adjustments of plans and behavior would be made according to given preferences and technologies, so that profits/utility could be maximized (at least theoretically). This aspect of behavior could be described mathematically in the usual analytic way. Prices and/or quantities of certain traded goods would rise and others would fall.

The "driving force" of biological evolution is that each species produces a number of offspring which is greater than the number of units which have a chance to survive. We may therefore say that evolution deals essentially with scarcity, which is the fundamental problem of economics. The driving force of memetic economic evolution, it should be emphasized again, is the fact that consumers usually want to consume more units than supplied and that firms usually want to produce more than demanded. This also explains for the fact that for a system of partial equilibria a tendency towards general equilibrium is unlikely. Not every possibility of arbitrage will be utilized by agents whose rationality is different from neoclassical economists' standard notion of economic rationality; it will be difficult to find a general notion of utility/profit maximizing behavior in such a world (see [17]).

Another example may explain why standard economic reasoning and the evolutionary view do not fully contradict each other; in part they are complementary rather than substitutive. In nature, scarcity is a disadvantage: if a species has only a few members its chance of survival will be small. To some extent this is also true for commodities. But markets provide an additional mechanism that does not occur in nature: prices rise if commodities are scarce. For that reason and under certain circumstances (concerning elasticities of demand and supply, for example) this may ensure the "survival" of a certain type of commodities. We see that standard microeconomics, which describes the price mechanism and assumes the workability of economic selection, and evolutionary economics are not contradicting each other in every case. A more pluralistic view of economic theorizing leads to the observation that various approaches restrict the others' generality.

The fact that consumers imitate other consumers can be interpreted as a cultural process on the memetic level. On the supply side, there may be direct imitation by certain firms (i.e. the Schumpeterian case). On the level of cultural interaction, memes are passed on, changed, and developed by means of direct communication between individuals and organizations (being interactors in this case). Transmission takes place via teaching and learning, talking within families, among friends, via books, mass media, etc. What we learn, read, and see on TV influences our consumption behavior, and the discussion in scientific journals will have an impact on new technologies introduced in the market. This triggers not only certain reactions to a changed environment, but rather a complex process of actions and reactions. "Change within a given structure or set of equations is considered to be Newtonian or mechanistic in nature, and not Darwinian or evolutionary in nature" ([26], p. 655). The interesting point is the explanation of the process itself and not of reactions according to an assumed mechanism. It becomes clear that this point is closely related to the institutionalist research program: Someone who is interested in changes in the state of order cannot get past changes in institutions. We have to deal with open systems that "are defined by their interactivities with their environment, which means in practice (since the latter cannot in general be uniquely specified and controlled) that determinate solutions to pre-specified problems are no longer possible. Instead, units of analysis (say business firms or biological organisms) are conceived according to a relationship with the environment ... " (Norman Clark, [27], p. 519). The more interrelations we take into account, the more important becomes the fact that non-equilibrium transactions take place, the unplanned outcomes of which influence each other at certain moments of historical time.

4 Conclusions

Summarizing the previous discussions it should be emphasised once more that the approaches discussed here are seen as complementary rather that substitutive and that they contradict the contemporary economic mainstream only as long as the latter is seen as a general device, sufficient to explain any economic phenomenon. The language used here may sound unfamiliar to many an economist, but in order to promote transdisciplinary dialogues we have to look for a common language so that any partner in such a process should try to make some steps in the others' direction.

Some of the restrictions of contemporary economics can be removed if evolutionary and self-organization approaches can successfully be built in. Neural nets can be used for developing further the inductive method of describing human decision-making. System dynamics simulation as well as certain mathematical approaches of synergetics could lead us to a better insight into how individual behaviors relate to each other and how collective phenomena are the results of individual action (though not necessarily of human design). Darwinian evolution, last but not least, may help us to understand better what goes on in an economy based on individual decision and competition. All these problems have a long tradition within economics as a science and contemporary answers are often far from satisfying. Much more interdisciplinary work will be needed in order improve at least some of them.

Acknowledgements

Encouragement and helpful comments on earlier versions of this and related papers by Stefano Bartolini, Michael Common, Giovanni Dosi, Hans-Georg Petersen and Ulrich Witt are appreciated. The work was financed by the Deutsche Forschungsgemeinschaft.

References

1. F.A. von Hayek: The Results of Human Action but not of Human Design, in: Studies in Philosophy, Politics and Economics. London (Routledge & Kegan Paul), 1967.
2. A. Smith: An Inquiry into the Nature and Cause of the Wealth of Nations, London (W. Strahan and T. Cadell), 1776.
3. E. Ullmann-Margalit: Invisible Hand Explanations, Synthese 39, pp. 263-291.
4. G.S. Becker: The economic approach to human behavior. Chicago (Chicago University Press), 1976.
5. Ph. Mirowski: More heat than light: economics as social physics and physics as nature's economics. Cambridge, New York (Cambridge University Press), 1989.
6. I. Prigogine and I. Stengers: Order out of Chaos. Man's New Dialogue with Nature, New York (Bantam Books).
7. U. Witt: Individualistische Grundlagen der evolutorischen Ökonomik, Tübingen (J.C.B. Mohr (Paul Siebeck)), 1987.
8. K. Dopfer: The Complexity of Economic Phenomena: Reply to Tinbergen and Beyond, Journal of Economic Issues 25, 1991, pp. 39-76.
9. G. Schiepek: Grundprinzipien der Selbstorganisation, in: Grundprinzipien der Selbstorganisation, ed. by K.W. Kratky and F. Wallner, Darmstadt (Wissenschaftliche Buchgesellschaft), 1980, pp. 182 - 200.
10. E.J. Schlicht: Ökonomische Theorie, speziell auch Verteilungstheorie, und Synergetik, in: Selbstorganisation. Die Entstehung von Ordnung in Natur und Gesellschaft, ed. by A. Dress, H. Hendrichs and G.Küppers, München, Zürich(Piper), pp. 219 - 227.
11. K. Lorenz: Die Rückseite des Spiegels, (Behind the Mirror: a Search for a Natural History of Human Knowledge), München (Piper), 1977.
12. D. North: Institutions. Journal of Economic Perspectives 5, 1991, pp. 97-112.
13. E. Baum: Neural Nets for Economists, in: The Economy as an Evolving Complex System, ed. by Ph.W. Andersen, K.J. Arrow and D. Pines, Redwood (Addison-Wesley), 1988, pp. 33-48.
14. I. Steedman: Economic theory and intrinsically non-autonomous preferences and beliefs, in: From Exploitation to Altruism, (Polity Press), 1989.

15. M.J. Radzicki: Institutional Dynamics, Deterministic Chaos, and Self-Organizing Systems. Journal of Economic Issues 24, 1990, pp. 57-102.
16. H. Haken: Synergetics. Berlin, Heidelberg (Springer), 1977.
17. F. Hinterberger: A Note on Sociobiology. Georgescu-Roegen, Schumpeter and Beyond, in: Entropy and Bioeconomics, ed. by J.Martinez-Alier and E.Seifert, Roma (Nagard), 1993.
18. F. Hinterberger: Biological, Cultural and Economic Evolution and the Economy-Ecology-Relationship, in: Concepts, methods and policy for sustainable development: Critiques and new approaches, ed. by J.C.J.M. van den Bergh and J. van der Straaten, Washington, Covelo (Island Press), 1993.
19. M. Rothschild: Bionomics. Economy as Ecosystem, New York (Henry Holt), 1990.
20. R. Dawkins: The Selfish Gene, Oxford (Oxford University Press), 1976.
21. K. Boulding: Evolutionary Economics. Beverly Hills, London (Sage), 1981.
22. D.J. Depew and B.H. Weber (Eds.): Evolution at the Crossroads: The New Biology and the New Philosophy of Science. Cambridge/Mass, London (MIT Press), 1985.
23. M. Dardi: The Concept and Role of the Individual in Marshallian Economics, in: Alfred Marshall's "Principles of Economics" 1890 – 1990, ed. by M. Dardi, M. Gallegati and E. Pesciarelli (Quaderni di Storia dell' economia politica), 1991.
24. J. Foster, The Thermodynamical Approach to Economic Science: Marshall Revisited and Prigogine Reassessed. Paper prepared for the third annual conference of the E.A.E.P.E., Wien, 1991.
25. R.H. Day: Disequilibrium Dynamics. A Post-Schumpeterian Contribution. Journal of Economic Behavior and Organization 5, 1984, pp. 57-76.
26. M.J. Radzicki: Institutional Dynamics: An Extension of the Institutionalist Approach to Socioeconomic Analysis. Journal of Economic Issues 22, 1988, pp 633 - 665.
27. N. Clark: Some New Approaches to Evolutionary Economics. Journal of Economic Issues 22, 1988, pp. 511 - 531.

Can Synergetics Serve as a Bridge Between the Natural and Social Sciences?

Hermann Haken

Abstract

This article first recapitulates some general concepts of synergetics and illustrates them by means of applications to physics (lasers), biology (human finger movements), computer sciences (parallel networks for pattern recognition), and, in particular, sociology. As is shown, a great variety of phenomena may be described by means of general concepts, such as stability and instability, order parameters, and the slaving principle. Based on such common principles, a bridge may be formed between the natural and social sciences.

1 What is Synergetics About?

For a long time it has seemed that science is being split into more and more disciplines, and that there would be no unifying principles at all. Over the last one or two decades, however, this trend has been changing. A number of attempts are being made that aim to build bridges between the different sciences. Synergetics may be considered as one of these bridges. I coined the term "Synergetics" some twenty years ago [1] in order to characterize an interdisciplinary field of research that did not exist at that time. Its basic ideas may be characterized as follows: When we browse through different scientific disciplines, we will find that quite often they deal with the following problem: The objects of research are composed of individual parts which by their cooperation may produce spatial, temporal, or functional structures. Let us consider a few simple examples. In physics, molecules may form a liquid which may exhibit different kinds of motions. In chemistry, specific kinds of molecules may undergo reactions by which macroscopic patterns, e.g. in the form of spirals or concentric rings, are formed. In biology, the individual cells constituting organisms cooperate in a highly organized fashion. High cooperation can also be found in animal societies, in human economy, or in human society. It is important to stress that these structures are not imposed from the outside, but that they are fully organized by the system itself, i.e. in other words, we are considering *self-organizing* systems. As far as physics, chemistry, and biology are concerned, synergetics focuses its attention

Springer Series in Synergetics, Vol. 61   **On Self-Organization**
Eds.: R.K. Mishra, D. Maaß, E. Zwierlein   © Springer-Verlag Berlin Heidelberg 1994

on open systems whose functioning or organization is maintained by a more or less continuous input of energy and/or matter into the system.

The question I asked some twenty years ago was whether there are general principles governing self-organization irrespective of the nature of the individual parts of a system. At least at that time, this question must have sounded rather far-fetched because the subsystems may be as diverse as atoms or molecules in physics and chemistry, cells in biology, or animals or humans in a society. But actually this question could be answered in the affirmative for large classes of systems. As one may guess, a price had to be paid for the generality of the applicability of these principles. The price is that we have to focus our attention on those situations in which the macroscopic properties of a system change qualitatively. What is meant by this statement will become clear by means of the examples provided below.

The principles governing self-organization have been revealed in a rather comprehensive, rigorous mathematical theory which has been published in some of my books [2], [3], [4]. It is not my intention to describe these mathematical theories here. All I want to use here is the fact that the statements made are statements on structural relationships. It is important to stress at the very beginning that I am not advocating any physicalism in which we try to transfer concepts of physics to other disciplines, say sociology. Rather what we are doing here is to grasp the mathematical relationships by means of simple examples, some of them belonging to physics.

2 The Laser Paradigm

Let us begin with an example of the general principles in the field of physics. This example has an advantage and a disadvantage. Its advantage is that the concepts can be deduced in every detail, and the predictions made on account of the theory have been checked experimentally. The disadvantage will be that, say, a sociologist not familiar with physics will not appreciate the stringency of the conclusions; rather, he may even argue that a society is extremely complex and a rather soft system. For this reason, we shall discuss these points later on in more detail. But let us start with our simple example and show why it is so pertinent for the spontaneous formation of structures, or, in other words, for self-organization.

A simple example of a laser is provided by a gas laser which has the following constituents: A glass tube is filled with a gas of atoms. At the end faces of the glass tube two mirrors are mounted. They serve the purpose of reflecting light waves propagating in the axial direction very often so that these kinds of light waves can interact strongly with the atoms of the gas in a way which we shall describe below. By means of an electric current sent through the gas, it becomes possible to energetically excite the individual atoms. After an excitation, an individual atom acts as a miniature radio antenna by emitting a wave train of light (instead of a radio wave). If the current is weak, only a small percentage of the atoms become excited. Each of them emits an individual wave train which can be compared to water waves that emerge when a pebble is thrown into a pond. When several atoms are excited, it is as if we were throwing a handful of pebbles into the water, and a wildly excited water surface will emerge. However, when we increase the current, more and more atoms will become excited. Suddenly a new phenomenon occurs: Instead of the many independent wave trains, a practically continuous giant wave emerges. In other words, the microscopic chaos of the original emission of light is replaced by macroscopic order. How is this achieved? As was shown by Einstein [5] at the beginning of this century, an excited atom may not only spontaneously emit a wave train, but it can also be forced by a wave train impinging on it to add its own energy to that wave train so that the latter is enhanced. When several excited atoms are hit one after the other by a wave train, it is clear that a light avalanche will be generated. A subtle point must be considered, however. It turns out that there are different kinds of wave trains that act differently on the atoms. A particular wave train is more efficient in forcing an atom to enhance its strength than other wave trains. In this way, a competititon between different avalanches occurs and one specific wave being amplified wins the competition. Some kind of Darwinism (of the non-intentional kind) of the inanimate world is at work here.

And now the important concepts of synergetics come in. Once a particuar light wave has won the competition, it forces all the atoms to deliver their energy to it. At the same time, the electrons in the atoms are forced to oscillate in a highly ordered fashion prescribed by the emerging light wave. Thus, the light wave that evolves describes both the order in the system and gives order to the individual atoms, i.e. to the individual parts. This is why we call this quantity the order parameter. At the same time, we recognize the existence of a circular causality. The order parameter enslaves the individual atoms, whereas the individual atoms support the order parameter. The behavior of one subsystem (atom) conditions the behavior of the others. When we disturb the order parameter "light wave", it turns out that it can return to its former state only after a rather long period of time. The subsystems, namely the atoms, on the other hand, relax after any perturbation very quickly. Order parameters and enslaved sybsystems are thus distinguished by the different time scales of their individual adjustments.

This will be an important criterion for the applicability of the concept of the order parameter and the slaving principle. As the mathematical theory reveals, the transition of the microscopic state to the highly ordered state of laser light can be described in the following terms: The order parameter behaves as if it is a ball moving in a landscape. If the electric current is small enough, the landscape has the form as shown in Fig. 1.

After each emission of a light wave train, the ball will relax towards its equilibrium value, i.e. the order parameter relaxes to zero and shows only fluctuations around its zero value. However, when the current exceeds a critical value, the landscape is deformed into that of Fig. 2, which has apparently two minima. (Actually in the laser case, the situation is still more complicated, but for our purposes it will be sufficient to treat the present case.) Quite evidently, the former value "zero" of the order parameter has become unstable and is replaced by

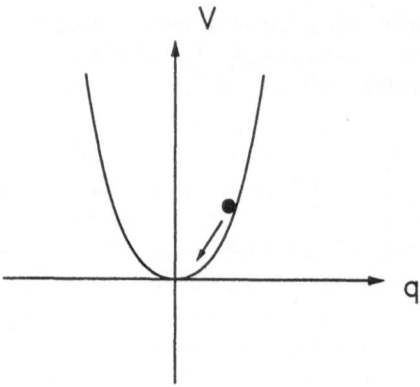

Fig. 1. Visualization of the behavior of an order parameter q by means of the position of a ball moving in a hilly landscape with one valley

two new stable equilibrium points at the bottoms of the valleys. Of course, the system can go only to one of the two valleys, i.e. it has to break the symmetry. Now a very important but subtle point comes into play, namely the question what causes the system to go to one or the other minimum. This is achieved by an initial spontaneous emission of a wave train which, according to quantum theory, cannot be predicted. Thus a chance event at a microscopic level decides which course the system will take on the macroscopic, observable level.

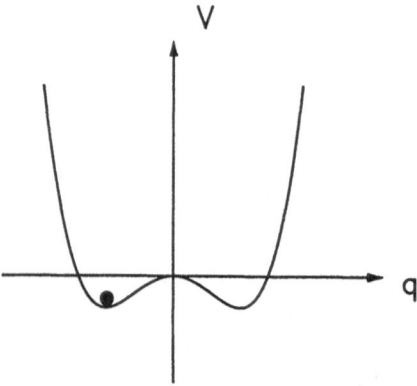

Fig. 2. Same as Fig. 1, but the landscape now has two valleys with two bottoms

Another phenomenon will turn out to be of fundamental importance, namely when the current is increased from below to above its critical value, the curve of Fig. 3 becomes very flat close to the equilibrium point. But as we have seen, the ball is still subject to fluctuations. Because the restoring force in such a flat potential is extremely small, the ball will strongly feel the fluctuations to which it is exposed. Thus its amplitude will oscillate strongly; we are dealing with so-called critical fluctuations. When the ball is pushed away from its equilibrium point, because of the very small restoring force, it will relax very, very slowly to its equilibrium value. This phenomenon is called critical slowing down.

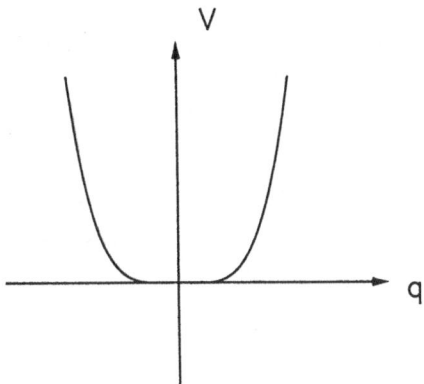

Fig. 3. When a control parameter is changed, there is an intermediate situation between the Figs. 1 and 2 where there is still only one valley, but one which has a very broad bottom

When the electric current is increased more and more, the previously established ordered light wave may become unstable and may be replaced by other phenomena, e.g. by regular light flashes or by what is called deterministic chaos. In these cases not only one but several order parameters occur, and their interplay determines the total behavior of the laser. This example allows us to formulate the results of the abstract mathematical theory of self-organizing systems in the following way: A change of rather unspecific conditions, in the laser case the power of the electric current, may cause the system to undergo a qualitative change on a macroscopic scale. In technical terms, the old state, e.g. the microscopic chaotic state, becomes unstable and is replaced by a new state, in our case the laser light state. At the instability point, one or a few order parameters occur. They enslave the individual parts of the system and thus cause a specific structure within the system. At instability points, in general, the system has the choice between several possibilities; which one is realized depends on microscopic fluctuations. In the transition region, critical slowing down and critical fluctuations occur. These concepts and the corresponding mathematical tools have been applied to either explain or predict a variety of phenomena in physics, such as structure formation in fluids, in plasmas, in semiconductors, and so on. But this is not our concern here; we rather want to proceed to biology, psychology, the computer sciences, and sociology.

3 Biology: The Finger Movement Paradigm of Kelso

As we all know, man and higher animals are composed of billions of cells of different types, such as muscle cells, nerve cells, tissue cells, and others. They have to cooperate in a highly organized fashion, so as to produce morphogenesis, locomotion, movements, feeling, heart beat, and blood circulation; and quite clearly, such a highly organized cooperation must go on at the cognitive level. What are the principles behind this high coordination? An experiment done by Scott Kelso [6] may serve as a fundamental paradigm. A few years ago, Scott Kelso visited me and told me about his following experiment: He asked test persons to move their fingers in parallel and then asked them to move their fingers more and more quickly. Suddenly, the finger movement changed quite involuntarily from the parallel to the antiparallel, i.e. symmetric configuration (Fig. 4). Quite clearly, what happens here is a qualitative change of a system on a macroscopic level, or, in other words, a phase transition.

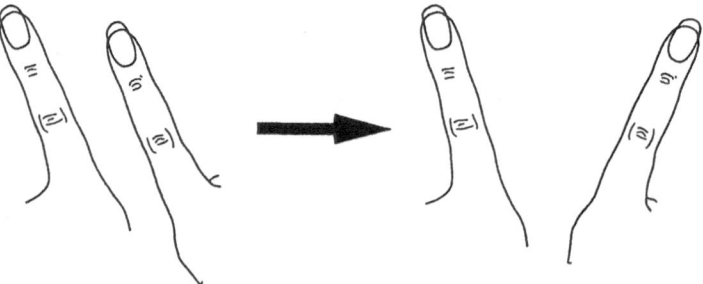

Fig. 4. Transition from a parallel finger movement to an antiparallel symmetric one

Can we apply the concepts of synergetics to this experiment, and can we model its features? Quite clearly, the relative position of the fingers or, in more technical terms, the relative phase between the oscillating fingers, suggests itself as an adequate order parameter. In the simple case of a single order parameter, one may try to construct a landscape describing its movement. Such a landscape can be easily devised by simple arguments that I will not repeat here [7]. As it turns out, the landscape has the form shown in Fig. 5 and undergoes a series of slight deformations when the speed of the finger movement is increased from the left upper to the right lower corner. From the qualitative point of view, a number of predictions can already be made: When the situation of the middle row, right-hand side of Fig. 5, is reached, the position corresponding to the parallel finger movement becomes unstable; the ball will fall down to the absolute minimum and stay there. This corresponds to the symmetric finger movement. When a person moves his or her fingers quickly in the symmetric mode, and is then asked to slow the finger movement down, the ball will, of course, stay in the absolute minimum. This prediction could easily be checked by Kelso and verified. This is the effect of hysteresis that is well-known in physics for example. In the case of hysteresis, the state of a system depends on its past history. For instance, when a ferromagnet is subjected to an external magnetic field, the magnetization may become parallel to the magnetic field at a specific field strength. When we reverse the field, the magnetization will switch again, but at a different field strength than before. In other words, the ferromagnet has retained some kind of memory of what has happened to it before.

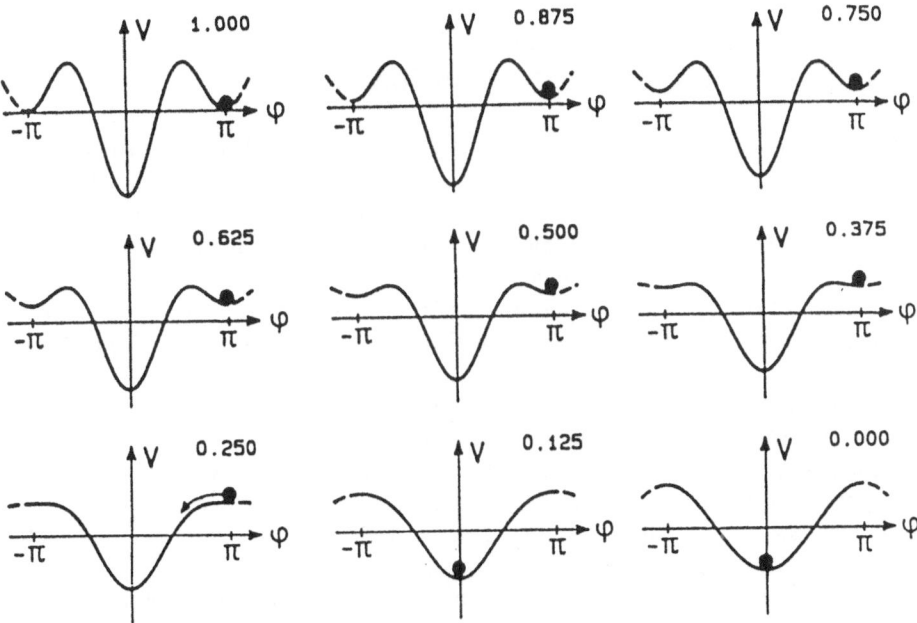

Fig. 5. The landscape describing the behavior of the relative phase in the finger movement experiment. In the upper left corner, the finger movement is still slow, and the system is in a stable state in the upper right valley, as indicated by the ball. With increasing speed of finger movement, the landscape is deformed until in the middle row, r.h.s., an unstable situation is reached and a little push now can cause the ball to fall down to the absolute minimum corresponding to antiparallel finger movement. With still higher speed of finger movement, the lower minimum is the only minimum available

But as we have seen before, close to the instability point, critical fluctuations and critical slowing down must be expected [8]. By careful measurements, Kelso could show that the relative phase undergoes pronounced fluctuations in the transition region, and that it can also show the phenomenon of critical slowing down [9] once the finger movement is disturbed. Our model can be cast in a mathematical form so that these statements can be made quantitative ones. This, however, is not our concern here. What is important is the following statement: It has often been argued that our brain is a computer which by specific programs steers the motion of our extremities and other functions. However, the picture we are drawing here is quite different. It strongly points to the fact that a biological system is a self-organizing one when coordinating the movements of its extremities. The concept of a computer program could not explain how critical slowing down and critical fluctuations arise. Rather these features are typical for self-organizing systems. A variety of experiments presently being undertaken shows that this interpretation of biological coordination holds in a variety of cases.

4 The Application to Computer Science: The Synergetic Computer for Pattern Recognition

How far can we take the concepts of order parameters and slaving principle and so on when we are dealing with complicated patterns? Remember that the slaving principle allows us to introduce a considerable compression of information in complex systems, because it allows one to express the behavior of the numerous components of a system by means of just a few quantities, namely the order parameters. To demonstrate the power of the order parameter concept, we constructed the "synergetic computer", which is based on principles of synergetics and is capable of pattern recognition. Since this computer has been described elsewhere in great detail [10], we shall not dwell on any details here, but rather stress the salient features. The basic idea is as follows: A specific pattern is described by its order parameter and the slaving principle by which the system is brought into the state prescribed by the order parameter. Once a set of features, e.g. the eyes and the nose, are given, the system generates its order parameter which competes with order parameters to which other noses or eyes might belong. This competition is eventually won by the order parameter connected with the presented nose and eyes. Once the order parameter has won the competition, it is able to restore the whole pattern. The original relationship between each order parameter and its pattern is determined by a learning procedure. While Fig. 6 shows some examples of faces learnt by the synergetic computer, Fig. 7 shows how it may restore the full face, if necessary including the family name, so that the computer can recognize faces out of individual features.

The order parameter concept is thus a highly powerful tool in constructing a new type of computer which, actually, can be fully parallelized and thus acts as a competitor to the presently discussed neural computers. It may be worthwhile to elucidate similar features and in particular differences between neural and synergetic computers. Both concepts aim at realizations of computers by computer architectures in which computations go on in specific basic cells in parallel. In neural computers these cells have only two internal states, namely on and off. A switch from the off-state to the on-state occurs when the input signals from the other cells exceed a certain threshold. The cells of the synergetic computer may occupy a continuum of states and their activity depends smoothly and in a nonlinear fashion on the input from the other cells. In contrast to neural computers there exists a complete learning theory of synergetic computers and their performance leads to unique results, whereas in neural computers there are

Fig. 6. Examples for stored prototype patterns

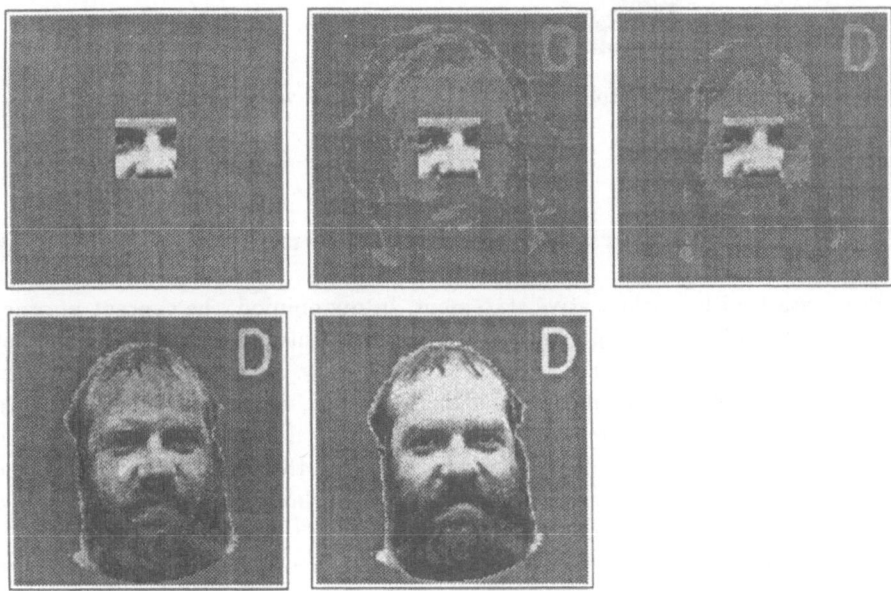

Fig. 7. Example of how the computer may reconstruct a full face with its family name encoded by a letter out of an initially given section including, for instance, the eyes and some part of the nose

still difficulties to be overcome, for instance the appearance of so-called spurious states.

We only mention here that the order parameter concept allows us to make connection with Gestalt theory originally formulated by Wertheimer and Köhler [11], and that it allows us to describe a number of detailed experiments in the field of psychophysics on ambiguous figures, as shown in Figs. 8-10. Now that we have found some confidence in the concepts of order parameters and the slaving principle, let us turn to the so-called softer sciences which, actually, will produce hard problems. I will take the example of sociology, though a number of remarks will also apply to economy and, possibly, ecology.

Fig. 8. A well-known example of an ambivalent picture: vase or faces?

5 Order Parameters and the Slaving Principle in Sociology

Let me start with a provocative statement. In my opinion, the concept of order parameters and the slaving principle will become central issues in sociology. Let us consider some typical order parameters in society. A language certainly lives longer than any individual. When a baby is born, he or she is subjected to the language of his or her parents and later on to that of other people of the same nation. When grown up, the individual carries the language further. In the technical vocabulary of synergetics, the baby is enslaved by the language. How much we are confined in our thinking by our mother tongue becomes clear when we go to other countries. There is by no means a one-to-one correspondence between words in different languages, and there is a great variety of subtleties that cannot be translated properly. This becomes particularly evident when we try to translate poetry. At the same time, language provides us with a powerful means of communication between members of the same nation. How much we are confined by a language becomes also evident when we look at different accents.

Fig. 9. Another example: old or young woman?

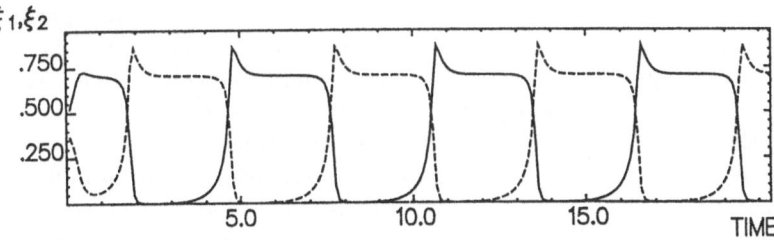

Fig. 10. Results from a computer simulation of the perception of ambiguous figures. Along the abscissa time is plotted, whereas the ordinates ξ_1, ξ_2 represent the size of the order parameters corresponding, say, to the vase or the faces. The oscillation between these two percepts can clearly be seen

For instance in the United States, one can distinguish not only between people from the North and South, but a skilled person can even tell, by listening to somebody from New York in which part of New York he or she has grown up. Generally speaking, language establishes an identity among a certain class of people and, at the same time, serves to distinguish them from other people. Clearly then, language is a typical order parameter.

Another order parameter is national character. Whether or not one may define a national character was beautifully discussed by Gregory Bateson some decades ago [12]. It is still worth reading his article. As he points out, a national character is brought about by the evolution of attitudes of people interacting with each other. He characterizes a national character not by single properties, but by dual properties, such as dominance or obedience. Bateson was, of course, not aware of synergetics which came into existence only later. From his definition of national character, one can easily deduce that it has all the properties of an order parameter.

Another order parameter is "ritual". Rituals serve for the identification of a group and to discriminate between that group and others. The external form of a ritual is not at all important. All that counts is the fact that the ritual is done within a specific group of people only. In it, a newly born baby or a youngster becomes a member of that group by means of the ritual. Another order parameter is the kind of state, such as democracy, dictatorship, and, possibly, other forms. The set of laws constitutes an order parameter. Law is the outcome of a typical collective effect. Quite a number of laws can be interpreted as means to resolve conflicts originally at an individual level by a collective regulation. Just to give a simple example: When a couple marries, it is not obvious at all whether the husband and wife will adopt the husband's or the wife's name. This solution can be negotiated at a personal level between the two married people, but the conflict can be resolved by a law stating that the family has to carry the name of the husband, for instance. Looking at some less serious issues, we may state that fashion is an order parameter. Another order parameter is public opinion, though here there is a complex interplay among individuals, mass media, government, and so on.

When I use the term "slaving principle" or "enslavement" for relationships between individuals and the order parameters, sociologists are quite often shocked. They tell me that individuals have a free will and may, certainly, choose, so to speak, their order parameter quite freely. A friend of mine, Mr. Balck, has suggested that the word "enslavement" be replaced by "consensualization". Since with this issue we are coming to the heart of the relationships between individuals and society, let us dwell somewhat longer on it. First of all, one cannot deny that there are strong cases for an influence of order parameters on individuals. Such a case is certainly prescribed by language, possibly also by rituals. It is certainly true that joining a religious movement is a voluntary decision. However, we know quite well of recent examples how strong the pressure of a group can become on an individual, even leading to collective suicide in extreme cases. Another reason why I wish to retain the word "enslavement" is based on the following observation: One may ask how it was possible that Germany became so easily prey to Nazism. In my opinion this was again a collective effect, where people just looked at each other, and each individual just followed what the other did. This is a typical source of instability which may lead anything – even to criminal acts. I think it will be very important to make people aware of these collective effects with their enslavement mechanisms such that, where necessary, they can react early enough against these mechanisms, which are linked up, for instance with national character, such as obedience.

Let me list a few other order parameters in companies, for instance. One of them is corporate identity; another one is social climate within a company. Why do we believe that the concept of the order parameter so important? This concept points to the way we may or may not change an order parameter. Let us consider social climate within a company. One cannot give a command to the individuals like be friendly to each other. The social climate has grown, and when newcomers enter the company, they are subjected to the social climate and will be enslaved by it. As we have learnt in synergetics, control parameters cannot be changed by the action of an individual (subsystem), but rather by the change of external conditions, in the present case by a change of the general frame of working conditions.

Before we discuss the mechanisms of changes in more detail, let us list a few further order parameters from other fields, namely economy and science. The important role of slowly varying variables (in synergetic terms, the order parameters) and fast varying variables was clearly stated in the important book by Paul Samuelson [13]. This distinction also plays an important role in a recent work by Wei-Bin Zhang [14]. In his well-known book *The Structure of Scientific Revolutions*, 1970, Thomas S. Kuhn [15] introduced the term "paradigm" and "changes of paradigms". Scrutinizing his book readily shows us that his paradigms are also the order parameters of synergetics.

Let us now discuss "mechanism of changes". Let us ask the question: Can an individual change the order parameter, or under what condition can he or she change it? Let us take a concrete example. Can an individual start a revolution? In my opinion, there is a clear prerequisite, namely the whole situation must be close to an instability. In politics, such instabilities may be caused by long-lasting economic depressions, persistent acts of terrorism, loss of credibility of a specific form of state, and so on. Only under such a condition can a revolution start. However, as the mathematical theory of synergetics illustrates, beyond the instability point, there will, in general, be different possible realizations. Which one actually occurs depends on small fluctuations or, in the present case, on a decisive group of individuals who drive the destabilized system into the new state. From this point of view, a revolution is a two-step process consisting of destabilization and then the decisive fluctuation driving the system into a specific new stable state. During such a transition, typical phenomena known in synergetics, namely critical fluctuation and critical slowing down, occur. These are phenomena which we can clearly watch now in the decay of the Soviet Union. But, from general principles of synergetics, it follows that the further course of this system is not at all evident. One must not expect that, for instance, democracy will be established now automatically in all parts of that decayed union. What new kind of state emerges in each case is a very subtle question which depends on individuals but which may also be hopefully influenced by external means, e.g., by the kind of economic support given.

An important issue is, of course, whether one can do experiments on the formation of order parameters in society. In this respect, an early experiment by Solomon Asch [16] seems to be relevant and very important. Solomon Asch made the following test: He had about ten test persons of whom, however, only one was a real test person, the others being his helpers. But that fact was not known to the real test person. Now he showed the group three lines of different lengths and asked the group: "Which one is the longest line?" Then his helpers gave the correct answer, and the test person too. Then he repeated the experiment, and his helpers gave a specific wrong answer. In this case, about 60 per cent of the test persons changed their opinion and agreed to the wrong answer of the helpers. For me, this was a clear indication of how easily people can be influenced with respect to their opinion formation. In my opinion, such formation of a collective opinion (an order parameter) is still more pronounced if a situation becomes truly difficult and complex. When I discussed these results of my interpretation with a sociologist, to my surprise he denied the applicability of the results of Asch's experiments to social issues. He said: "In the situation caused by Asch, the test person wanted to be sociable. If there were, however, really important issues, then the test person would develop his or her own opinion." I quote my conversation because it shows the limits of the possibility to have an interdisciplinary talk. What seemed to me quite obvious was entirely denied by the other person.

I hope that the above points illustrate the limits of predictability. When a system is destabilized, we cannot predict, at least not in general, to which new stable state it will go. This depends on very subtle events. Nevertheless, before such an instability point is reached, there are indications of its occurrence, namely critical fluctuations and critical slowing down. In retrospect, these critical fluctuations could be clearly observed before the German Democratic Republic broke down, for instance mass demonstrations, and so on.

6 The Role of Chaos Theory

More recently it has become fashionable to apply chaos theory to events in society and economy, to mention just two examples. First of all, we must bear in mind that chaos theory applies only when the so-called dimension of a system is small, or if there are only a few degrees of freedom which are decisive. But how can a complex system, such as a society, or a market exhibit only few degrees of freedom if there are so many individuals involved? The answer is provided by synergetics, according to which the degrees of freedom are the order parameters. But, in general, these degrees of freedom are not explicitly known in chaos theory; rather we have only one or few indicators, e.g. the gross national product. A number of mathematical methods have been developed to reconstruct the dynamics of a chaotic system once the time evolution of one variable is known. In such a case, one may determine so-called fractal dimensions and Lyapunov exponents, the latter being a measure of the length of time over which we may predict the development of such a system. But there is the following crucial point: In order to derive the above mentioned quantities, the observed time series must be long enough. There is no case known to me in which this condition is fulfilled in economy, society, or ever in weather forecasting. Thus chaos theory can give us some qualitative hints, but I think a search for underlying models giving quantitative agreement with observed data will be in vain.

I am very sceptical about quantitative predictions for complex systems, such as economic and social ones. While on the one hand it seems necessary to make such predictions, we must be also aware of their great limitations. The only way out is to steer a system continuously and softly by setting again and again conditions so that it can smoothly self-organize into a hopefully optimal state. This then leads to the question of control parameters, which in economy may be interest rates, the amount of cash flow, and so on. But the discussion of these issues will be beyond the scope of this article.

7 Concluding Remarks

I have tried to show how the same principles can be applied to quite a variety of disciplines that are dealing with self-organizing processes in complex systems. The systems treated ranged from physics to sociology. Because of the analogies thus unearthed, it becomes possible to model phenomena in complex systems, for instance by computer procedures. It should be stressed that I restricted myself in this article to the conceptual level; the level at which concrete models are produced has been described in several other volumes of the Springer Series in Synergetics.

References

[1]   H. Haken, Lectures at Stuttgart University (1969), H. Haken and R. Graham, Umschau 6, 191 (1971)
[2]   H. Haken, Synergetics. An Introduction, 3rd. ed., (Springer, Berlin 1983)
[3]   H. Haken, Advanced Synergetics, (Springer, Berlin 1983)
[4]   H. Haken, Information and Self-Organization, (Springer, Berlin 1988)
[5]   A. Einstein, Physik Zeitschr. 18, 121 (1917)
[6]   J.A.S. Kelso, American Journal of Physiology: Regulatory, Integrative and Comparative Physiology, 15, R 1000 - R 1004 (1989)
[7]   H. Haken, J.A.S. Kelso, H. Bunz, Biol. Cybern. 51, 347-356 (1985)
[8]   G. Schöner, H. Haken, J.A.S. Kelso, Biol. Cybernetics, 53, 442 (1986)
[9]   G. Schöner, H. Haken, J.A.S. Kelso, J.P. Scholz, Physica Scripta 35, 79-87 (1987)
[10] H. Haken, Synergetic Computers and Cognition, (Springer, Berlin 1990)
[11] W. Köhler, Die physischen Gestalten in Ruhe und im stationären Zustand, Vieweg, Braunschweig (1920)
[12] G. Bateson, Morale and National Character, In Civilian Morale. Society for the Psychological Study of Social Issues, Second Yearbook. Ed. by Goodwin Watson, pp. 71-91, Boston: (Houghton Mifflin Co. for Regual and Hitchcock, New York) (1944)
[13] Paul Samuelson, Foundations of Economic Analysis, (Harvard University Press, Cambridge MA 1947)
[14] Wei-Bin Zhang, Synergetic Economics, (Springer, Berlin 1991)
[15] Thomas S. Kuhn, The Structure of Scientific Revolutions, (University of Chicago Press, Chicago 1970)
[16] Solomon E. Asch, Group Forces in the Modification and Distortion of Judgements, in: Social Psychology, Prentice Hall Inc. New York, p. 452 (1952)

Modelling Concepts of Synergetics with Application to Transitions Between Totalitarian and Liberal Political Ideologies

Wolfgang Weidlich

1 Introduction

Synergetics is a new branch of science dealing with the universal laws of the dynamic macro-structures which are generated in multi-component systems through the interactions between their elements.

Human society is such a multi-component system with a manifold of material and mental interactions between its elements, the individuals. Therefore, synergetics should also be applicable to the society, that means to the modelling of social processes!

In the following such a modelling framework is presented. It consists in the combination of concepts taking into account the special nature of social systems with concepts taken from statistical physics and synergetics, which are universally applicable to stochastic multi-component systems (A more comprehensive presentation of the general modelling concepts is given in [1] and [2]. Applications to migration processes of human populations including empirical evaluations are presented in [3].)

The purpose of this modelling procedure is the development of an integrated concept of theory construction for the quantitative description of collective evolutions in the society. The following partial purposes are included:

1. The formulation of the interrelation between the microlevel of individual decisions and the macrolevel of dynamical collective processes in the society.
2. The derivation of a probabilistic description of the macro-process including stochastic fluctuations, and the derivation of a quasi-deterministic description, in which the fluctuations are neglected.
3. The investigation of model solutions by analytical methods, e.g. the exact or approximate solution of master equations or meanvalue equations, or by numerical simulation of characteristic scenarios.
4. The evaluation of empirical systems, including field inquiries on the microlevel, as well as regression analysis on the macrolevel for the determination of model parameters, and forecasting of future evolutions by model simulation.

Springer Series in Synergetics, Vol. 61
Eds.: R.K. Mishra, D. Maaß, E. Zwierlein

The potential domains of application belong to different sectors of social science, namely to

- *Sociology* (for instance socio-political opinion formation, the example treated below in this article)
- *Demography* (for instance migration of populations)
- *Regional Science* (for instance formation of settlements and urban dynamics)
- *Economics* (for instance nonlinear models for business cycles and market instabilities)

We finish this introduction with a scheme of the general conceptual framework for the quantitative modelling of socio-dynamics in synergetics. The blocks constituting this scheme are explained in more detail in the next two sections.

2 Characterization of the Social System

2.1 The state of the Society

a) Microvariables: The Social Role of the Individual The individual possesses an *"attitude vector"*, that means a multiple

$$\mathbf{i} = (i_1, i_2, \ldots, i_a, \ldots, i_A) \tag{1}$$

of publicly exhibited *"external"* attitudes (e.g. opinions, activities, behavioral modes) with respect to A different aspects $a = 1, 2, \ldots, A$. These attitudes are conditioned by internal inclinations and by external social constraints.

Furthermore, the individual may possess an *internal propensity* with respect to its own publicly exhibited attitude i. This propensity is described by the *trendparameter* ϑ_i assuming integer values only for simplicity. The "internal" propensity may be in agreement with or in opposition to the "external" attitude i.

$$
\begin{aligned}
&\text{Values } \vartheta_i > 0 \text{ describe} &&\textit{internal affirmation} \text{ of attitude } i \\
&\text{The value } \vartheta_i = 0 \text{ describes} &&\textit{internal neutrality} \text{ to attitude } i \\
&\text{Values } \vartheta_i < 0 \text{ describe} &&\textit{internal opposition} \text{ to attitude } i
\end{aligned}
\tag{2}
$$

b) The Macrovariables of the Society There exist *subpopulations* $\mathcal{P}_\alpha, \alpha = 1, 2, \ldots, P$, each consisting of individuals of the same social background (e.g. social "classes"). A *coarse-grained* model uses one or few subpopulations $\mathcal{P}_\alpha$ only, but a *fine-grained* model uses many differentiated subpopulations $\mathcal{P}_\alpha$.

A central macrovariable is the *socioconfiguration*. It describes the distribution of attitudes (opinions, activities, behaviours) among the subpopulations $\mathcal{P}_\alpha$ of the society and thus characterizes the macrostate of the society. The *socioconfiguration* consists of a multiple of integers

$$\mathbf{n} = \{n_1^1, \ldots, n_i^\alpha, \ldots, n_C^P\} \tag{3}$$

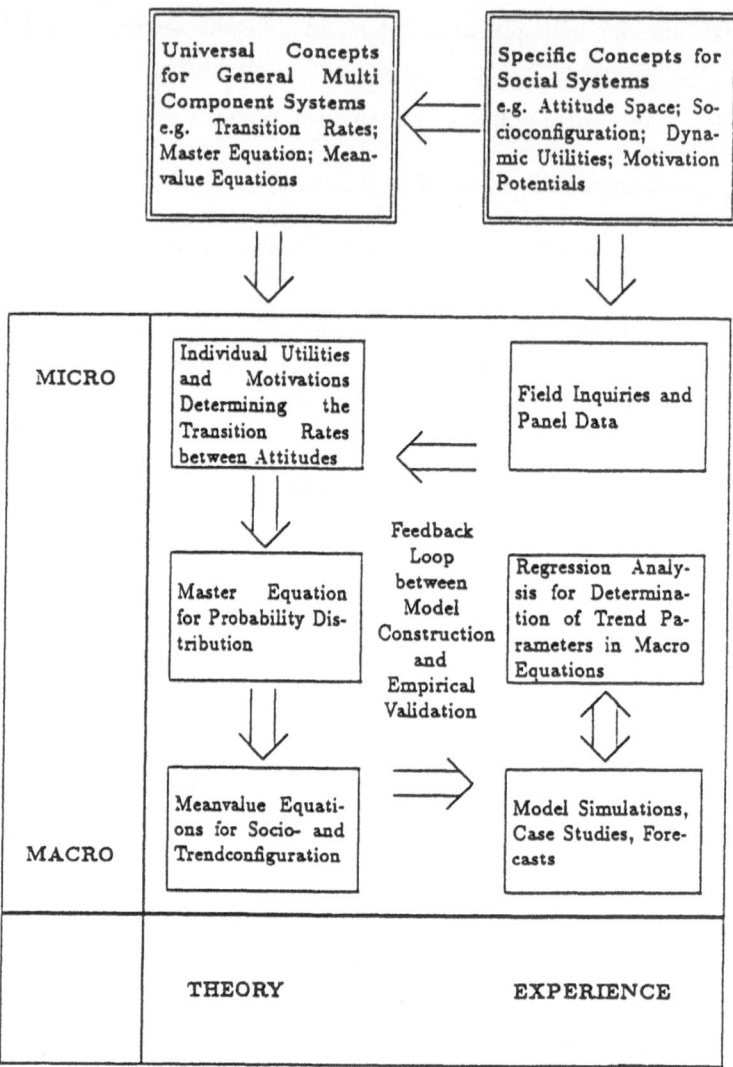

Fig. 1. The general conceptual framework for the quantitative modelling of socio-dynamics

where n_i^α is the number of individuals of subpopulation $\mathcal{P}_\alpha$ having the attitude i.

If *trendparameters* ϑ_i^α are assigned to the members of $\mathcal{P}_\alpha$ who have the attitude i, then also the *trendconfiguration*

$$\boldsymbol{\vartheta} = \{\vartheta_1^1, \ldots, \vartheta_i^\alpha, \ldots, \vartheta_C^P\} \qquad (4)$$

can be introduced as a further set of macrovariables in addition to the socioconfiguration. It is assumed that the integer value ϑ_i^α varies between a minimal amplitude $-\Theta$ and a maximal amplitude $+\Theta$.

2.2 Elements of Socio-Dynamics

a) Utilities and Motivation Potentials The *utility* u_i^α is defined as a measure of the usefulness of the adoption of the external attitude i (namely the publicly exhibited opinion, activity, behaviour) for a member of subpopulation $\mathcal{P}_\alpha$. The utility u_i^α is a real number: $-\infty < u_i^\alpha < +\infty$. It may be a function of the socioconfiguration $\mathbf{n}$ and the trendconfiguration $\boldsymbol{\vartheta}$ as follows

$$u_i^\alpha(\mathbf{n}, \boldsymbol{\vartheta}) \;=\; g_i^\alpha(\mathbf{n}) + h_i^\alpha(\mathbf{n})\vartheta_i^\alpha$$
$$\text{with } h_i^\alpha(\mathbf{n}) \geq 0 \tag{5}$$

Formula (5) indicates, that the utility u_i^α of attitude i will in general depend on the collective *external* social situation by the term $g_i^\alpha(\mathbf{n})$. On the other hand this external influence will be modified by the term $h_i^\alpha(\mathbf{n})\vartheta_i^\alpha$ in (5) containing the *internal* trend ϑ_i^α of a member of population $\mathcal{P}_\alpha$ with attitude i: An affirmative trend $\vartheta_i^\alpha > 0$ will enhance the utility u_i^α, whereas an opposing trend $\vartheta_i^\alpha < 0$ will diminish the utility u_i^α.

The *motivation potential* v_i^α is defined as a measure of the internal psychological satisfaction of a member of $\mathcal{P}_\alpha$ being in attitude i, with his own internal trend ϑ_i^α. Putting

$$v_i^\alpha = b_i^\alpha(\mathbf{n})\vartheta_i^\alpha, \text{ where } b_i^\alpha(\mathbf{n}) \gtrless 0 \tag{6}$$

the satisfaction is increasing for growing trend ϑ_i^α, if $b_i^\alpha(\mathbf{n}) > 0$, but decreasing for growing ϑ_i^α, if $b_i^\alpha(\mathbf{n}) < 0$. Hence, positive b_i^α describe a high psychological satisfaction with positive affirmative values of ϑ_i^α, whereas negative b_i^α describe a high internal satisfaction with negative, dissident values of ϑ_i^α.

b) Utility- and Motivation-Guided Individual and Configurational Probability Transition Rates Our application of the utility concept differs from that utilized in conventional economics in two respects: In classical economics the utility concept is applied in *static* situations to the *deterministic* behaviour of economic individuals. Instead, we make use of utilities and motivation potentials in order to describe the *dynamic* evolution of situations by modelling the *probabilistic* behaviour of homogenous ensembles (subpopulations) of individuals.

The following key-concepts are introduced:

a) The changes of publicly exhibited attitudes $i \Rightarrow j$ and of internal trends $\vartheta_i^\alpha \Rightarrow \vartheta_i^\alpha \pm 1$ of individuals of population $\mathcal{P}_\alpha$ are the *elementary dynamic processes* on the *microlevel* of the society. They induce corresponding changes

$$\mathbf{n} = \{n_1^1 \ldots n_j^\alpha, \ldots, n_i^\alpha \ldots n_C^P\}$$
$$\Rightarrow \mathbf{n}_{(ji)}^\alpha = \{n_1^1 \ldots (n_j^\alpha + 1), \ldots, (n_i^\alpha - 1), \ldots n_C^P\} \tag{7}$$

and

$$\boldsymbol{\vartheta} = \{\vartheta_1^1 \ldots \vartheta_i^\alpha, \ldots, \vartheta_C^P\}$$
$$\Rightarrow \boldsymbol{\vartheta}_\pm^\alpha = \{\vartheta_1^1, \ldots, (\vartheta_i^\alpha \pm 1), \ldots, \vartheta_C^P\} \tag{8}$$

of the socioconfiguration and trendconfiguration, that means of the macrovariables.

b) It is assumed, that the difference between the utility or motivation potential of an origin state and a destination state plays the role of a "driving force" for a transition of an individual between these states.

Therefore, the functional form of the individual and configurational *probability transition rates from origin to destination state* is constructed in terms of these driving forces. The following exponential ansatz proves convenient and plausible as well, because it guarantees the positive-definiteness and simultaneously the factorization into a push and a pull term of the transition rates. Hence we put

Individual rate from attitude i to attitude j for a member of $\mathcal{P}_\alpha$:

$$p_{ji}^\alpha(\mathbf{n}, \boldsymbol{\vartheta}) = \nu \exp[u_j^\alpha(\mathbf{n}_{(ji)}^\alpha, \boldsymbol{\vartheta}) - u_i^\alpha(\mathbf{n}, \boldsymbol{\vartheta})] \tag{9}$$

Configurational rate for the transition (7):

$$w_{ji}^\alpha(\mathbf{n}, \boldsymbol{\vartheta}) = p_{ji}^\alpha(\mathbf{n}, \boldsymbol{\vartheta})n_i^\alpha \tag{10}$$

Rate for the transition $\vartheta_i^\alpha \Rightarrow (\vartheta_i^\alpha + 1)$ for members of $\mathcal{P}_\alpha$:

$$\begin{aligned} r_{i\uparrow}^\alpha(\mathbf{n}, \boldsymbol{\vartheta}) &= \mu(\Theta - \vartheta_i^\alpha) \exp[v_i^\alpha(\mathbf{n}, \vartheta_i^\alpha + 1) - v_i^\alpha(\mathbf{n}, \vartheta_i^\alpha)] \\ &\equiv \mu(\Theta - \vartheta_i^\alpha) \exp[b_i^\alpha(\mathbf{n})] \end{aligned} \tag{11}$$

Rate for the transition $\vartheta_i^\alpha \Rightarrow (\vartheta_i^\alpha - 1)$ for members of $\mathcal{P}_\alpha$:

$$\begin{aligned} r_{i\downarrow}^\alpha(\mathbf{n}, \boldsymbol{\vartheta}) &= \mu(\Theta + \vartheta_i^\alpha) \exp[v_i^\alpha(\mathbf{n}, \vartheta_i^\alpha - 1) - v_i^\alpha(\mathbf{n}, \vartheta_i^\alpha)] \\ &\equiv \mu(\Theta + \vartheta_i^\alpha) \exp[-b_i^\alpha(\mathbf{n})] \end{aligned} \tag{12}$$

3 Equations of Motion for the Socio-Dynamics

3.1 The Stochastic Level of Description

We consider the probability $P(\mathbf{n}, \boldsymbol{\vartheta}; t)$ to find at time t the socioconfiguration $\mathbf{n}$ and the trend configuration $\boldsymbol{\vartheta}$ in the society. This probability distribution obeys the following fundamental *master equation*;

$$\begin{aligned} \frac{dP(\mathbf{n}, \boldsymbol{\vartheta}; t)}{dt} &= \left[\sum_{j,i,\alpha} w_{ji}^\alpha(\mathbf{n}_{ij}^\alpha, \boldsymbol{\vartheta})P(\mathbf{n}_{(ij)}^\alpha, \boldsymbol{\vartheta}; t) - \sum_{j,i,\alpha} w_{ji}^\alpha(\mathbf{n}, \boldsymbol{\vartheta})P(\mathbf{n}, \boldsymbol{\vartheta}; t) \right] \\ &+ \left[\sum_{i,\alpha} r_{i\uparrow}^\alpha(\mathbf{n}, \vartheta_{i-}^\alpha)P(\mathbf{n}, \vartheta_{i-}^\alpha; t) - \sum_{i,\alpha} r_{i\uparrow}^\alpha(\mathbf{n}, \boldsymbol{\vartheta})P(\mathbf{n}, \boldsymbol{\vartheta}; t) \right] \\ &+ \left[\sum_{i,\alpha} r_{i\downarrow}^\alpha(\mathbf{n}, \vartheta_{i+}^\alpha)P(\mathbf{n}, \vartheta_{i+}^\alpha; t) - \sum_{i,\alpha} r_{i\downarrow}^\alpha(\mathbf{n}, \boldsymbol{\vartheta})P(\mathbf{n}, \boldsymbol{\vartheta}; t) \right] \end{aligned} \tag{13}$$

The solution $P(\mathbf{n}, \vartheta; t)$ of eq. (13) describes not only the mean trajectory of a society in the configuration space, but also the probability of stochastic deviations from this mean behaviour.

The right hand side of the probability evolution equation (13) consists of three terms. The first term describes the change of the probability of the configuration $(\mathbf{n}, \vartheta)$ by probability inflows from neighbouring states $(\mathbf{n}_{(ij)}^{\alpha}, \vartheta)$ and probability outflows from state $(\mathbf{n}, \vartheta)$. Similarly, the second and third term describe changes of the probability of $(\mathbf{n}, \vartheta)$ by enhancement or diminution processes of the internal trend, respectively.

3.2 The Quasi-Deterministic Level of Description

A description of the macro-evolution of the society in terms of expectation values (meanvalue) is indicated, if one is only interested in the mean behaviour and not in probabilistic fluctuations.

The expectation values of the components of the socio- and trend-configuration are defined as follows

$$n_k^{\beta}(t) = \sum_{\mathbf{n}, \vartheta} n_k^{\beta} P(\mathbf{n}, \vartheta; t) \tag{14}$$

$$\vartheta_k^{\beta}(t) = \sum_{\mathbf{n}, \vartheta} \vartheta_k^{\beta} P(\mathbf{n}, \vartheta; t) \tag{15}$$

In the case of unimodal (or appropriately truncated) probability distributions the following approximate equations of motion can be derived for $n_k^{\beta}(t)$ and $\vartheta_k^{\beta}(t)$, making use of the master equation (13):

$$\frac{dn_k^{\beta}(t)}{dt} = \sum_i w_{ki}^{\beta}(\mathbf{n}(t), \vartheta(t)) - \sum_j w_{jk}^{\beta}(\mathbf{n}(t), \vartheta(t)) \tag{16}$$

$$\frac{d\vartheta_k^{\beta}(t)}{dt} = r_{k\uparrow}^{\beta}(\mathbf{n}(t), \vartheta(t)) - r_{k\downarrow}^{\beta}(\mathbf{n}(t), \vartheta(t)) \tag{17}$$

The equations (16) and (17) form a set of coupled selfcontained, autonomous, in general nonlinear differential equations for $n_k^{\beta}(t)$ and $\vartheta_k^{\beta}(t)$, with $k = 1, 2, \ldots, C$ and $\beta = 1, 2, \ldots, P$. The standard methods of nonlinear analysis can be applied to this dynamical system.

4 Example: A Dynamical Model of Collective Political Opinion Formation

4.1 The Components of the Model

We consider the simplest version of such a model by assuming only two competing political opinions $i = +$ and $-$ (that means two parties or two ideologies) and only one homogenous population (this means, that the index $\alpha = 1$ can be skipped). The *socioconfiguration* now consists of

$$\mathbf{n} = \{n_+; n_-\};\tag{18}$$

with the total population number

$$2N = n_+ + n_-\tag{19}$$

and

$$n_+ = N + n; \quad n_- = N - n\tag{20}$$

after introducing the *majority variable*

$$n = \frac{1}{2}(n_+ - n_-); \quad -N \leq n \leq +N\tag{21}$$

The *trend configuration* is given by

$$\vartheta = \{\vartheta_+, \vartheta_-\}; \quad -\Theta \leq \vartheta_\pm \leq +\Theta\tag{22}$$

The form of the *utility functions* is a special case of (5):

$$u_+(n_+, \vartheta_+) = g(n_+) + h(n_+)\vartheta_+$$
$$u_-(n_-, \vartheta_-) = g(n_-) + h(n_-)\vartheta_-\tag{23}$$

with the simplest nontrivial ansatz for g and h:

$$g(n_\pm) = \frac{1}{2}\kappa n_\pm; \quad h(n_\pm) = \frac{1}{2}\gamma\tag{24}$$

Similarly the *motivation potentials* are given by (see (6)):

$$v_+(n_+, \vartheta_+) = b_+(n_+)\vartheta_+$$
$$v_-(n_-, \vartheta_-) = b_-(n_-)\vartheta_-\tag{25}$$

with the simplest nontrivial form of $b_\pm(n_\pm)$:

$$b_\pm(n_\pm) = \beta(n_\pm - N) = \pm\beta n\tag{26}$$

The meaning of (23) ...(26) becomes evident, if the *transition rates* are constructed according to the general rules (9) ...(12), with the result:

$$\begin{aligned}
\text{a) } & w_{+-}(n, \vartheta) = \nu(N - n)\exp[\kappa n + \gamma\vartheta]\\
\text{b) } & w_{-+}(n, \vartheta) = \nu(N + n)\exp[-(\kappa n + \gamma\vartheta)]\\
\text{c) } & r_\uparrow(n, \vartheta) = \mu(\Theta - \vartheta)\exp[\beta n]\\
\text{d) } & r_\downarrow(n, \vartheta) = \mu(\Theta + \vartheta)\exp[-\beta n]
\end{aligned}\tag{27}$$

It is plausible here to put

$$\vartheta_+ = -\vartheta_- \equiv \vartheta\tag{28}$$

and to characterize the socio- and trend configuration (18) and (22) by the two variables $\{n, \vartheta\}$ only. The transition rates (27) then induce the following next neighbour transitions:

$$a) \{n\vartheta\} \Rightarrow \{n+1, \vartheta\}$$
$$b) \{n\vartheta\} \Rightarrow \{n-1, \vartheta\}$$
$$c) \{n\vartheta\} \Rightarrow \{n, \vartheta+1\}$$
$$d) \{n\vartheta\} \Rightarrow \{n, \vartheta-1\} \tag{29}$$

The name and the interpretation of the paramters $\nu, \mu, \Theta, \kappa, \gamma, \beta$ follows as a consequence of the form of the transition rates in terms of the macrovariables $\{n, \vartheta\}$ of the society with respect to its external opinion (n) and internal trend (ϑ) state.

$$\nu = \text{opinion evolution speed parameter}$$
$$\mu = \text{trend evolution speed paramter}$$
$$\Theta = \text{maximal trend amplitude}$$
$$\kappa = \text{opinion pressure parameter}$$
$$\gamma = \text{trend influence parameter}$$
$$\beta > 0 \,(\text{or } \beta < 0) = \text{propensity parameter}$$

for developing an affirmative (or a dissident) internal trend in relation to the external majority opinion.

4.2 The Equations of Motion of the Model

The master equation for the probability $P(n, \vartheta; t)$ to find the configuration $\{n, \vartheta\}$ at time t is a special case of eq. (13). It reads:

$$\frac{dP(n, \vartheta; t)}{dt} =$$

$$[w_{+-}(n-1, \vartheta)P(n-1, \vartheta; t) + w_{-+}(n+1, \vartheta)P(n+1, \vartheta; t)$$
$$-w_{+-}(n, \vartheta)P(n, \vartheta; t) - w_{-+}(n, \vartheta)P(n, \vartheta; t)]$$
$$+[r_{\uparrow}(n, \vartheta-1)P(n, \vartheta-1; t) - r_{\uparrow}(n, \vartheta)P(n, \vartheta; t)]$$
$$+[r_{\downarrow}(n, \vartheta+1)P(n, \vartheta+1; t) - r_{\downarrow}(n, \vartheta)P(n, \vartheta; t)] \tag{30}$$

Of course, the explicit form (27) of the transition rates must be inserted.

The equations of motion for the expectation values $n(t)$ and $\vartheta(t)$ follow from the general form (16), (17) and read

$$\frac{dn}{dt} = w_{+-}(n, \vartheta) - w_{-+}(n, \vartheta)$$
$$= 2\nu\{N \sinh(\kappa n + \gamma\vartheta) - n \cosh(\kappa n + \gamma\vartheta)\} \tag{31}$$

$$\frac{d\vartheta}{dt} = r_\uparrow(n, \vartheta) - r_\downarrow(n, \vartheta)$$
$$= \mu\{(\Theta - \vartheta)\exp(\beta n) - (\Theta + \vartheta)\exp(-\beta n)\} \tag{32}$$

Introducing the scaled variables

$$x = \frac{\vartheta}{\Theta}; \; -1 \leq x \leq +1; \quad y = \frac{n}{N}; \; -1 \leq y \leq +1$$

$$\tau = 2\nu t; \; \tilde{\mu} = \frac{\mu}{\nu}; \; \tilde{\kappa} = N\kappa; \; \tilde{\gamma} = \Theta\gamma; \; \tilde{\beta} = N\beta \tag{33}$$

one obtains the scaled form of the meanvalue equations:

$$\frac{dy}{d\tau} = \{\sinh(\tilde{\kappa}y + \tilde{\gamma}x) - y\cosh(\tilde{\kappa}y + \tilde{\gamma}x)\} \tag{34}$$

$$\frac{dx}{d\tau} = \tilde{\mu}\{\sinh(\tilde{\beta}y) - x\cosh(\tilde{\beta}y)\} \tag{35}$$

4.3 Simulation of Characteristic Scenarios and their Interpretation

The model equations are now solved numerically for concrete parameter sets, which correspond to characteristic scenarios of political behaviour. In all cases we exhibit the fluxlines of the meanvalue equations (34), (35) and the corresponding stationary solution of the master equation (30). Throughout all simulations we choose $\tilde{\mu} = 2$ and $2N = 20$. The very small value of $2N$ has been chosen for illustrative purposes, because it yields broad probability distributions, whereas the scaled meanvalue equations do not explicitly depend on N.

We distinguish two groups of scenarios: the group A with affirmative trend-dynamics (i.e. with $\beta > 0$) and the group D with dissident trenddynamics (i.e. with $\beta < 0$). In both groups we vary the value of the parameter $\tilde{\beta}$ (i.e. the strength of the propensity to go in case A to affirmative or in case D to opposing internal trends). Furthermore we vary the opinion pressure parameter $\tilde{\kappa}$. A small (large) $\tilde{\kappa}$ means a liberal (totalitarian) society with a small (high) pressure on the individual to adapt his external opinion to the majority opinion.

Group A: Affirmative Trend Dynamics ($\tilde{\beta} > 0$)

Case A.1, shown in Fig. 2, corresponds to a liberal society without opinion pressure and with weak affirmation propensity. The origin $(\hat{y}, \hat{x}) = (0, 0)$, which corresponds to a balanced opinion and trend situation, is a stable fixed point, (see Fig. 2a) around which probabilistic fluctuations occur (see Fig. 2b).

Case A.2, shown in Fig. 3, corresponds to a liberal society without opinion pressure but with a strong affirmation propensity. The balanced opinion / trend situation is now unstable, because the strong inclination for affirmation leads to the self-stabilization of stable majority opinions $\hat{n} > 0$ or $\hat{n} < 0$ (see Fig. 3a) and to fluctuations around these stable situations (see Fig. 3b).

Case A.3, shown in Fig. 4, corresponds to a society with considerable opinion pressure, that means with totalitarian tendencies, and with intermediate affirmation propensity. We find again stable opinion majorities (see Fig. 4a) with

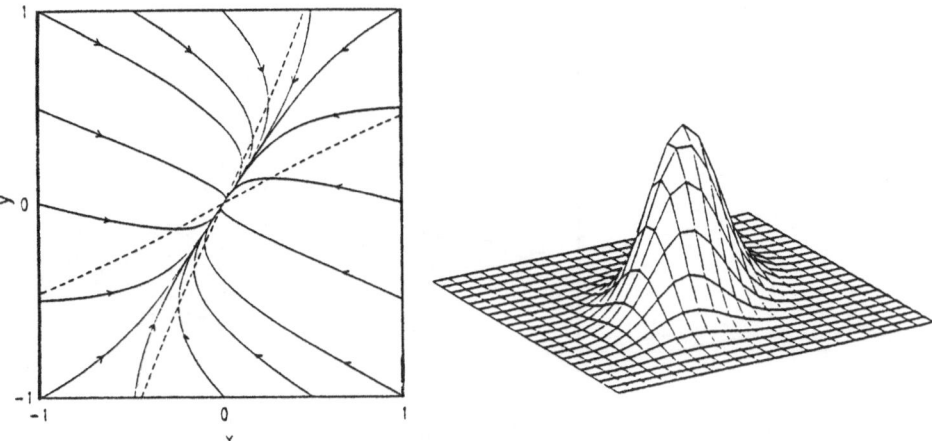

Fig. 2. Case A.1   Parameters: $\tilde{\kappa} = 0$; $\tilde{\beta} = \tilde{\gamma} = 0.5$

No opinion pressure; weak affirmation strength

a) The flux-lines approach the stable balanced opinion situation $(\hat{y}, \hat{x}) = (0,0)$.

b) Unimodal stationary probability distribution peaked around the origin $(0,0)$.

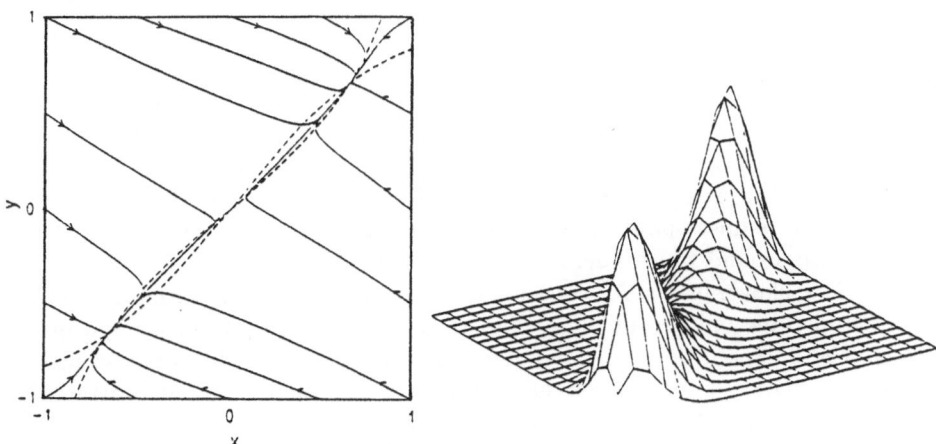

Fig. 3. Case A.2   Parameters: $\tilde{\kappa} = 0$; $\tilde{\beta} = \tilde{\gamma} = 1.2$

No opinion pressure; strong affirmation strength

a) Unstable balanced opinion situation $(0,0)$. The flux-lines approach unbalanced opinion majority situations stabilized by affirmation.

b) Bimodal stationary probability distribution peaked around the stable opinion majority situations.

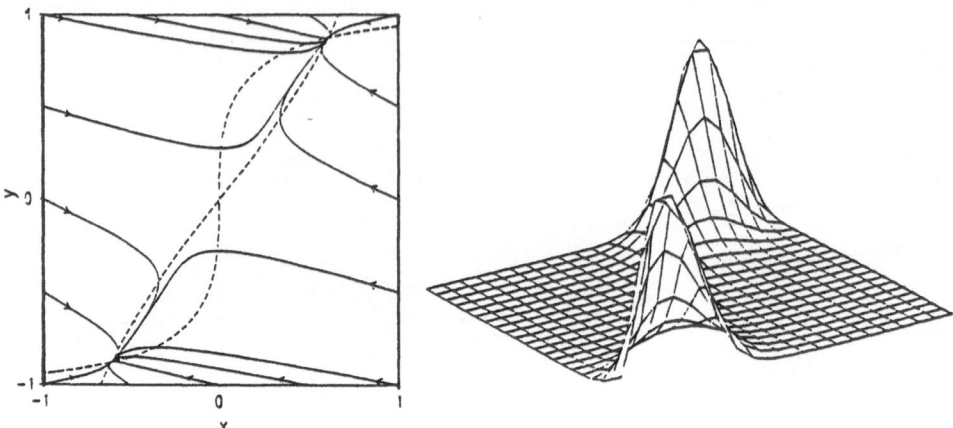

Fig. 4. Case A.3   Parameters: $\tilde{\kappa} = 1$; $\tilde{\beta} = \tilde{\gamma} = 0.8$
Considerable opinion pressure; moderate affirmation strength
a) Unstable balanced opinion situation $(0, 0)$. The flux-lines approach unbalanced opinion majority situations stabilized by affirmation and opinion pressure.
b) Bimodal stationary probability distribution peaked around the stable opinion majority situations.

fluctuations around them (see Fig. 4b). However, the opinion majority is now partially stabilized by an affirmative internal trend, and partially by the effect of opinion pressure.

Group D: Dissident Trend Dynamics $(\tilde{\beta} < 0)$

Case D.1, shown in Fig. 5, corresponds to a liberal society without opinion pressure but a strong propensity to develop an opposing internal trend. Therefore, any existing external opinion majorities are soon removed by opposing trends. The flux-lines spiral into the balanced opinion / trend situation $(0, 0)$ (see Fig. 5a), around which fluctuations occur (see Fig. 5b).

Case D.2, shown in Fig. 6, corresponds to a society with totalitarian tendencies (considerable opinion pressure) and weak inclination to develop opposing trends. The balanced opinion / trend situation is still stable, but opinion majorities ($n > 0$ or $n < 0$), once established, have a long lifetime as demonstrated by the broad probability distribution of Fig. 6b.

Case D.3, shown in Fig. 7, describes a totalitarian society with strongly developed opinion pressure and weak propensity to develop dissident trends. The society stabilizes in a state of high external opinion majority induced by opinion pressure, and of simultaneous weak opposing internal trend (see Fig. 7a). The bimodal probability distribution (see Fig. 7b) describes the probabilistic fluctuations around the stable fixed points.

Case D.4, shown in Fig. 8, corresponds to a totalitarian society with fully developed opinion pressure, but equally strong propensity to develop a dissident internal trend. A dramatic revolutionary dynamics arises from the antagonistic

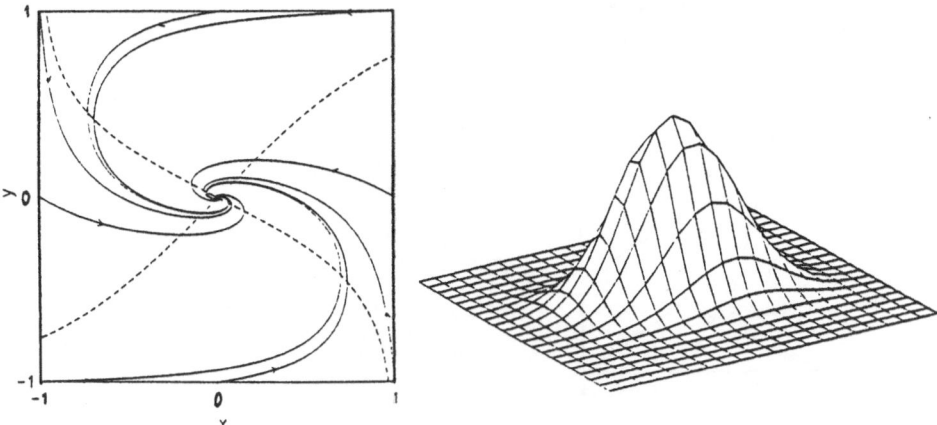

Fig. 5. Case D.1   Parameters: $\tilde{\kappa} = 0$; $\tilde{\beta} = -2$; $\tilde{\gamma} = 1$
No opinion pressure; strong dissidence strength
a) Stable balanced opinion situation $(0,0)$. The flux-lines spiral into the origin $(0,0)$.
b) Unimodal stationary probability distribution peaked around the balanced opinion situation $(0,0)$.

competition between the effects of opinion pressure and dissidence strength. The evolution approaches a limit cycle (Fig. 8a) whose phases can be interpreted as follows.

Starting somewhere in the first quadrant, transient states with affirmative trend are quickly traversed, until in the second quadrant a long-living metastable state builds up. Its lifetime is proportional to the stationary probability (see Fig. 8b). In this state there exists a strong antagonism between the external opinion majority ($n \lesssim N$) sustained by opinion pressure, and the simultaneous opposing internal trend ($\vartheta \gtrsim -\Theta$). The metastable state finally breaks down by the "victory" of the opposing trend. Thereupon intermediate transient states are traversed in the third quadrant until there stabilizes in the fourth quadrant another metastable state with the opposite opinion majority ($n \gtrsim -N$) and a trend ($\vartheta \lesssim +\Theta$) again in dissidence to this new opinion majority. This metastable state has again a high perseverance probability (see Fig. 8b), but will finally break down, too.

This result is a consequence of the oversimplified model assumption, that the parameters $\tilde{\kappa}, \tilde{\beta}, \tilde{\gamma}$ of political psychology remain constant during the evolution. It is however highly plausible (but not yet included in the model equations) that the revolutionary breakdown of the metastable state also leads to the evolution of other psychological trend parameters $\tilde{\kappa}, \tilde{\beta}, \tilde{\gamma}$. If, for instance, the opinion pressure $\tilde{\kappa}$ and the dissidence strength $\tilde{\beta}$ would decrease after the breakdown of the metastable totalitarian state, the further evolution could approach the balanced opinion / trend situation instead of the opposite totalitarian metastable state.

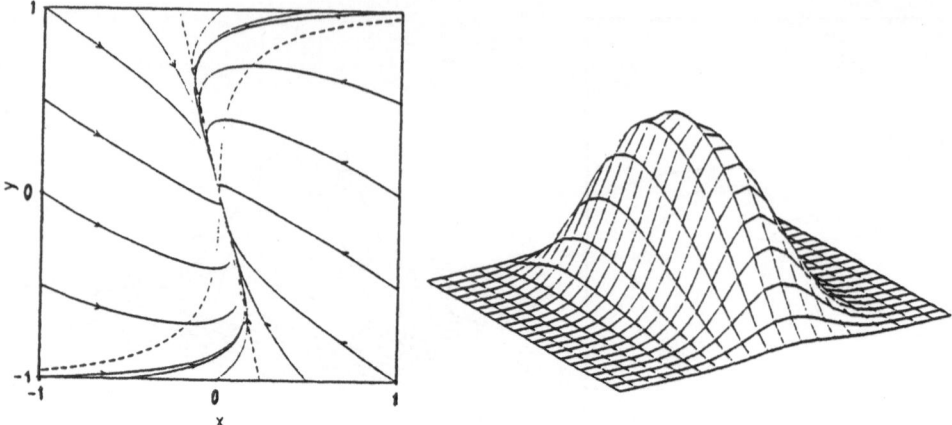

Fig. 6. Case D.2   Parameters: $\tilde{\kappa} = 1$; $\tilde{\beta} = -\frac{1}{4}$; $\tilde{\gamma} = 1$
Considerable opinion pressure; weak dissidence strength
a) Still (marginally) stable balanced opinion situation $(0,0)$. The flux-lines approach the origin via strongly unbalanced opinion situations.
b) Broad but still unimodal stationary probability distribution with large variance around balanced opinion situation $(0,0)$.

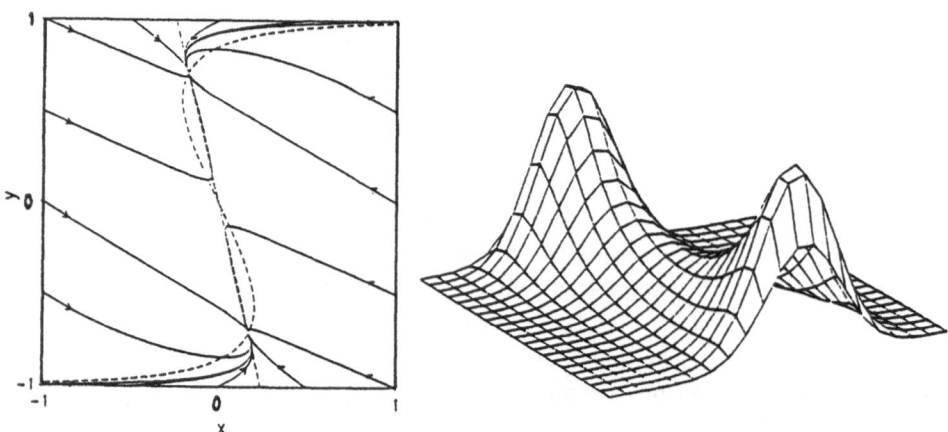

Fig. 7. Case D.3   Parameters: $\tilde{\kappa} = 1.5$; $\tilde{\beta} = -\frac{1}{4}$; $\tilde{\gamma} = 1$
Strong "totalitarian" opinion pressure; weak dissidence strength
a) Unstable balanced opinion situation $(0,0)$. The flux-lines approach stable unbalanced opinion majority situations, in which the opinion pressure outweighs the dissidence strength. b) Bimodal stationary probability distribution peaked around the stable opinion majority situations.

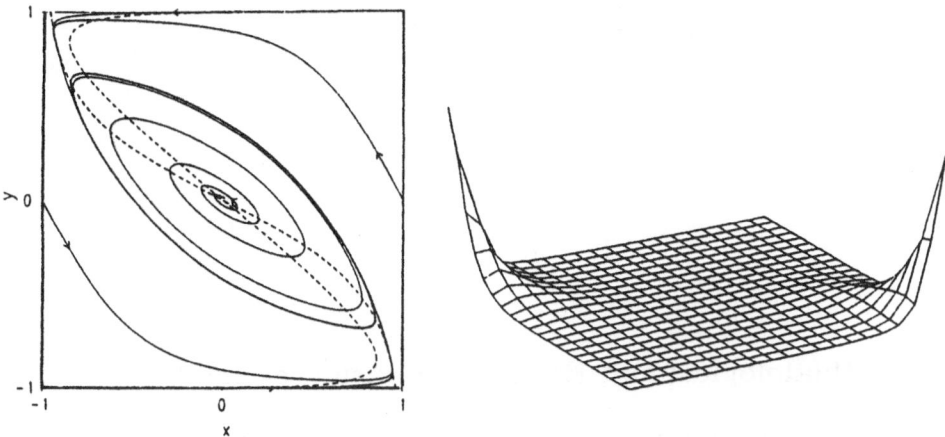

Fig. 8. Case D.4a and D.4b   Parameters: $\tilde{\kappa} = 3.5$; $\tilde{\beta} = -2$; $\tilde{\gamma} = 2$
Very strong "totalitarian" opinion pressure and strong dissidence strength
a) and b) No stable stationary situations. The flux-lines approach a limit cycle comprising metastable pronounced opinion majority situations sustained by very strong opinion pressure but finally destabilized by the strong dissidence trend.

Summarizing, it seems that the model, in particular its scenarios D, can provide semiquantiative insights into the dynamics of recent historical events, even in spite of the assumed oversimplifications.

References

1. Weidlich, W., Haag, G.: Concepts and Models of a Quantitative Sociology. Springer, Berlin, Heidelberg, New York (1983)
2. Weidlich, W.: Physics and Social Science – The Approach of Synergetics. Physics Reports 204, pp. 1 - 163, (1991)
3. Weidlich, W., Haag, G.: Interregional Migration – Dynamic Theory and Comparative Analysis. Springer, Berlin, Heidelberg, New York (1988).

Self-Organization, Artificial Intelligence and Connectionism

Michael M. Richter

1 Methodological and Historical Considerations

In the decade after 1945 Artificial Intelligence and connectionistic ideas were not separated. The term 'Artificial Intelligence' was not even born at that time. One of the aims at that time was to understand the working of the human brain and to describe the process of thinking in a formal and mathematical way. Different aspects came together and evolved into what researchers like Norbert Wiener called *cybernetics*. It was based on a long history and contained important ingredients from mathematical logic as well as from statistical mechanics. In the following we will isolate three periods in an attempt to structure human science. The connectionistic paradigm denotes in this light a new and fourth historical period.

The first three phases have been described by M. Foucault in his book[2] 'Les mots et les choses'. He describes the development of sciences from the 17th to the 19th century and distinguishes three periods. It is not a historical description in the usual sense but a systematic approach to discover common methods in different disciplines. These common elements can be regarded as fundamental codes of a culture; at the corresponding times one was not explicitly aware of them.

In the first phase thinking was dominated by what Foucault called *similarities*, today one would call it *associative knowledge*. A typical piece of knowledge of that form was a relation between a certain disease and its medical treatment. Such a relation was not explained and therefore in some sense of a magic character. The world seemed to be an accumulation of such knowledge units and the aim of research was to collect as much knowledge as possible in this way.

Nevertheless there was a desire to structure the knowledge and this led to the second phase which can be characterized by the term *taxinomia*. The main task of research was to structure and to classify knowledge. A prominent example is the classification introduced by Linnaeus into botany. Associative knowledge was still present but it was now subject to the aspect of ordering. One was still not able to explain the underlying reasons for observed behaviour but it was an important step in this direction. The third and Foucault's final period can be described by the term *modelling*. The intention of a model is to describe a machinery which is able to simulate behaviour and to predict observations. For instance, in economy the taxonomy of the price structure was complemented by the introduction of product models. The primary aspect was now not only the search for taxonomic attributes but more the analysis of dynamic events.

It is interesting to observe that these three periods of together 300 years have their direct counterparts in the the 30 years evolution of knowledge representation and expert system technology in artificial intelligence after 1960. The first period was characterized by the more or less exclusive use of facts and rules in expert systems. Prominent examples are seen in MYCIN, XCON and others. Such systems were difficult to handle when they exceeded a certain size. Similar effects were known from the experience with large programming packages in classical computer science. The answer was the development of software engineering techniques, in our context the most important paradigm was *object oriented programming*. This combines two principles, information hiding and taxonomic orderings. The principle of information hiding results in the encapsulating of complex data structures which then form closed words similar to the monads of Leibniz. The objects however communicate with each other and can be structured into hierarchies. In artificial intelligence the development of semantic nets and frames cultivated these ideas. In particular, the old paradigm of facts and rules could be incorporated into the new approach. There were still many unsatisfactory aspects present, however. The explanatory character of such systems could not be developed convincingly enough, simulations were impossible and future effects could not be predicted. This lead to the development of *model-based systems*. In a model all interesting aspects are formally described. For a machine a model contains not only the collection of its parts but also all interconnections, causal relations, intended functionalities etc.. The application of this technique leads to model-based diagnosis, model-based configuration, model-based planning and more.

Complexity arguments, however, also show the boundaries of the paradigm of models. If an organism becomes too complex, it is no longer possible to describe it formally in all detail in a reasonable time. This applies in particular to biological structures but also to complex artificial objects. This observation was made in some generality already around 1950 as well as thirty five years later in artificial intelligence again.

The main new aspect here is that one has to give up the main feature of a classical algorithm or of a logical calculus. The old idea here was to have every step made by a system under individual and perfect control. A formula is either true or false, an algorithm either produces a certain step or not and the question of termination is totally determined (although in general undecidable). When one wants to give up the deterministic view some other principles have to replace it. There are two main principles involved here. The first is the principle of self-organization; this means that the system itself determines the steps to make which was formerly regulated from the outside. The second principle is of statistical nature. The external world still plays a role for the system but in a more indirect way. The external world gives rewards and penalties according to the actually existing context. From the viewpoint of the system this is of probablistic nature because it has only little knowledge about the outside. This finally leads to the notion of evolution which not only applies to nature but also to artificial systems as long as they are guided by some principle of self-organization.

This way of handling knowledge can be regarded as a new period complementing those three introduced by Foucault. There are very many possible candidates for applications ranging from pattern recognition through administrative problems to very complex political dimensions. An example of the latter is summarized in the customary term *European Nervous System* (ENS). It is based on the observation that a unified Europe cannot be the result of a few political decisions but needs the adjustment and coordination of a huge amount of small actions. The only promising perspective is to use some kind of self-organization. The political problem here is to determine rewards and penalties in such a way that desirable developments are favoured. A connectionistic model of some sort seems to be unavoidable.

2 Machine Learning, Numerical Methods and Neural Nets

2.1 Basic Approaches

Neural nets [6] describe one possibility for the formalization of the connectionistic paradigm. Their main feature is that they not only produce outputs but also change their internal structure and behaviour. Such a change is usually referred to as a learning step. In neural nets connecting weights can be changed in such a way that the output is optimized in a desired way. The aspect of learning in a formal way is, however, by no means novel nor even exclusively elaborated in neural nets. The concept of learning has a long history and was investigated in many branches.

We will first make a basic distinction between analytic-numerical learning methods and logic-symbolic techniques. This distinction grasps in a nutshell the two main developments and methodological approaches which have taken place separately with almost no overlap. Nevertheless, we still have in principle only one problem area and one large class of concepts, methods and tools which are formulated in different terminologies, realised and implemented in different ways and which very often can be translated into each other.

In symbolic machine learning, supervised learning has dominated almost exclusively. At first sign this is somewhat surprising because in the philosophy of sciences inductive learning was most important in the development of empirical sciences where one never knows the truth absolutely and tries to approximate unknown laws. The explanation is probably that the supervised methods of inference are closer to the classical calculi of formal logic while clustering algorithms are somewhat unnatural in this context.

The main area where unsupervised learning takes places is explorative data analysis in statistics. Many algorithms, in particular for cluster analysis, have been developed and applied successfully. In neural networks both supervised and unsupervised learning are represented. The backpropagation algorithm is the most prominent representative of supervised learning. This has two phases. In the forward phase the input is propagated through the net and a teacher estimates or determines the error of the output. In the backward phase the weights of the net are adjusted in such a way that for the error function the minimum is approximated using the gradient method of steepest descent.

The father of all methods for unsupervised learning is competitive learning [7]. The advice of the teacher has to be replaced by some internal decisions and actions. This means that neurons have to be selected which get a reward and others which get a penalty. The only information available is the amount of input the neurons obtain and therefore it is natural to make this responsible for the reward or penalty. In competitive learning only the winner (i.e. the neuron with the largest input) is rewarded while in the Kohonen net the reward goes down gradually in a neighborhood around the winning neuron.

A typical example is the learning of concepts [4]. Suppose we are given a universe U of φ which usually is finite and a formula $\varphi(x)$ of first order predicate logic which is defined over U. $\varphi(x)$ is unknown and has to be learned by a learning system L. For this purpose a training sequence $a_1, a_2, \ldots$ of elements is given to L. Let $CB_n = \{a_1, \ldots, a_n\}$ be the case base at time n.

In supervised learning an external teacher tells L for each a_i whether $\varphi(a_i)$ holds or is false, in unsupervised learning no such information is provided.

In symbolic learning logical inference steps generate formulae which may eventually result in the correct formula $\varphi(x)$. The inference steps are, however, not neccessarily logically correct, i.e. truth-preserving, but rather of an inductive character.

The most common φ algorithm for supervised machine learning is the *version space algorithm*. The version space is the set of all formulae $\psi(x)$ which are still candidates for $\varphi(x)$ when the case base CB_n has been seen. A step from CB_n to CB_{n+1} leads to an updating and in particular to a shrinking of the version space. Under certain assumptions it can be shown that the version space converges to the unit element set $\{\varphi(x)\}$ in which case the unknown formula has been learned succesfully.

In order to apply numerical methods all participating objects like data, predicates or functions have to be translated to points, regions or functions of an n-dimensional real space. An inference step is now called an iteration step which tries to approximate a known or unknown magnitude. The counterpart of the termination of a logical procedure is now the convergence of a numerical algorithm.

The problem of learning such an unknown formula is approached by numerical methods with the use of distance measures (these are like pseudo-metrics but may violate the triangle inequality); equivalently one could use similarity measures. Suppose our objects are represented as vectors of the R^n; the universe is hence some finite set $U \subseteq R^n$. Furthermore we are given a distance measure $d(x, y)$ and a case base, i.e. a set $CB \subseteq U$ the elements of which are already correctly classified.

The *nearest neighbour method* now makes an attempt to classify also the unseen elements from $U \backslash CB$: For each $a \in U \backslash CB$ the class of a is determined by the class of $y \in CB$ where $d(a, y) = min(d(a, z) \mid z \in CB)$. The (unknown) formula $\varphi(x)$ splits U into two classes, one for $\varphi(x)$ and one for $\neg\varphi(x)$ and the aim is to determine them correctly using (d, CB). The analogue of the version space algorithm is now an algorithm that tries to learn a distance measure which classifies correctly [3]. This view is slightly misleading because the correctness of d also depends on the set CB; CB is the set of objects seen in the training sequence so far or a subset of it.

The nearest neighbour method can be regarded as a special case of analogical reasoning. Using analogy one tries to modify the known solution of an old problem in order to solve a new problem as seen in figure 1; for the nearest neighbour method the modification reduces to the identity.

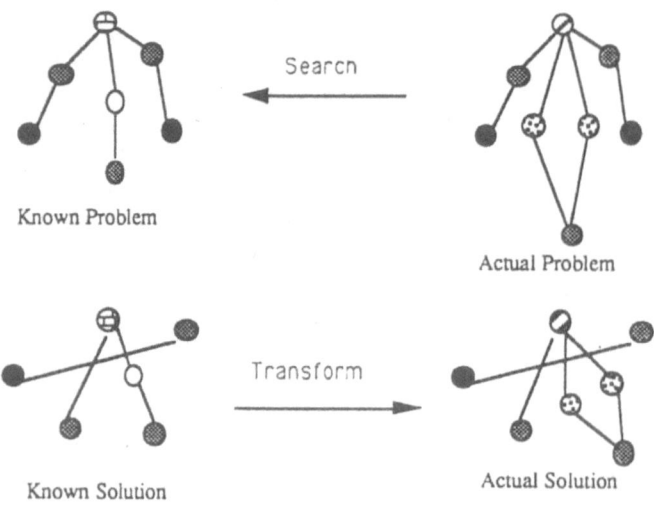

Fig. 1. Analogy

This technique is of course only useful if the modification is easier to perform than the solution of the actual new problem itself. This means that the two problems have to be similar in some way and if one has a case base available one is interested in retrieving the particular case which is in this sense most similar to the actual problem.

2.2 Aspects of Similarity

The notion of similarity is central to case-based reasoning. It has many interesting features and is presently discussed extensively. In order to define similarity it is plausible to use a distance measure. There are however different possibilities to represent similarity which we will now introduce:

1. a binary predicate $SIM(x, y) \subseteq U^2$ meaning "x and y are similar";
2. a binary predicate $DISSIM(x, y) \subseteq U^2$ meaning "x and y are not similar";
3. a ternary relation $S(x, y, z) \subseteq U^3$ meaning "y is at least as similar to x as z is to x";
4. a quaternary relation $R(x, y, u, v) \subseteq U^4$ meaning "y is at least as similar to x as v is to u";
5. a function $sim(x, y) : U^2 \to [0, 1]$ measuring the degree of similarity between x and y;
6. a function $d(x, y) : U^2 \to \mathrm{I\!R}$ measuring the distance between x and y.

The obvious questions which arise here are:

(i) How can one axiomatize these concepts, i.e. which laws govern them?
(ii) How are the concepts correlated, which are the basic ones and which can be defined in terms of others?
(iii) How useful are the concepts for the classification task?

Some interrelations between the introduced concepts are immediate. If d is a distance measure and sim a similarity measure then we define

$$R_d(x, y, u, v) : \Longleftrightarrow d(x, y) \leq d(u, v)$$
$$R_{sim}(x, y, u, v) : \Longleftrightarrow sim(x, y) \geq sim(u, v)$$

and

$$S_d(x, y, z) : \Longleftrightarrow R_d(x, y, x, z)$$
$$S_{sim}(x, y, z) : \Longleftrightarrow R_{sim}(x, y, x, z).$$

We say that d and sim are compatible , iff

$$R_d(x, y, u, v) \Longleftrightarrow R_{sim}(x, y, u, v) ;$$

compatibility is ensured by $d \equiv_f sim$ for some f.
As usual in topology the measures also define a neighborhood concept. For $\epsilon > 0$ we put

$$V_\epsilon(x) := V_{d,\epsilon}(x) := \{y | d(x, y) \leq \epsilon\},$$

and analogously $V_{sim,\epsilon}(x)$ is defined; if d is a metric then these sets are ordinary closed neighbourhoods. $S_d(x, y, z)$ expresses the fact that each neighbourhood of x which contains z also contains y.

The relation S allows one to define the concept y is most similar to x' : For some set $M \subseteq U$ some $y \in M$ is called *most similar* to x with respect to M iff

$$(\forall z \in M)S(x, y, z)$$

This notion is essential in case-based reasoning.
There are various ways to define specific measures or distance functions. It is convenient to assume that the objects are given as vectors of some R^n where each coordinate corresponds to an attribute and the entries will represent the value of that attribute. For most of our considerations is suffices to assume Boolean

attributes which have only the values 0 and 1. The simplest similarity measure comes from the well known Hamming distance. A much more refined measure is given by the Tversky Contrast Model [8]:

For two objects x and y we put

$A :=$ the set of all attributes which have equal values for x and y ;

$B :=$ the set of all attributes which have value 1 for x and 0 for y ;

$C :=$ the set of all attributes which have value 1 for y and 0 for x ;

The general form of a Tversky distance is

$$T(x,y) = \alpha \cdot f(A) - \beta \cdot f(B) - \gamma \cdot f(C)$$

where α, β and γ are positive real numbers. Most of the other possible measures are located between the Hamming and the Tversky measure with respect to the information which they can contain.

It is clear that the power of the measures depends on the choice of the constants α, β and γ and the specific function f. These magnitudes contain the knowledge in the measure.

The basic question is how to define a distance measure in such a way that the classification is at least sufficiently correct. This requires that some knowledge about the solution is encoded into the distance measure which at a first glance leads to some kind of a cycle: In order to find the unknown solution of a problem we have already to know it in some sense. In other words, the similarity of two objects cannot be detected a priori but only a posteriori after we have seen the classes they belong to.

There are two possible ways to deal with this problem. The first is to encode as much knowledge as possible into the measure and the second is to improve its capacity by a learning process.

The learning of distance measures will be discussed below in section 4. We will first be concerned with the comparison of the two approaches sketched above, a question which has been entirely neglected in the literature. Under moderate assumptions the following is the case:

a) There is a one-to-one correspondence between the formulae $\psi(x)$ in the version space after CB_n has been presented and the distance measures d which classify correctly each $a \in CB_n$ using $CB_n \backslash \{a\}$.

b) Each updating step of the version space corresponds to an updating of the set of distance measures. This equivalence in prinziple does not mean that there are no differences. The most important ones are:

(i) A predicate classifies absolutely while a measure introduces a scaling factor which results in a graduated classification. This classification has an ordinal as well as a cardinal aspect.

(ii) A measure makes a global assertion about the universe; it puts every two objects into a certain relation. It is important to observe in this context that the global statements of classical logic and statistics or analysis are of a very different character. The only global statements of logic are those provided by universal or existential quantifiers and they are of an absolute character while the analytical methods introduce probability distributions, measures etc. There is of course the attempt to extend logic in order to cover uncertainty reasoning using uncertainty factors, probability logic or related methods. From a foundational point of view these approaches have not yet been fully convincing.

(iii) The nearest-neighbour methods can be directly applied to noisy data while logical methods require (as mentioned above) an additional mechanism for dealing with uncertainty. The same is the case if the objects of the universe are only partially known.

(iv) In logic, additional knowledge can be formulated, added and changed in a natural and easily controllable way. When distance measures are used, knowledge has to be encoded and is no longer explicitly available. It is hidden in the parameters of the distance measure d and the cases which are selected for the case base CB.

3 The Integration Problem

As we have outlined, the symbolic approach to learning by artificial intelligence and the connectionistic approach using numerical techniques both have their advantages and disadvantages. The two main points which speak in favour of symbolic learning are on the one hand the large number of methods and concepts developed in logic and on the other hand the possibility to represent knowledge in an explicit form. The strengths of the connectionistic approach consist in the way of dealing with uncertainties, the fact that procedures need not terminate but can come to an equilibrium state and finally the possibility of expressing self-organization easily. There is of course a desire to integrate both techniques into one amalgamated system. Such a system should ideally have only the advantages of both kinds of methods.

From the theory of programming languages similar integration problems are known. An example is the integration of logical and functional languages, which has been attacked in various ways and can still not be regarded as finally solved. The underlying difficulty in this case is that the two paradigmas have an entirely different semantics; logic programming is essentially first order logic while functional programming is essentially recursion theory and therefore inherently of higher order. The problem of semantics is also basic for the integration of the different learning paradigms. The semantics for symbolic machine learning is mainly that of predicate logic and therefore essentially clearly defined. This is not so much the case in connectionism. As semantics has never been formally defined, it is doubtful that there is a semantics which is valid for all the different variations occurring in connectionism. What is clear is that such a semantics should incorporate elements of probability theory and we have mentioned above that it is not trivial to make it compatible with classical logic.

An additional difficulty lies in the fact that the symbolic methods generally use supervised learning. In principle, there is of course no visible argument which would forbid the use of self-organization also in symbolic-learning; the point here is more that such methods have not yet been developed.

When one has to integrate a method m_1 and a method m_2 then there are essentially two different ways to perform this task. One way is to embed one system into the other one, say e.g. m_1 into m_2. In this way m_1 becomes a subsystem of m_2. The problem reduces to the requirement that the embedding of m_1 into m_2 should not lose the advantages of m_2 by a too complex transformation. This approach is used by simulating symbolic manipulations in neural networks. A different technique is to embed both m_1 and m_2, into a common system m_3. This approach was successfully used in the integration of some programming languages. In the area of learning no such attempt can be seen. In the next section we will discuss an approach which takes elements from symbolic and connectionistic learning and uses ideas from supervised learning and self-organization as well; the approach is still in its infancy and is to some extent only experimental.

4 How to Combine Explicit Knowledge and Self-Organisation: The PATDEX System as a First Approach

The PATDEX system [5],[9] was developed at the University of Kaiserslautern as a part of the MOLTKE system [1] which is concerned with the fault diagnosis of complex technical systems like Computerized-Numerical-Controlled machining centers. It contains elements from traditional artificial intelligence as well as from the connectionistic paradigm. PATDEX is concerned with the diagnostic process as a whole, using such machining centers as its leading example. This process deals with situations of incomplete information which have to be at least partially completed by new measurements or observations until a situation is reached where a sufficiently correct diagnosis can be made. The principal computational model of PATDEX is shown in figure 2.

PATDEX has two types of case bases, the first contains the strategy cases and the second the diagnostic cases. The diagnostic cases are used for the classification task, i.e. the final diagnosis. The purpose of the strategy cases is to guard the diagnostic process, i.e. to select the next observation to make in order to complete the information. In our consideration here we will restrict ourselves to the diagnostic cases, i.e. we deal with classification.

We assume that there are n possible observations where the outcome of the information i ranges over a finite domain r_i. We define an information vector as a vector $(y_1 ..., y_n)$ where for each i either $y_i \in r_i$ or $y_i = x_i$ where x is a variable. A variable indicates that the information provided by observation i is not yet available. The universe U is the set of all such information vectors.

The measure which PATDEX uses is a special form of the Tversky-measure, cf. section 2.2. The function f is simplified to the cardinality of its argument. The main addition is that the information vectors may contain unknown values, i.e. variables. We give up the symmetry condition of the measure because a vector x from the case base and an actual given vector x^{act} have a different significance.

We will first introduce the necessary notation:

We put

$x_{act} = (w_{i_1}, ..., w_{i_k}), \quad x = (v_{r_1}, ..., v_{r_j})$; here we list only the coordinates with a known value.

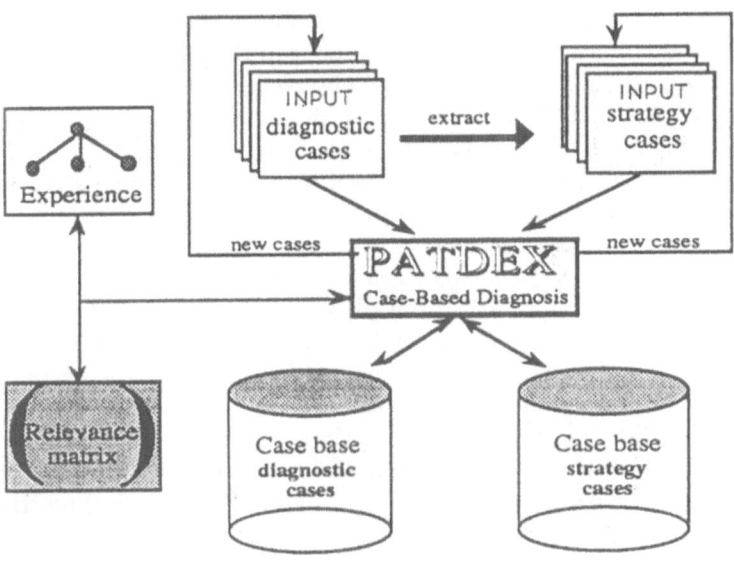

Fig. 2. Computational Model of the PATDEX System

$H = \{i_1, ..., i_k\}$,

$K = \{r_1, ..., r_j\}$;

$E = \{i | i \in H \cap K, w_i = v_i\}$, the set of attributes where the values agree;

$C = \{i | i \in H \cap K, w_i \neq v_i\}$, the set of attributes with conflicting values;

$U = H \setminus K$, the set of attributes with a known value for x but unknown value for the actual object;

$R = K \setminus H$, the set of attributes with a redundant value for x_{act}.

The measure used is of the form

$$sim_{PAT}(x_{act}, x) = \frac{\alpha \cdot |E|}{\alpha \cdot |E| + \beta \cdot |C| + \gamma \cdot |U| + \eta \cdot |R|} .$$

The parameters α, β, γ and δ can be chosen. For exemple, for

$$\alpha = 1, \qquad \beta = 2, \qquad \eta = 1/2, \qquad \gamma = 1/2 ;$$

we obtain

$$sim_{PAT}(x_{act}, x) = \frac{|E|}{|E| + 2 \cdot |C| + 1/2 \cdot |U| + 1/2 \cdot |R|} .$$

Next we are concerned about four different types of domain knowledge which should be respected by the similarity measure.

a) Importance of an attribute: Some attributes or attribute values may have a higher significance then others.

b) Functional dependencies: Certain attribute values may be totally determined by others and therefore they should not count in the measure.

c) Redundant values: Some values may be observed in the actual case but
not in the case from the base. If these redundant values are normal they should
be neglected but abnormal values should still play a role.

d) Unknown values: If these values are known for the case of the base then
one has different possibilities. They are mainly of statistical character, depending
on the probability of the unknown value.

Whether one of these situations really occurs may depend on the context, i.e.
on the values of certain attributes, and the available knowledge should be strong
enough to analyse this. As a consequence, the measure should be changed in such
a way that it reflects this knowledge; in the PATDEX system this is partially
implemented. The purpose of the subsymbolic part of PATDEX is to extract
knowledge from the cases which are not explicitly represented. The main concept here is the notion of the relevance matrix. If n is the number of attributes
and m is the number of classes then the relevance matrix R is an (n, m) matrix.
Its entries w_{ij} have to satisfy (i) For all i and j $0 \leq w_{ij} \leq 1$ holds;
(ii) For all j we have $\sum_{i=1}^{n} w_{ij} = 1$.

The idea is for the weights to be an estimate of the probability that a certain
attribute value a_i determines a class C_j. The initial setting of the relevance
matrix is given by simply counting the frequencies in the case base. Another
feature takes care of attributes with more then two possible values. For such
attributes an additional similarity function w is introduced which describes the
similarity between values of that object. The user can be choose a threshold λ
in order to distinguish between sufficiently and not sufficiently similar values.
This leads to a redefinition of the similarity measure. For the similarity between
values the user chooses a threshold λ. We put
$x_{act} = (w_{i_1}, ..., w_{i_k})$, $x = (v_{r_1}, ..., v_{r_j})$; here we list only the coordinates with a
known value or where the value in x_{act} can be predicted with probability $\geq \theta$.
H= $\{i_1, ..., i_k\}$,
K= $\{r_1, ..., r_j\}$;
$E' = \{i|\omega(w_i, v_i) \geq \lambda\}$, the set of attributes with sufficiently similar values;
$C' = \{i|\omega(w_i, v_i) < \lambda\}$, the set of attributes with not sufficiently similar values;
$U' = H \setminus K$, the set of attributes with a known or estimated value for x but
unknown value for x_{act}.
$R' = K \setminus H$, the set of attributes with a redundant and classifying value for x_{act}.

Using this we define

$$E_0 = \sum_{i \in E'} w_{ij} \cdot \omega(w_i, v_i);$$

$$C_0 = \sum_{i \in C'} w_{ij} \cdot (1 - \omega(w_i, v_i));$$

$$R_0 = |R'|;$$

$$U_0 = \sum_{i \in U'} w_{ij} \quad .$$

This leads finally to the measure of PATDEX/2:

$$sim_{PAT/2}(x_{act}, x) = \frac{\alpha \cdot |E_0|}{\alpha \cdot |E_0| + \beta \cdot |C_0| + \gamma \cdot |U_0| + \eta \cdot |R_0|}$$

α, β, γ and η can be chosen as before.

Unfortunately it turns out that this similarity measure does not even classifiy the cases from the case bases correctly. Therefore the similarity measure is improved by an adaptive learning process. This process uses the case base as the training set. It turned out that the best choice for the updating rule in the learning phase is the *winner take all* rule from competitive learning. In this way an element of self-organisation is introduced here.

This is only a very small step towards the integration of symbolic and subsymbolic reasoning. Humans are able to combine these two types of reasoning in a very efficient way. It is certainly a big deficiency of the present situation that all attempts to reproduce this in artificial systems are still not convincing. The integration of symbolic and subsymbolic reasoning and self-organisation principles seems to be a major task in order to make progress in artificial intelligence and connectionism as well.

References

1. Althoff K.D.: Eine fallbasierte Lernkomponente als integrierter Bestandteil der Moltke Workbank zur Diagnose technischer Systeme. Doctoral Dissertation, University of Kaiserslautern, 1992
2. Foucault, M.: Les Mots et les Choses. Editions Gallimard 1966.
3. Globig, Chr.: Symbolisches und fallbasiertes Lernen, In preparation, Kaiserslautern, 1993
4. Michalski, R.S: A Theory and Methodology of Inductive Learning. In: Machine Learning: An Artificial Intelligence Approach, Vol.1, p.83-134, Tioga Publ.Comp. 1983
5. Richter, M.M., Wess, S.: Similarity, Uncertainty and Case-Based Reasoning in PATDEX, in: *Automated Reasoning, Essays in Honor of Woody Bledsoe* (ed. R.S. Boyer), Kluwer Acad. Publ., 1991.
6. Rumelhart, J.L. McClelland: Parallel Distributed Processing, Vol.1. MIT Press Cambridge 1986.
7. Rumelhart, D.E., Zipser, D.: Feature Discovery by Competitive Learning, *Cognitive Science* 9, 75- 112, 1985
8. Tversky, A. : Features of Similarity, *Psychological Review* 84,p.327-352, 1977
9. Wess, S.: Inkrementelle und wissensbasierte Verbesserung von Ähnlichkeitsurteilen in der fallbasierten Diagnostik. Proc. 2. Deutsche Tagung Expertensystem XPS-93. Springer-Verlag 1993

Speedup of Self-Organization Through Quantum Mechanical Parallelism

Michael Conrad

Abstract

Organisms act under real time constraints. The capabilities for information processing and control ultimately depend on how much molecular pattern recognition work can be performed subject to these constraints. The speed of complex formation is all important. The model developed in this article shows that the parallelism inherent in the electronic wave function speeds up the self-organization of the complex. The analysis suggests that all modes of biological self-organization – evolutionary, dynamical, and algorithmic – draw on the quantum speedup effect through their dependence on macromolecular specificity.

1 Four Modes of Self-Organization

Life is a process that self-organizes from many constituent self-organizing processes. The variety of individual mechanisms and dynamical structures is stunning. Four general classes of processes stand out, however.

i. *Energy-based.* These are processes such as polymolecular complex formation that are largely governed by energy minimization. Enzyme–substrate complex formation and the principle of macromolecular self-assembly are of particular importance.

ii. *Entropy-based.* These are processes in which organization is associated with the export of heat. The spatial and temporal order exhibited by nonlinear dynamical systems falls into this category. Attractors, for example, should be a property of gradient (or dissipative) systems.

iii. *Natural selection.* This is self-organization through the Darwin-Wallace mechanism of variation and selection. The process is self-organizing since the selection arises endogenously from the interactions among organisms and between organisms and the environment.

iv. *Rule-based (or algorithmic).* This is self-organization through procedural mechanism, similar to the algorithms used in artificial intelligence programs.

These classes are closely interleaved in modern biological systems, and may occur in different mechanistic and material guises at different levels of organization. For example, complex macromolecules presumably arose through an entropy-driven process early in the history of the universe. Then self-assembly and catalytic processes largely driven by energy became possible, yielding reaction-level dynamics that supply dynamical order of the dissipative structure (or synergetic) type [1, 2]. These processes are molded for specific function through variation-selection mechanisms. As the evolution proceeds both continuous and discrete aspects of the dynamics become more sophisticated. Sometimes it appears possible to abstract rule-like behavior from these dynamics, as in the coding rules involving the translation of DNA sequences to amino acid sequences [3], or in the various forms of rule-like behavior associated with intelligence and language [4]. In the case of coding rules at the genetic level it is clear that the rule-like description is very incomplete, since the amino acid sequences fold, and since it is the specificity of the folded structures that does the coding. Energy- and entropy-driven mechanisms mix into the genetic process in a way that is entirely foreign to the manner in which digital computers execute programs. Analogous considerations suggest that the rule-like aspects of human intelligence are also closely intertwined with energy, entropy, and evolution [5].

Any one of the above general forms of self-organization could be taken as a starting point for tracing out the manner in which the different forms interact with each other. My focus in this paper will be on energy-driven self-organization at the molecular level, in particular enzyme-substrate complex formation and self-assembly. I will argue that electronic-conformational interactions allow these molecular level processes to draw on the most fundamental feature of the physical world – the quantum mechanical superposition principle – as a vast source of self-organizing power. *The basic idea is that the parallelism inherent in the electronic wave function of the interacting macromolecules speeds up complex formation, and that this speedup enormously magnifies the information processing and control capabilities of cells and organisms, including capabilities that on the surface appear to have a dynamical or algorithmic character.*

2 Enzymatic Processing

Biological information processing is largely based on the recognition capabilities of macromolecules such as proteins and nucleic acids. Molecular shape plays an important role in these recognition processes. The role of shape and self-organization is illustrated by the interaction between an enzyme (or protein catalyst) and the substrate (or object molecule) on which it acts. The general scheme is:

$$E + S \rightleftharpoons ES \rightleftharpoons E + P$$

The classical picture is that the enzyme and substrate first explore each other through diffusive (Brownian) search. If the dynamical and structural features of the two molecules are complementary (yield a very specific fit) the complex self organizes and a catalytic action is taken. The complex then breaks up to yield a product molecule and the regenerated enzyme.

Enzymatic reactions are ordinarily quite sensitive to milieu. Whether the enzyme recognizes the substrate and acts on it may depend on a variety of physiochemical influences. These may include control molecules that also complex with the enzyme. Some enzymes are reported to respond to twenty known control molecules [6].

The context sensitivity of enzymes is enormously important for control and information processing (see Fig. 1). Seed germination provides a useful metaphor, due to the fact that seeds often germinate only when some set of conditions is satisfied. Similarly, an enzyme takes an action when a variety of milieu influences are present. The enzyme not only recognizes the molecular object on which it acts, but also recognizes the pattern of influences acting on it. How it acts may also be modulated by context.

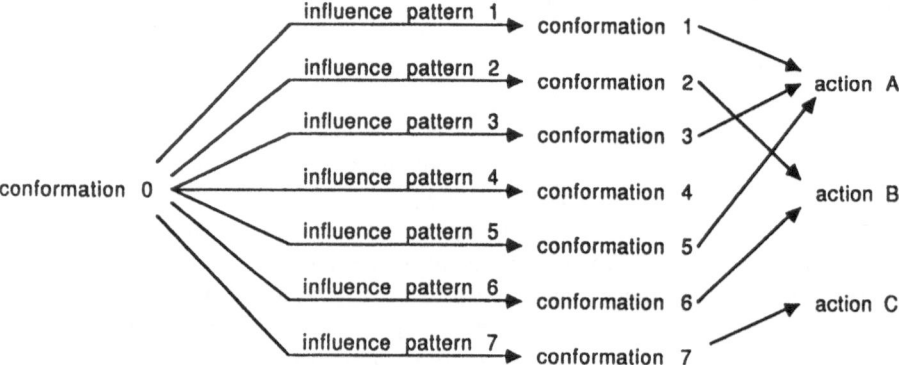

Fig. 1. Seed germination model of enzyme catalysis. "Conformation 0" is the initial conformation. The influence patterns (or contexts) are sets of milieu features, including physiochemical features and control molecules, along with the time intervals during which these features are present. Distinct influence patterns lead to different conformational forms of the complex, each characterized by some collection of shape features. The different conformations lead to different actions (including nonactions). If two conformations share a shape feature they can lead to the same action. Thus in the diagram conformations 1, 3, and 5 all lead to catalytic action A. This will happen if the common shape feature triggers the action of an enzyme that performs A. In this way the complex groups influence patterns into different categories. Catalytic action B might be similiar to action A, but with different rate constants. Action C could be a nonaction (no catalysis). The number of distinct influence patterns would in general be very much larger than indicated in the diagram.

This local context depends on the external inputs impinging on the cell and on endogenous cellular activities. In general the local pattern of influence reflects the external input and internal state in only a very partial manner. Each enzyme is potentially an information processing node that controls the internal or externally directed activities of the cell in a manner that reflects the global situation to the extent that the local situation does. This model of the cell as a collection of highly context-sensitive nodes is obviously very different than the

model of it as a network of switching elements that respond to a small number
of definitively specifiable influences.

This picture extends to conformational processing by polymacromolecular
(quaternary) complexes. Impinging patterns of influence should produce "conformational ripples" in such complexes analogous to the conformational changes
in a single protein. The effector action of the complex is controlled by these
conformational features. All that is necessary is for some effector enzyme to be
controlled by one of these features. This effector enzyme will then respond to
(categorize) all the input patterns that give rise to this feature. The number of
categories that the complex can recognize depends on the number of recognizable
shape features that appear in response to distinct groupings of milieu contexts
and on the repertoire of effector enzymes.

In principle each different context imposes different initial and boundary
conditions on the dynamical behavior of the complex. Different groupings of
milieu contexts occur because the atoms in each different surface region of the
complex interact with the remaining atoms of the whole complex in a different way. Different interactions will be significant (or insignificant) for different
regions. Consequently each different surface region may exhibit particular conformational features, perhaps transiently, in response to different groupings of
milieu influences.

Self-assembly of the complex can itself mediate a "perception-action" function at the molecular level. Suppose that signals impinging on different regions
of the cell surface trigger the release of specifically shaped macromolecules. The
pattern of inputs is then represented as a collection of molecular shapes that can
self-assemble, somewhat in the fashion of a jigsaw puzzle, to form a supermolecular mosaic. In this case it is quite clear that the collection of shape features of
the complex must depend on the input pattern, since the constituent molecules
will be different for different input patterns. Particular shape features may still
be common to different input patterns, however.

3 The Free Energy Surface

The enzyme-substrate complex must form rapidly and then break up rapidly.
How can this speed be compatible with high specificity? Formation of the complex is invariably accompanied by conformational changes. It is reasonable to
suppose that the complex draws the substrate in an active manner from many
different initial contacts. This requires the formation of the complex to be associated with a greater reduction in free energy than would be the case in a
more passive (purely diffusive) docking process. The lowering of the free energy
should reduce the speed of complex decomposition, however, and therefore lead
to a conflict between specificity and speed.

This conflict is inherent in a static free energy surface. The static picture
is artificial, however. The formation of the complex involves dynamical motions
that are nonthermal in character. Consequently, the true potential energy is time
dependent, and a surface that represents the energy (or a linear combination

of energy and entropy) should also be time dependent. Pathways connecting different minima on the surface transiently appear and disappear (the minima are transiently converted into saddlepoints).

The time dependence is schematically illustrated in Fig. 2. The term "energy loan" is used to describe this surface, since the nonthermal shift from the surface that is active during formation to the surface that is active during decomposition is due to an exchange between potential and kinetic energy [7, 8]. The potential energy surface of a bouncing ball would exhibit similar behavior. Even a classical model of enzyme dynamics could exhibit the energy loan phenomenon. However, we shall show that the effect is greatly amplified in quantum systems due to the exchange of energy between electronic and nuclear coordinates. Note that the energy loan merely serves to speed a process. It does not alter the equilibrium of substrates and products, and hence cannot lead to violation of macroscopic thermodynamics.

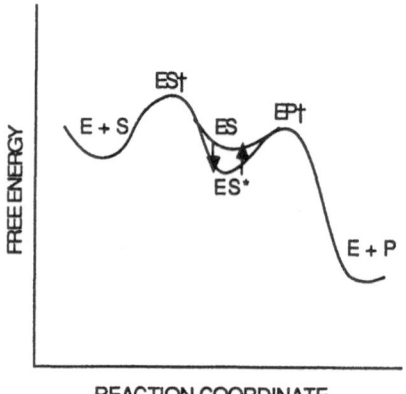

Fig. 2. Dynamic free energy surface of the energy loan model. In order for the enzyme to be specific the barriers to complex formation should be high, otherwise the enzyme could complex with any substrate ($ES\dagger$ and $EP\dagger$ are the activation free energies). The recognition process can nevertheless be fast if the enzyme actively draws in the substrate. This means that the free energy of the complex, ES*, must be comparatively low as determined by direct energy measurements. The free energy as determined by kinetic measurements, ES, would be higher. The reason is that the complex is inherently unstable, due to a nonthermal exchange of kinetic and potential energy between the nuclear and electronic coordinates. Roughly speaking, the molecules fall into the complexed state like a rubber ball might fall to the earth due to the force of gravity. Then they bounce out of the complexed state, just as the ball bounces off the earth. The bounce is represented by the switch between the two free energy surfaces. Such energy loans can contribute to complex formation, and could allow for the dynamic appearance and disappearance of activation barriers. Stable self-assembly of quaternary structures occurs if the aggregate state is stable; if all aggregated states are unstable all complexed forms are transient. This is the case in enzyme catalysis.

Fast decomposition is not necessary in self-assembly. The transient appearance of pathways between minima will speed up complex formation in this case. Self-assembly occurs when the complex is basically stable. Enzyme–substrate complex formation occurs when the energy loan itself facilitates the breakup of the complex. In this case the complex is inherently unstable. Such instability is critical to the real-time capabilities of organisms. Classical systems with nonlinear dynamics can of course exhibit unstable modes; we will show that the quantum features of the electronic-conformational interaction can serve to greatly amplify instabilities.

The above discussion emphasizes energy. The problem of recognizing a pattern of influences (signals) is converted to a free energy minimization problem. As noted earlier, energy minimization always involves some dissipation. For example, crystallization is accompanied by entropy production, but it is energy controlled because the crystal is a minimum energy structure. The important question for complex formation is whether the direction of the process is more controlled by energy minimization or by an increase in configurational energy (number of possible alternative structures of the molecule). Energy should play the more important role in specific complex formation, since the end product involves a very specific (complementary) fit between or among the molecules. Many weak interactions contribute to the form of the complexed state. Additively these are significant relative to thermal energies. Thus crystallization is a good analogy. However, because of the intricate structure of the complexed state, it is quite likely that the pathways of formation involve the opening ("unfolding") and closing ("folding") of molecular structures. Entropy may play a role here. But for the purposes of the discussion to follow we will focus on the energy aspect, partly because it is the more important aspect and partly because mixing classical ensemble considerations into the discussion would obscure the role of the superposition principle.

4 Quantum Parallelism and Quantum Speedup

The conversion of the pattern recognition problem to an energy minimization problem has a simple implication for the real time control capabilities of organisms. Any physical-dynamical mechanism that speeds up complex formation can increase either the rate of processing of a given amount of information or the amount of information that can be processed in a given time period. Thus, it would be possible for an enzyme or supermolecular complex to perform a given amount of pattern recognition work faster, or to increase the size of pattern recognition problems that it solves in a given amount of time. The same amount of pattern recognition work is performed faster if the self-organization of the complex is sped up. In this case a given free energy surface is searched more rapidly. The amount of pattern recognition work performed per unit time is increased if the number of molecules or shape features that contribute to complex formation increases. The complexity of the free energy surface that can be explored in a given amount of time increases in this case.

We can now restate the manner in which the quantum mechanical superposition principle contributes to self-organizing power, and hence to information processing and control. According to the superposition principle the time evolution of a quantum system is controlled by the interference of possible states of different energy. Our hypothesis is that the parallelism of possible electronic states of a self-assembling complex is converted to speed of complex formation – that is, to the speed with which the free energy minimum corresponding to the complex can be discovered. The important electrons are those that are not rigidly bound to atomic nuclei (i.e., electrons for which the Born-Oppenheimer approximation is not accurate). These electrons will make transitions to states of different energy due to the perturbing effects of the radiation field or due to either thermal or nonthermal motions of the nuclei. The electronic charge density will therefore change in time due to constructive and destructive interference among electronic states of different energy. The nuclei undergo a bobbing motion in response to this changing charge density that can serve as a self-amplifying source of perturbation to the electronic structure and that can open up new pathways on the potential surface.

There are two possible courses of development. The electronic and nuclear wave functions can be self-consistent (have a self-stabilizing or negative feedback relation) or be inconsistent (have a self-destabilizing or positive feedback relation). In the picture to be developed the interacting molecules do not initially satisfy consistency conditions. The self-organization of the complex corresponds to the evolution of a self-consistent form of organization. The self-consistent state may correspond to a permanent complex (as in self-assembly) or to a dissociated state (as in enzyme catalysis). The difference between enzyme–substrate complex formation and self-assembly is that no self-consistent form of the complex exists in the former case.

5 Self-Consistent Field Theory

Hartree type self-consistent field theory provides a general scheme for describing the approach to a consistent form. The idea is to assume initial potentials governing the electronic and nuclear systems, then to obtain the electronic and nuclear systems by solving the Schrödinger equations for the electronic and nuclear systems given these initial potentials, and then to complete the cycle by using the wave functions to update the potentials. The process is repeated until the wave functions and potentials are consistent, where consistency either corresponds to a self-assembled complex or to a dissociated state.

Self-consistent field schemes of this sort are ordinarily used as approximation methods that start off with a trial potential that has no physical significance. Here we will view the scheme as representing the self-organization of the complex, and not as an approximation scheme (in principle it could also be an approximation scheme, but in practice this is infeasible for a problem as difficult as macromolecular complex formation). The conceptual difference is that the potential functions at each stage are viewed as representing the physically relevant features of the true time independent potential that controls the time development of the system. This true potential is not very useful, however, since in principle it controls not only the time development of the interacting molecules but also that of the heat bath as a whole. Information must be discarded to obtain the relevant features of the system at any particular stage. The relevant features of the potential change as initially distant components of the self-organizing complex come into proximity or initially adjacent components become too separated to exert much effect on each other.

Suppose, as an extreme idealization, that we describe the self-assembling complex in terms of two quasi-independent wave functions that become correlated through mutual interactions. The wave functions could not in reality be independent, since the potential energies (and hence the Hamiltonians) of the electronic and nuclear subsystems are clearly not additive. Our assumption is that the interaction between the electrons and nuclei is sufficiently limited that we can write down independent (notational) wave functions for the electronic and nuclear subsystems. If we can elicit a speedup effect under this condition, we could reasonably conclude that the effect would even be stronger if the interaction between electrons and nuclei is too strong to allow for a quasi-independence assumption.

Let us denote the electronic wave function of the complex by $\Psi_e(x_e, t)$, where x_e represents the electronic coordinates. Similarly we can denote the nuclear wave function by $\Psi_n(x_n, t)$, where x_n represents the nuclear coordinates. If the potential energies (or Hamiltonians) of two systems were additive we could write

$$\Psi(x_e, x_n, t) = \Psi_e(x_e, t)\Psi_n(x_n, t) \tag{1}$$

As noted above this is not possible in a system in which the potentials share coordinate dependencies. The time scales of electronic and nuclear transitions are very different, however. Our idealization assumption is that the difference is so great that we can treat the wave functions as being independently updated. We assume that the time development is linearizable between the updating points and that the potentials do not significantly depend on the same coordinates between updates. All the nonlinearity is thus concentrated into the updating points.

Apart from these points the potential function can then to a reasonable approximation be split into two parts

$$V(x_e, x_n, t) = V_n(x_e, x_n, t) + V_e(x_e, x_n, t) \tag{2}$$

Here V_n is the potential controlling the behavior of the electronic system, V_e is the potential controlling the behavior of the nuclear system, and the temporal separation of coordinate dependencies is not at this point explicitly specified. The reversal of subscripts relative to those of the corresponding wave functions (e.g. V_n is the potential for Ψ_e) is intended to indicate that the nuclei make a more significant contribution to the potential governing the time evolution of the electronic system than do interactions among the electrons and that the electronic system makes a bigger contribution to the potential governing the time evolution of the nuclear configuration than do the interactions among the nuclei.

The updating process may be expressed notationally by writing

$$V_e(x_e, x_n, t) = V_e(\Psi_e(x_e, t)), \Psi_n(x_n, t)) \tag{3a}$$

$$V_n(x_e, x_n, t) = V_n(\Psi_e(x_e, t)), \Psi_n(x_n, t)) \tag{3b}$$

where the functional notation expresses the fact that V_e and V_n depend on features of both the electronic and nuclear wave functions. The updated functional forms utilize the probabilities $p(x_e, t) = \Psi_e^*(x_e, t)\Psi_e(x_e, t)$ and $p(x_n, t) = \Psi_n^*(x_n, t)\Psi_n(x_n, t)$. But of course some procedure would have to exist for obtaining adequate potentials from these probabilities; for example, interactions may be considered irrelevant if the likely coordinate values are such that the particles are more than some specified distance apart, or some averaging may be done beyond a certain distance. For the purposes of constructing a theoretical demonstration of the speedup effect it is only necessary to assume that such a procedure exists in principle, however. Also note that the construction of the potentials does not imply that measurement type wave function collapse contributes to the self-organization of the complex (just as the approximative use of Hartree-Fock does not imply that measurement plays a role in maintaining the electronic structure of atoms and molecules).

The wave functions themselves are solutions of the Schrödinger equations

$$-\sum_r \frac{\hbar^2}{2m_e}\nabla_r^2\Psi_e(x_e, t) + V_n(x_e, x_n, t)\Psi_e(x_e, t) = i\hbar\frac{\partial\Psi_e(x_e, t)}{\partial t} \tag{4a}$$

$$-\sum_r \frac{\hbar^2}{2M_s}\nabla^2\Psi_n(x_n, t) + V_e(x_e, x_n, t)\Psi_n(x_n, t) = i\hbar\frac{\partial\Psi_n(x_n, t)}{\partial t} \tag{4b}$$

where m_e is the mass of the electron, ∇_r^2 is the Laplacian operator of the rth electron, M_s is the mass of the sth nucleus, and ∇^2 is the Laplacian operator of the sth nucleus.

Each cycle of the iteration scheme calls on Eqs. (4a) and (4b) for some period of time over which the respective potentials are applicable, and then updates the potentials. For the purpose of formally writing down the scheme it is convenient to abbreviate Eqs. (4a) and (4b) as

$$\Psi(x, t + \tau_b) = \Psi(V(x, t + \tau_a), t + \tau_b) \tag{5}$$

Here $\Psi(x, t + \tau_b)$ is the (electronic or nuclear) wave function at time $t + \tau_b$ and $\tau_b - \tau_a$ is the time interval over which the (nuclear or electronic) potential controlling the time evolution of this wave function remains applicable.

The assumption that time dependent perturbation theory is adequate for dealing with any changes in the potential functions during the intervals means that the potentials take the form

$$V(x,t) = V'(x) + v(x,t) \tag{6}$$

in a piecewise continuous manner, where $V'(x)$ is a time independent unperturbed potential applicable during the interval $\tau_b - \tau_a$ and $v(x,t)$ is a small time dependent perturbation that develops over this interval. The solution of the Schrödinger equation for $V'(x)$ may then be separated into time independent and time dependent components

$$\Psi'(x,t) = \psi'(x)\phi(t) \tag{7}$$

in a piecewise continuous manner. This assumption is more restrictive than necessary, but by eliciting the vibratory terms it serves to exhibit the role of the superposition principle in an explicit way.

Let the time interval $t_{u+1} - t_u$ denote the uth cycle of nuclear-electronic interactions. The electronic system undergoes an evolution over a time interval $\tau_{e1(u)}$ and the nuclear system follows over a time interval $\tau_{n(u)} - \tau_{e1(u)}$. The cycle is completed when the electronic system makes a fast adjustment to the potential defined by the new nuclear wave function over a time interval $\tau_{e2(u)} - \tau_{n(u)}$. The subscript (u) associates these intervals with the start time of the cycle over which they are valid.

With the above notational conventions, each cycle of nuclear-electronic interactions may be formally represented by the iteration scheme:

$$\Psi_e(x_e, t_u + \tau_{e1(u)}) = \Psi_e(V_n(x_e, x_n, t_u), t_u + \tau_{e1(u)})$$
$$= \Psi_e(V_n(\Psi_e(x_e, t_u), \Psi_n(x_n, t_u)), t_u + \tau_{e1(u)}) \tag{8a}$$

$$\Psi_n(x_n, t_u + \tau_{n(u)}) = \Psi_n(V_e(x_e, x_n, t_u + \tau_{e1(u)}), t_u + t_{n(u)})$$
$$= \Psi_n(V_e(\Psi_e(x_e, t_u + \tau_{e1(u)}), \Psi_n(x_n, t_u)), t_u + \tau_{n(u)}) \tag{8b}$$

$$\Psi_e(x_e, t_u + t_{n(u)} + \tau_{e2(u)}) = \Psi_e(V_n(x_e, x_n, t_u + \tau_{n(u)}), t_u + \tau_{n(u)} + \tau_{e2(u)})$$
$$= \Psi_e(V_n(\Psi_e(x_e, t_u + \tau_{n(u)}), \Psi_n(x_n, t_u + \tau_{n(u)})), t_u +$$
$$\tau_{n(u)} + \tau_{e2(u)}) \tag{8c}$$

Here $t_{u+1} = t_u + \tau_{n(u)} + \tau_{e2(u)}$ and t_0 can be taken as the time of initial contact. The nuclear wave function $\Psi_n(x_n, t_u)$ is used in Eq. (8b) rather than $\Psi_n(x_n, t_u + \tau_{e1(u)})$ since $\tau_{n(u)} > \tau_{e1(u)}$, but actually it would be more precise to use the latter. Similarly $\Psi_n(x_n, t_u + \tau_{n(u)})$ could be used for $\Psi_n(x_n, t + \tau_{n(u)} + \tau_{e2(u)})$ when updating V_n (and setting $u = u + 1$) for re-entering Eq.(8a).

Actually three interacting systems should be considered, since the de Broglie wavelengths of hydrogen bonds is large enough to allow for some quantum behavior (e.g. some tunneling). More importantly, hydrogen bonds provide preferred pathways for electron tunneling [9]. Consequently they can serve to mediate some of the interactions between the electrons and the heavier nuclei. They will respond more strongly to a changing electronic charge density than the heavier nuclei, and can have a strong effect on the electrons by virtue of their facilitating effect on tunneling. The iteration scheme can be expanded to include the mediating effect of hydrogen bonds, but this greatly complicates the picture and would complicate our in principle demonstration that the superposition of electronic states speeds up complex formation.

6 Divergent and Convergent Superpositions

For simplicity suppose that the complex has two accessible families of nuclear configurations, labeled by I and II, and two accessible families of electronic states, also labeled by I and II. The electronic wave functions can then be expressed as a superposition of the form

$$\Psi_e(x_e, t) = \sum_{r \in I} a_r(t) e_r(x_e) \exp(-iE_{e(r)}t/\hbar) +$$
$$\sum_{s \in II} a_s(t) e_s(x_e) \exp(-iE_{e(s)}t/\hbar) \tag{9}$$

where x_e denotes the electronic coordinates, the e_r are eigenfunctions of the electronic system that are most naturally associated nuclear configuration I, the e_s are eigenfunctions of the true potential most naturally associated with nuclear configuration II and the a_r and a_s are expansion coefficients. The $E_{e(r)}$ and the $E_{e(s)}$ are the electronic energies associated with the eigenfunctions e_r and e_s. The eigenfunctions of the true potential are of course unknown, but conceptually they exist and this is sufficient for demonstrating the speedup effect. Some of the eigenfunctions might not be assignable in an unambiguous way to either conformation I or conformation II. This class of eigenfunctions would contribute to speedup, due to the fact that they extend over a larger region of the potential surface. But for the purpose of demonstrating the speedup effect we can safely ignore the ambiguous eigenfunctions, since doing so strengthens the argument.

Similarly the wave function for the collection of nuclei may be expressed as

$$\Psi_n(x_n, t) = \sum_{r \in I} b_r(t) n_r(x_n) \exp(-iE_{n(r)}t/\hbar) +$$
$$\sum_{s \in II} b_s(t) n_s(x_n) \exp(-iE_{n(s)}t/\hbar) \tag{10}$$

where x_n denotes the nuclear coordinates, the n_r and the n_s are the eigenfunctions of the nuclear system when it is in conformation I and II, and the $E_{n(r)}$ and $E_{n(s)}$ are the energies associated with these eigenfunctions. However, the nuclear system, because of its large mass, should with high probability be assigned to a definite state. Thus if it is initially in conformation I all the b_s can for all practical purposes be set to zero. To make matters as unfavorable as possible to the speedup effect we can also assume that the nuclear configuration is initially in a single eigenstate, so that all the b_r apart from one can initially be set to zero. Practically speaking this is reasonable since the average values associated with the different n_r should peak very sharply due to the large mass.

For the electrons, however, it is not possible to ignore superposition effects. An electron, because of its small mass, might with reasonable probability be found in a location more naturally associated with II than with I even if the nuclei are in the latter conformation. In order not to beg the question, let us assume that initially all the a_s are negligible. Also, to make matters as unfavorable as possible, we can assume that the electronic system is initially in a single eigenstate, so that only one of the a_r is initially nonnegligible.

Recall that if some of the electrons are not tightly bound to the nuclei, it will be possible for perturbations to induce transitions to eigenfunctions with energies close to that of the energy of the initial eigenfunction. The probability distribution of electronic coordinates will then be given by

$$p(x_e, t) = \sum_r a_r^*(t) a_r(t) e_r^*(x_e) e_r(x_e) +$$

$$\sum_r \sum_{r'} a_r^*(t) a_{r'}(t) e_r^*(x_e) e_{r'}(x_e) \exp\left(\frac{(-iE_{e(r)} - iE_{e(r')})t}{\hbar}\right) \quad (11)$$

where $r \neq r'$. The explicit appearance of the vibratory terms makes it clear that the probability distribution of the electrons is time dependent, due to constructive and destructive interference between eigenfunctions of different energy [10].

What happens when the electronic superposition (Eq. 9) is run through the iteration scheme (Eqs. 8a-c)? The a_r and a_s change with time, due to the time dependence of the potentials. During the linearizable part of each cycle the expansion coefficients change gradually, due to the fact that the perturbations are adiabatic (by virtue of our simplifying assumptions). The expansion coefficients can change suddenly at the updating points, but only if the potentials change significantly at these points. Two basic types of development are possible:

i. *Divergence*. The a_r change with time, due to the small (initially radiative or thermal) perturbation $v_n(x_e, x_n, t)$, leading to a time dependence of the charge density. The nuclei undergo a slight bobbing motion in this changing charge density. This adds another contribution to $v_n(x_e, x_n, t)$, thereby further spreading out the a_r. This further increases the bobbing motion, which further spreads out the a_r, and so on. The effect is self-amplifying, due to the positive feedback between the a_r and the b_s. The bobbing motion will either rapidly or slowly trigger a switch in the nuclear conformation, that is, a switch from the the b_r to the b_s. This switch will drag the electronic system along with it, corresponding to a major decrease in the a_r and a major increase in the a_s. The electronic system should then rapidly fall to the lowest energy state of the new nuclear conformation (apart from perturbations that again spread it out or that initiate a new divergent process).

ii. *Convergence.* In this case bobbing will occur, but it is insufficient to trigger a switch in the nuclear conformation from I to II. Slight changes in the conformation (really configurational changes) may occur, with associated slight change in the electronic wave function. The dynamics are convergent (i.e. the electronic and nuclear wave functions are basically consistent) as long as the complex cycles within a given family of nuclear and electronic states. Strictly speaking conformation I then corresponds to a family of subconformations.

The self-consistent state could either be a global or a local minimum. If it is a local minimum it will at most be metastable, due to the fact that values of the electronic coordinates that extend beyond the boundaries of self-consistency will always have nonzero values of $p(x_e, t)$. When these values occur the system may either switch to the new conformation directly, or a self-amplifying divergence may be initiated. When the complex does discover the global minimum it will remain in it, apart from thermal fluctuations or quantum fluctuations that could not raise the energy permanently.

As noted previously, it is also possible for the global minimum to be inherently unstable. This will be the case when pathways for decomposition open up dynamically after the complex is formed. The term "pseudo global minimum" is thus more appropriate than "global minimum".

7 Predictions of the Model

Molecular structures that speed the self-organization or disassembly as a result of divergent dynamics are probably a minority of all possible structures. Biological macromolecules are also a very small, highly evolved subset of all macromolecular forms. Let us briefly consider some of the physical properties that would have to be present to exploit the superposition effect for self-organization and that could be used to test the model.

i. *Significance of infrared radiation.* Biological macromolecules are colorless, apart from chromophores. The main absorption is in the infrared. Ordinarily this is attributed to nuclear rotations and vibrations. However, a small part should also be attributed to the non-Born-Oppenheimer electrons, due to the fact that these are not rigidly attached to the atomic nuclei. These electrons will exhibit accelerations relative to the nuclei, and consequently some component of the infrared spectrum of a self-organizing complex should be attributed to these electrons. The attribution can be treated as a matter of convention in a stable structure undergoing periodic motions, but physically it is not a matter of convention since it is connected to closely spaced energy levels of the delocalized electrons (including tunneling electrons and surface electrons that may have extended states). If it were not for this feature the electronic expansion coefficients, a_s, could not spread out.

ii. *Electronic versus diffusive scanning.* The spreading out effect provides an electronic analog of Brownian search. The usual picture, as noted earlier, is that macromolecules "use" heat motion to explore each other's surface features. According to the quantum speedup model superposition effects in the electronic system provide an additional search mechanism. Ultimately this electronic search also derives from heat motion (infrared radiation), but it differs from diffusive search in two ways. First it is channeled through the electronic system. Second it affords a more active form of search, due to the fact that the changing phase relations among the electronic states of different energy induce nonthermal motions of the nuclei. Furthermore, the effect can be self-amplifying.

iii. *Effect of low frequency radiation.* Sustained application of low frequency (microwave) radiation should affect the rate of complex formation, or could have an altering effect on the conformation properties of the complex. Low frequency radiation would not initially have a very significant spreading out effect on the electronic expansion coefficients; but sustained dosage would allow for the effect to build up through self-amplification.

iv. *Effect of introducing color groups.* If color groups were intimately integrated with biological macromolecules the complex formation process would be sensitive to ambient light conditions.

v. *Effect of hydrogen isotopes.* As already noted, hydrogen bonds should play an instrumental role in the self-amplification effect. This is due to the fact that the spreading out of the electronic expansion coefficients will have a greater effect on the hydrogen bonds than on the heavier nuclei and due to the fact that the hydrogen bonds, as facilitators of electron tunneling, will have a greater altering effect on the electronic expansion than the heavy nuclei. In this way hydrogen bonds should serve to "kindle" the self-amplification effect. Replacement of hydrogen bonds by deuterium or tritium should have a substantial retarding effect on the rate of complex formation.

vi. *Unstable electron pairing.* The speedup effect is linked to unstable dynamics. All stages of complex formation are inherently unstable except for the final stage (the dissociated stage in the case of enzyme catalysis). Many different mechanisms could contribute to this instability. Previously the author proposed that the weak interactions holding the complex together could push parallel spin electrons closer together than is energetically favorable [7, 8, 11]. Ordinarily this would result in spin flipping. However, over very short periods of time, in the presence of downward fluctuations in thermal energy, transient pairing is possible (since it is energetically more favorable for the electrons to undergo anticorrelated motions). Such pairs are unstable. But for the short period of time during which they can exist the electronic ground state of the complex will be lowered. This can provide a source of energy to the nuclei, thereby inducing a conformational change. This is an energy loan, since the inherent instability of the lowered electronic ground state pulls the energy back from the nuclear system (hence breaking up the pair).

The electronic system in the pairing model is in a superposition of normal and true ground states. The two states will interfere, yielding a changing charge density that triggers the nuclear motion. The electronic ground state of the new nuclear conformation is higher than the unstable ground state. The unstable electron pairing mechanism has all the features that enter into the general model described in this paper. However, it introduces two important variations. The first is the transient collapse of the normal ground state. The second is that downward thermal fluctuations contribute to the spreading out of the electronic expansion coefficients, since pair formation can occur only during periods of downward fluctuation. This adds an entirely novel source of search power, since only upward fluctuations contribute to diffusive exploration. The self-consistency scheme (Eqs. 8a-c) allows for upward fluctuations to contribute (through infrared spreading out of the electronic expansion coefficients) and also allows for downward fluctuations to contribute (through spreading out of the electronic expansion coefficients via transient destabilization of the normal ground state).

8 Connection to Other Modes of Self-Organization

We have demonstrated that the superposition principle can yield increased organizational power in systems whose time development is controlled by energy. Additionally, the system must allow for an interaction between components (the nuclei) that require a classical description and components (the electrons) that require a quantum description. Divergent dynamics, therefore instability, plays a central role in this mode of self-organization. A stable system would not be able to take advantage of the ability of the electrons to "scan" for new minima, since the nuclei would always lead and never follow the electrons.

Needless to say many mechanisms contribute to speed of complex formation. The self-consistency scheme outlined in this paper accommodates the processes for which a quantum analysis is necessary. Our hypothesis is that the quantum nature of matter plays the key role in controlling the rate of time development and in affording high specificity (including context sensitivity).

The energy (and superposition) category of self-organization has important relations to the other categories: Darwinian, entropy-driven, and algorithmic. The molecular structures that yield the speedup effect must satisfy special organizational requirements. These structures are molded by natural selection. Reciprocally, the genetic mechanism that makes variation and natural selection possible is enabled by the specificity of macromolecules.

The entropy-driven mode of self-organization (flow of high grade energy into low grade energy) is required for the synthesis of macromolecules. It almost certainly played a role in their prebiological origin. Reaction systems catalyzed by enzymes can exhibit self-organization in the form of dynamical attractors: limit cycles, complex spatial and temporal structures, and chaos. A dynamical system will exhibit attractors whenever perturbations to it are canceled out by a countervailing dissipative process, at least within a certain range. The nonlinearities required to support biochemical reaction networks with basins of attraction are rooted in the specificity of enzymes, hence in the quantum speedup effect. The connection is deeper than this, however. Attractors are generic structures without personality. Organisms are always particulars [12]. The context sensitive pattern recognition activities of enzymes can yield classes of attractors whose generic properties depend on how much pattern recognition power the enzymes have. The particularities of the attractor–what actual sequence of biochemical changes occur or what actual pattern of genes is turned on–depends on the particular patterns recognized, however. It is these details that determine the specific response of cells and organisms to environmental inputs, or that determine the particular ways in which they develop. The generic properties of the dynamics are a higher level of abstraction.

The algorithmic mode of self-organization is also rooted in the quantum speedup effect. It is of course possible to implement algorithmic behavior in simple switching networks. This is how digital computers work. These networks consist of switches with extremely simple, context independent pattern recognition capabilities (they recognize simple patterns of 0's and 1's that arrive in a clock-controlled manner). It is always possible to implement the same rules executed by such switching networks in more efficient fashion by replacing subnetworks of simple switches by individual molecular or cellular processors with equivalent input-output behavior [13]. In this way the power of the algorithmic mode of information processing is enhanced by the quantum speedup effect. Furthermore, precise clocking of the signals is not necessary, since the enhanced pattern recognition power of the processors makes it possible for them to recognize input patterns whose temporal and spatial structure is less constrained. This is of great importance, since the human brain is clearly capable of algorithmic behavior, yet has no known mechanism for imposing a clock time on nerve impulses.

Acknowledgement

This research was supported in part by the U.S. National Science Foundation (Grant ECS-9109860).

References

1. Nicolis, G. and Prigogine, I., Self-Organization in Nonequilibrium Systems. (Wiley-Interscience, New York, 1977).
2. Haken, H., Synergetics. (Springer-Verlag, Berlin and New York, 1978).
3. Pattee, H.H., Cell psychology: an evolutionary approach to the symbol-matter problem, Cognition and Brain Theory 5(4), 325-341 (1982).
4. Conrad, D., Rules, laws, and reductionism, to appear in Applied Mathematics and Computation.
5. Conrad, M., The brain-machine disanalogy, BioSystems 22, 197-213 (1989).
6. Marijuán, P. and Westley, J., Enzymes as molecular automata: a reflection on some numerical and philosophical aspects of the hypothesis, BioSystems 27, 97-113 (1992).
7. Conrad, M., Unstable electron pairing and the energy loan model of enzyme catalysis, J. Theoret. Biol. 79, 137-156 (1979).
8. Conrad, M., Electronic instabilities in biological information processing, in: Molecular Electronics, P.I.Lazarev (ed.), (Kluwer Academic Publishers, Amsterdam, 1991), pp.41-50.

 9. Beratan, D. N., Onuchic, J. H., Betts, H.N., Bowler, B.E., and Gray, H.B, Electron-tunneling pathways in ruthenated proteins, J. Am. Chem Soc. 112, 7915-7921 (1990).

10. Bohm, D., Quantum Theory (Prentice-Hall, Englewood Cliffs, N.J., 1951).

11. Conrad, M., Electron pairing as a source of cyclic instabilities in enzyme catalysis, Physics Letters A, vol. 68A, no. 1, 127-130 (1978).

12. Mae Wan Ho, personal communication.

13. M. Conrad, Advances in Computers, M.C. Yovits, Ed. (Academic Press, Boston, 1990).

Living State for Self-Organization
A Plea

R.K. Mishra

Perhaps the most outstanding example of self-organization is the living organism. While it is moving or not-moving, sensing, responding, growing, developing, ageing the processes resulting in the maintenance of the organized state are always at work. There is throughflow of mass and momentum. Time also "flows". Yet the identity of the organism is never in doubt. There go on the intrinisic programmed patterns of interaction and overall behaviour. Not only a body but what is most surprising is that the properties and strategies of self-organization are passed on even to a successor organism. Yet we know little of the fundamental rules that will create and sustain functional dynamism. In many cases of sudden death of the organism the structures may remain intact, the functionalism vanishes. The problem is so complex that it may seem beyond experiments and theories. The organism is indeed a physical body whether a virus, or a slime mold or man. We should thus expect the laws of physics to apply. In every system that is isolated out from it they do, to an approximation, yet there is no theory for a global systemic organization. Is this to tease or teach the physicist? Do we need new physics? A living body is made of molecules, but no molecules, indeed not all the molecules of the same body put together, would create a functioning organism. Chemistry of molecules is therefore no final answer. If all the organs are put together the body will not come alive. A body must necessarily self-organize. 50 years back the demand was for dealing with non-linear systems. Demand was made for new mathematics. Then it was realised that there was indeed the landmark work of Henri Poincaré. Following him then we had the puzzles of entropy, thermodynamics for near-equilibrium, far-from-equilibrium, dissipative structures, fluctuating systems, chaos, catastrophe, elasticity, synergetics, new geometries, computers, new reiteration-generated geometries. Words which should have been there from the beginnings of languages, like self-organization, self-similarity, self-reference, autopoiesis, autocriny, infrequently heard before, now hold the central stage of discussions, yet something is missing. We know what objects can do when pushed, what they would do cooperatively, what they would do when moved unequally, what individuals would do as numbers in population dynamics or in computer iterations of some particular functions, but to our dismay many such doings cannot come to pass in the living systems. At least not in a shape so as to serve as the basis of self-organization inspite of all the plenum. Every single system, if small enough and in durations small enough,

Springer Series in Synergetics, Vol. 61   **On Self-Organization**
Eds.: R.K. Mishra, D. Maaß, E. Zwierlein   © Springer-Verlag Berlin Heidelberg 1994

may succeed in manifesting this or that result of these formulations, but in the
end we recognise that they succeed only for some components of the system,
some triggers for small acts; nothing will create the evolving, reacting ensemble
of functions that the living body is. Study shows that it is not the lack of major
rules discovered but rather the state of structures on which to apply them. The
problem is so central and so vital and so widely applicable that one must attend
to it, maybe merely for a concept, and even at the risk of being correctable. It
has been proposed that the missing something is the anatomy of energy flows in
the entire body, with pumps, in a system of "loose structure", loose enough to
be relevant for bounds of energy, thermodynamically open and operationally in
non-equilibrium, – so to say the Living State [4, 7, 14, 15, 46]. It can be shown
that all necessary properties using formulations of self-organization [55]-[58] are
derivable from this essential structuration, effectively a pumped boson body.

1 The Central Issue: Functional Ensemble

The most exclusive and distinguishing feature of living systems is their capacity
to self-organize themselves as an ensemble of functions which enslave the subsystems. Perhaps the most remarkable example for this is the case of fragmentary
living systems which live in fragments, "saltatory life", as seen in metamorphosis. An ordered functionalism "dances" through outwardly dissimilar organisms.
Take the cestode intenstinal worm, *Dibothriocephalous latus*. Its eggs are passed
in the feces. Ciliated larva, coracidium escapes. This is ingested by cyclops, a
crustacean, where it becomes a procercoid. Ingested by a freshwater fish this
develops into a pleocercoid in its muscle. No apparent disease yet, but when the
fish is incompletely cooked and eaten it becomes the worm in the human and ful-
filling its destiny, causes symptoms of worm infestation. For this organism flow of
time and the habitats determine strikingly different forms. No amount of looking
at the worm or the procercoid will let us anticipate the future form and function. There is no consistent plan of the intermediate larvae in different animals.
Planula of a coelenterate, trocophore of an annelid, larva of starfish, nauplius of
a crustacean differ in external forms; the stage is recognised by its functionalism
and the locus in the trajectory of metamorphosis. Only the nymphs of the insect
metamorphosis give some clue to future form, but then the story is already out
and the speculation is redundant.

The duration of transformation can be controlled by chemical triggers like
tri-iodothyronine or thyroxine but not the stages of transformation. Such life-
cycles are usually baffling. The stages of the intra-RBC form of malaria plas-
modium deciphered by Ross are a marvel, but the identification of the extraery-
throcytic form had to await another half a century of investigation; it had not
been easy. Perhaps the most significant geometrical shaping due to programmed
transformation-trajectories is afforded by regeneration and wound healing. A
striking example of the former is the development of a full-grown functional
limb from an ablated one, through regeneration in the newt of a blastema or a
bud, in just a few days, remarkably stopping just where it should. Here a functional need is fulfilled, and no more, by microscopic triggers and biochemical

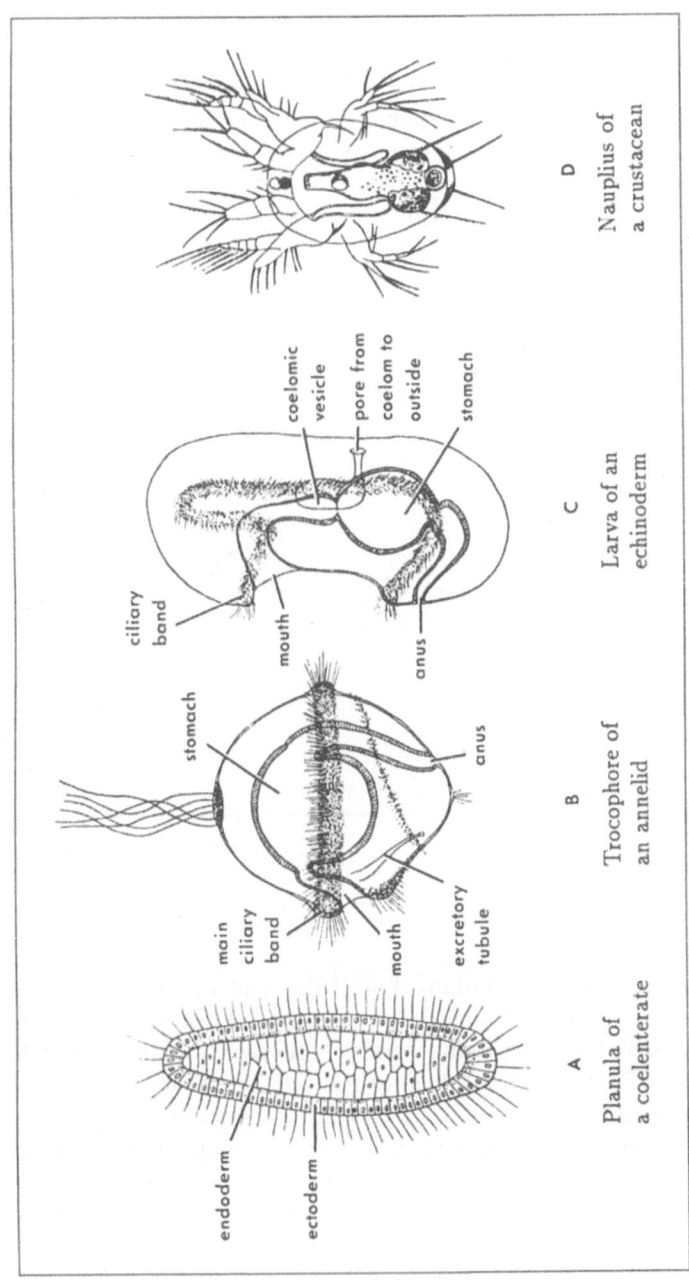

Fig. 1. Various types of larvae [10]

pathways invisible to the naked eyes. Repair of a wound with no loss or gain in morphology is an atavism of this regenerative potential. This phenomenon of epomorphosis may not be so puzzling as reorganization of a whole body from 1/200 of the original individual, e.g. morphallaxis or morphallactic regeneration in Hydra, a freshwater polyp. That is not mere modelling of a limb bud, but it is pursuit of form by processes of enslavement by a function. Sometimes the function is hidden as in autotomy (self-mutilation, shedding of a limb of prey, a crab, due to the seizure of a segment of it by a predator or a scientist). It shows inbuilt options which are exercised as a consequence of functional need, in this case development of a stricture to mark the site of dismemberment.

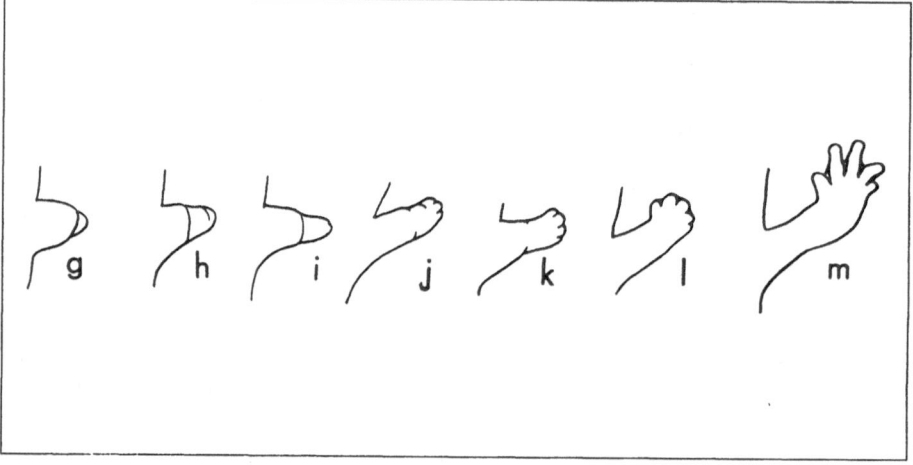

Fig. 2. Regeneration from a limb bud [10]

In the higher forms, including humans, both, plus and minus activity, are initiated at the minutest level of molecular recognition. This is indicated by terms so commonly used by an immunochemist - "nonself" distinguished by immune system from a "self" immunochemical object. These are outstanding examples of the targeting of functional activities by a functional trigger. These highlight extreme specifity for interactions. Other examples like hormones and biomodulators are such multifunctional determinants. For example, β-estradiol is active, α is not. These differences in orientation of hydrogen atoms bonded to a carbon at 4 position in the steroid skeleton are of the same order as if one could distinguish from earth two cats sitting side by side on the moon. Yet a few milligrams in an incomparably larger body, the young 50 kg female, would make a drastic transformation, working even quite contrarily, as regards growth of the body, glands, fat in the breast, or hair, colour, and so on. The hormone stops some activity, initiates some others. Adipose tissue grows in breast, but little or none at the waist. As Selye said, the aim of biology is to accomplish a function and not a structure; an estrogen would do everything needed to prepare the female body for reproduction [36]. Even the illusive is influenced – the psyche. Attractants may be phenomena like mimicry, display of colour, odour, male dominance in a crowd, pecking-order, etc. We mention these examples not to sound mysterious, or to bring esoterics into science, but to resolve the mystery to the extent possible, by enlarging the scope of the application of principles of physics, which the body as a tangible physical object ought to obey.

2 The Essential in Functionalism

The first lesson in life sciences, of any description, anywhere, starts with answering the question: What is the difference between living and nonliving? Of these movement, apprehending food, ingestion, digestion, biochemical modifications in metabolism, and excretion are secondary and excludable for our purposes. One can live without these for appreciable length of time. Given a living organism they are necessary for its optimal maintenance. Responsivity to environmental factors in order to maintain existence is the only central one, the only one relevant for defining the Living State, that by knowing which every other function can be derived or discovered. Even reproduction is not an obligatory necessity for the purposes of one organism, it may be necessary for another level of organization of the living forms namely, the species. In naming existence as the basic property perhaps we are repeating the law of action-reaction in the context of biology. Perhaps there is only difference in phraseology of its rules, the resistive reaction is not equal and opposite, but equal and "necessary". In this are subsumed attraction, repulsion, change, rates, magnitudes and signature of change, all that is necessary for adaptation in the bounded phenomenological environment of the basic unit. This basic unit may be a molecule or cell or the entire body at different levels of organization.

Kaganov and Lifshitz [11] discussing pseudoparticles emphasized that the 'units' of any one structure are manifold. For each evolution the unit may be different; it is actually the one that cannot be broken down further for the purposes of activity in question. Thus, it is molecules when one studies the specific heat of molecular gas at a not too high temperature. A small admixture of ions is neglected. But when electrical conductivity of a molecular gas is considered the elementary particles are indeed ions. In the same way one could recognise the unit of function of a system as a living system is a 'process structure', a dynamic structure of diverse ingredients of mass in motion, ordinarily unresolvable further, but distinct unavoidable trajectories of transformation for a manifest function.

This communication is addressed to the question: how can laws of self-organization be successful in open non-linear systems and unfold universally for this functional end in biological or living systems which are partitioned by organs, tissues, cells, organelles, membranes and integrated into structured multiphasic water? We examine a state of matter unaffected by these partitions, liquid like in mobility, yet providing necessary degree of spatial stability for anchoring functiuons. Bernal [52] visualised this in a remakable interjection in Faraday Society meeting nearly sixty years back pleading for liquid crystallinity in living systems. Far from equilibrium one requires ready change of states by subtle stimuli, molecular biology demands highly specified sites and domains to be held in a whole spectrum of spatio-temporal transformation, forces are weak, structures too massive. The nub of the solution is to go beyond partitions for a continuous edifice.

3 The Perturbing Influences in the Relevant Environment and the Sensitivity of Responding Systems

(i) The Bounds of Electromagnetic Fields [16, 17] [22] - [24] [30, 51, 53].

Table 1 illustrates the ranges in the environment encountered.

Table 1. Electric and magnetic fields within the biological system which represent energy of kT at 310 K [12]

Living system	Range of Cell Sizes	Egg	Man
Size of system	1 μm ... 100 μm	70 g	70 kg
E (min)	3 kV/m ... 3V/m	250 μV/m	8 μV/m
B (min)	100 μ T ... 100nT	9 pT	300 fT
Flux Quanta at B (min)	$\Phi_O/300...\Phi_O/3$	$8\Phi_O$	$25\Phi_O$
Poynting Vector	200 kW/m...80 mW/m	600 pW/m	1 pW/m

Sensitivity to above fields which are subtle have been clinically observed leading to widespread allergic response [12]. The relevant conclusion from the above is that very weak fields evoke wide response in a living body, as contrasted with those affecting non-living bodies.

The range of frequency to which responsive adaptations are made, from cosmic radiation to VLF and ELF, is vast, 10^{12} to 1 Hz. In fact frequency oscillation of charged surface, as in case of the heart, is about 1 Hz; and in case of exceptional individuals like Napoleon less than 1 Hz.

Summed surface potentials like EEG, EKG, EMG, EOG, ERG cover a range from few μV to several tens of mV, in some electric fishes a shock of even several hundred volts may result from the electric organ.

Magnetic fields in the central nervous system may be of strength of 10^{-12} T to 10^{-11} T which are in the frequency range of 8-12 Hz for alpharhythm in EEG and MEG.

For electrical polarisation a nerve membrane achieves a potential difference around 100 mV across a membrane only 100Å thick.

While the atmospheric pressure on land varies comparatively slowly the hydrostatic pressure in the sea varies by 101325 Pa per 10 meter depth. At a depth of 10 km it is around 10^8 Pa. Some fishes and whales which accelerate fast are exposed to millions of pascals in swift movement. By this metabolism and protein conformation may be affected. Barophilic bacteria are thus constituted to adapt to 5×10^7 Pa pressure. Internal osmotic pressure, turgor, corresponds to $10^5 - 10^6$ Pa. Quite clearly living systems in nature are adapted to very weak fields within and without. Some can stand up to very high fields. There is no doubt about variation in species, but the capability of living systems over a wide range is clear. Perhaps a stone can withstand them, but it would show no new dynamic structure in response.

(ii) Internal Fields.

When the organism is not exposed to external stimuli of the above types it continues to show fine variation in internal fields. EEG and similar activities have been cited above. The pacemaker potentials in myoneural tissue of the heart would fire and continue to do so from intrauterine life when the potential develops until the very end. Indeed prolonged absence of heart-beat in an electrical record conclusively establishes organismic death. Bursting pacemaker neurones in some molluscs keep firing every several seconds. This is due to underlying calcium and sodium movements. Miniature end-plate potentials are continuously present as shown by Fatt and Katz [13]. It was later shown that they arise by continuous quantal release of transmitter and their currents were assessed as per Poisson distribution. They were further shown to be actually binomially distributed where m quanta are released while n are releasable with probability p for each, thus $m = np$. Spontaneous activity of this nature is perhaps widespread. In the visual sense organs, or acoustic receptors, they are responsible for random noise. It is important to note that the ionic currents which are responsible for these are gated by ubiquitous albumen molecules as gate. These ports for sodium ion may cover nerve membrane surface less than 1 in 10,000 of the membrane area. The selectivity of Na^+ ion is due to size compatibility and spatial arrangement of oxygen atoms at the membrane which react with the hydration shells of ions. From the point of view of the present postulate it is important to note that the gate molecule for channel entrance is controlled by a field sensor.

The model of ionic channel has been inferred by studying noise spectra of the membrane currents. In the accepted model τ_m is the time constant for opening and τ_h for closing. If a potential is active there flows a random total current. The electric power density δ as a function of frequency can be calculated. The sprectrum has three parts - S_m, a part related to opening parameter m, S_h related to closing and the third proportional to l/f. The first two elements are as follows:

$$S_h(f) = \frac{4I_{Na}^2}{M_h}(1 - h)L(\tau_{hg}, f)$$

$$S_m(f) = \frac{4I_{Na}^2}{M_h} \sum (\frac{3}{b})(\frac{1-m}{m})^b L(\frac{\tau_m}{b,f})$$

$L(\tau, f)$ is a Lorentzian function τ $\quad [1 + (2\pi\tau f)^2]^{-1}$

The values obtained by voltage clamp technique agree very well with τ_m or τ_h in these equations.

It is interesting that single channel conductance in this is of the order of 5 to 8 pS with a rather large transference i.e. some 10 millions ions per second. Interesting even more than this is that the gate opening is accompanied by very small currents, which are of course swamped out by much larger ionic flows and are revealed by stopping them by specific ions like tetradotoxin for sodium or tetramethylamonium for chloride.

Since the organism is exposed to various unrecognised perturbations all receptors are simulated and give rise to receptor currents, which are not all-or-none action potentials, but are graded with reference to stimulus.

(iii) The Weakness of Stimulus.

The response of nerves to the strength of the stimulus, specially in receptors, is very surprising. In the visual photoreceptor a single photon of red light corresponding to 3×10^{-19} Joules leads to a receptor current of 5×10^{-14} Joules, an amplification of 1.7×10^5 times. Even this weak current is stronger than required - a receptor will respond to 4×10^{-21} Joules. Actually in a dark-adapted eye 5 to 10 photons if simultaneously given in a narrow area will generate a weak transient sensation of light.

In the acoustic sensing in normal 60 dB conversation the basilar membrane movement may be as small as 10^{-11} m; atomic lengths are involved [44]! Indeed the loudest sound that can be felt may produce a vibration of 1 μm only.

What is even more baffling is that centrally programmed action may result involving a complex multistep energy handling. In the case of a frog, it may not respond to a fly, but the movement of the fly across the visual field will cause a reflex, the frog sticks out its tongue.

Weak stimuli will activate signals in nerves in other receptor modalities. Nocturnal desert scorpions can sense movement of prey 10.5 m away. Distance and direction can both be determined at 15 cm. Its basitarsal vibration receptor senses displacement of the segment by less than 0.1 nm.

Sometimes sensation or modality is in response to merely a fine variation in the physical state or modality of the stimulus. Time-compensated celestial markers may condition response - so called clock-compass. This guides some bees and birds. Geomagnetic cues may be important. Honeybees when they return to hive after 100 meters or more communicate the location of source by waggle-dance inside the dark chamber in a vertical honeycomb. The angle between the axis of the dance and direction of gravity indicates the fight direction relative to the sun. This angle changes as the sun changes, between 0 and $\pm 15°$ and has relation to the magnetic field and orientation of the vertical honeycomb.

Navigational capability of sharks or rays is well known. Electroreceptor ampullae of Lorenzini respond to 10^{-8} V/cm. Infrasound can be a simulus in marine organisms, while 0.002° C temperature difference in the pit can be felt by the rattlesnakes i.e.; a rat 40 cm away. Equally or even more striking are the sensing abilities for chemoreception, including immune response. In the silk moth Bombyx mori, bomkykol, a female sex attractant pheromone, can be effective in a dilution of 10^{17} in air, i.e., sensed from several kms away.

(iv) Extraordinary Amplification.

While the stimuli are weak the response can be comparatively very strong. Across a membrane of a nerve, a potential can exist of nearly 100 kV/cm, and this may be reached at the speed of 1 kV/s. Again, a single muscle fiber may have very weak tension, but athletes can lift their body weight in a single jerk.

(v) Extreme Specificity.

We have already referred to this aspect of biological systems by mentioning β-estradiol. The following table by Schuster and Sigmund [64] illustrates this in connection with some aspects of evolution.

Studying the dependence of mutant distribution whithin a quasispecies on the chain length of the polynucleotide, the quality of single digit accuracy connoting quality is expressed as q_m and is taken as 0.99 and the value of $\sigma_m = 4$ is taken as the superiority parameter. $1 - q/m$ designates error rate per digit.

In summary it can be that the animal physiology is replete with unexpected data on weakness of stimulus, strength of response, cooperativity of units in response and well directed response in the presence of noise. Perhaps ionic noise is as necessary as a background radiation field for biological energy. The modalities perceived are electromagnetic fields, electric charge or potential, temperature, gravity, magnetic lines of force (independent of polarity), state of polarisation of light, flicker, movement of object, colour, diffuse light, chemical molecules in hormones, drugs, food, odorants, orientation in gravitations field, ambient fluid pressure, and magnets, concentration of ions and the thermodynamic activity of water. It is true that we have summed D.C. potentials like EEG, EGG or ERG but it is again of significance that most of these interactions are due to some common aspect of the cells. These cells are derivatives of embryonic ectoderm. The receptors at the target surfaces have the same model of epithelial structure. That the electronic stimulation is effective as also the chemical messengers may well highlight the electronic aspects of the transduction of simulus to response. This appreciation supports the view that the basic functional end is of adaptation and existence. In the responsive systems it is due to primary epithelial-related structures. If noise, common ionic population or ionic potentials are unavoidable intermediates, where then is the source of complexity that we observe or sense? A cue is provided by the fact that the cortical representation and the psychophysical perception present features of complexity for an organism which may be performing very simple tasks. The continuous visual

Table 2. Single digit accuracy, superiority, and error threshold in some chemical and biological systems.

Single digit accuracy q_m	Error rate per digit $1 - q_m$	Superiority σ_m	Maximum digit content v_{max}	Biological examples
0.95	5×10^{-2}	2 4 10 20	14 28 46 60	Enzyme-free RNA replication
0.99	1×10^{-2}	2 4 10 20	64 139 230 300	tRNA precursor
0.9995	5×10^{-4}	2 4 10 20	1386 2772 4606 5991	single stranded RNA replication via specific replicases Phage $Q\beta$, $v=4500$
0.999999	1×10^{-6}	2 4 10 20	$0.7\ 10^6$ $1.4\ 10^6$ $2.4\ 10^6$ $3.0\ 10^6$	DNA replication via polymerasen and proof-reading
0.9999999	1×10^{-7}	2 4 10 20	$6.9\ 10^6$ $13.9\ 10^6$ $23.0\ 10^6$ $30.0\ 10^6$	E. coli, $v = 4\ 10^6$

field that we see with all its fine modalities and details is not represented in area 17, the primary visual cortex. There are the orientation sensitive hypercolumns of cells which sense orientation; other cortical visual areas must contribute to the image. These having been recognised in monkeys [17].

It may be true that in case of perception "the most sweeping generalisation that we can make is that its complexity and variety of functions are manifestation of complexity and variety of neural circuits - and not of a large variety of different kinds of signals" [17]. But even this statement is not able to explain the high specificity, extraordinary stimulus-response factor, stimuli related to electromagnetic fields, gravity-like inputs, vibrations and tremendous coherence in time and of action synergistically and simultaneously evoked at innumerable sites which are far removed but the concerted activity of which leads to avoidance of destruction.

4 Possibilities

(i) Features Desired for a Basic State

If a global feature exists to co-ordinate sensation and adaptive response of a high order, it must be responsive to electromagnetic fields, gravity-like stimuli, electrical currents, molecular orbitals, and geomagnetic fields. It must be responsive to energy units of the order of 7 kilocal/mole and not much more than 12 kilocal/mole. It should be able to respond to logarithm of the intensity of a stimulus. Adaptation should be achieved by coherence of response. Since at any approaching danger to life billions of muscle spindles, reflexes from ligaments, tendons, joints, visual, postural, auditory reflex pathways and biochemical reactions are called into play, the communicating channels should be exceedingly fast. It is impossible to have data on this, but the refractive index of cellfluid during physiological activity, may vary by a digit in the 3rd or 4th place of decimal. Therefore the propagation velocity should not be very much slower than the velocity of light in the tissues.

(ii) Feature That Cannot Offer Satisfactory Explanation

With such requirements we can exclude the following as possible levels of networking and as
  (a) covalent bonds and strong atomic and molecular associations. These are too strong and immutable
  (b) long lived structures like organs, tissues, cells. These are two discrete, too massive and direction-insensitive for the purposes of adaptation
  (c) strong and long-lived oxidation-reduction states. These are too slowly mutable to be relevant for the speeds necessary

(iii) The Desired Properties in the Edifice for a Network?

1. The edifice must be able to use water in a way suitable for phase transitions and metastable structures.
2. It must allow smooth transition of force fields, possess amenability to transmission of electromagnetic fields and vibration and pressure waves suitable for conformational and inactive/active phase transitions.
3. It must allow for time ordered hierarchical states in which information, is communicated fast, if at all possible with about the average velocity of light in the tissues.
4. It must allow for cooperative coherence and amplification and entrainment of signals.

(iv) Possible Alternatives

One can easily delete several levels from consideration as a substratum for functionalism, a process structure, i.e. a process recognised as a structure: (i) Could it be an aspect of gross anatomy? This is hardly relevant for understanding many processes for example, perception of vibration. (ii) Body as a collection of cells. We understand some aspects of cellular pathology and physiology but there are not cellular systems which are general enough for the entire body say, photosynthesis in plants, this is not for all of plant physiology. (iii) We may ignore cells, consider body as a mixture of molecules. This approach can hardly explain why one gets irritated by a dripping faucet or why there exists an artist. (iv) Lastly, one can ignore atomic nuclei and regard body as a conglomerate of orbitals and reaction surfaces. We can perhaps explain some isolated reactions by examining potential surfaces and decide in instances what is possible or impossible. But then fluids like blood flowing in the body or large amounts of fluid water would imply large areas which are not related to special features. But, we can already see that one is now removed far from bulk properties and is at a microscopic level. So, lastly, one takes a step further and be concerned with elementary excitations only, like photons, phonons or excitons. It is proposed that this is the level at which coherence and cooperativity result due to global happenings in the biophysicists "field". This field can be affected by stimuli which are due to electrical, magnetic, vibrational or pressure changes. It would also be specific enough because of very specific features in energy and frequency domain. This is probably also the locus for resonant targets and fulfills the criteria as a basis for information generation and locking: "entelechy" (this word is used here in the sense of goal seeking).

Now we invoke a useful rule here: Nature universally prohibits abrupt change of force-fields. And with this we can consider a useful solution.

5 The Model

In this paper we study the dynamical behaviour of a biological system. We present here quite a general quantum field theory model which has hitherto not been applied to biology and in which three boson fields interact [18] with each other. We indentify these fields as follows:

(i) The metabolic energy (closely packed quantum levels of photons if in thermal range) which is fed into the system. This is one of the boson fields. This field can be identified with an internal pump mentioned earlier. Other pumps of the same entity would be actual optical photons out of exciplex radiation [19] - [21] or out of bioluminiscence of various origins. Electroluminescence by electron transfer in mitochondria has been demonstrated. Coherent emission occurs in some Thai fire flies. Examination of interhydrogen distances in enzymes [22] and other biological structures show they are so small that a non-linear interaction is most expected [23] - [25]. These distances would lead to excited states due to intermolecular interactions in entire assemblage, no matter whether intra- or intermolecular, as per London formulation for London – van der Waals interactions. Exchange of virtual photons is specifically conjectured [24]. Recent studies on electron velocity-loss when electrons in electron microscope pass through ribosomes [28] are to be interpreted as indicative of the presence of oscillations due to plasmons in ribosomes which consume a very substantial part of the energy of high energy electrons passing through them. This would be as per Bohm-Pines formulation for plasma oscillations in solids.

(ii) The second boson field corresponds to the quanta of energy of another boson (say excitons in sufficient density); as in retina following stimulation by light.

(iii) The third field is due to the longitudinal sound waves (phonons) arising due to vibrations of hydrogen bond and/or groups with degenerate energies like sequence of methylenic groups. The expressions for number and energy densities of each type of field are obtained and we notice that solitary wave solutions indeed arise in this model.

Because of its wide applications in nonlinear systems the three wave interaction model was studied in detail earlier at the classical as well as at the quantum level [54]. Ohkuma [18] obtained the thermodynamic properties of the quantum three wave interaction model. He concentrated on two of the fields as fermion fields. However, in our problem we consider all the fields as boson fields. We also treat the system under consideration in one space and one time (1+1) dimensions.

The Model and Its Dynamics

The dynamics of such a system is described by the following Hamiltonian [31, 32]

$$H = \int dx \left\{ \sum V_n B_n^+(x) \frac{1}{i} \frac{\delta}{\delta x} B_n(x) \ g \ [\ B_2^+ \ (x) B_3 \ (x) B_1(x) \right.$$
$$\left. + \ B_1^+ \ (x) B_3^+ \ (x) B_2(x)] \right\} \qquad (1)$$

where the v_n are constant velocities, g is the coupling constant, B_n^+ and B_n are the creation and annihilation boson operators. These operators obey the following equal time commutation relations:

$$[B_j \ (x,t), \ B_k^+ \ (y,t)] = \delta_{jk} \delta(x-y), \qquad (2a)$$
$$[B_j \ (x,t), \ B_k \ (y,t)] = [B_j(x,t), B_k(y,t)] = 0 \qquad (2b)$$

where j,k = 1, 2, 3. From Heisenberg's equation of motion

$$\frac{\delta}{\delta t} B_j(x,t) = i [H, \ B_j \ (x,t)], \qquad (3)$$

using commutation relations in (2), we obtain the following equations for the operators $B(x,t)$

$$\frac{\delta}{\delta t} B_1(x,t) + v_1 \frac{\delta}{\delta x} B_1(x,t) = -ig \ B_3^+(x,t) B_2(x,t), \qquad (4a)$$

$$\frac{\delta}{\delta t} B_2(x,t) + v_2 \frac{\delta}{\delta x} B_2(x,t) = -ig \ B_3(x,t) B_1(x,t), \qquad (4b)$$

and

$$\frac{\delta}{\delta t} B_3(x,t) + v_3 \frac{\delta}{\delta x} B_3(x,t) = -ig \, B_1^+(x,t) B_2(x,t), \tag{4c}$$

The structure of the Hamiltonian in (1) is such that the number of each type of particles is not conserved. Further, $B_2(x,t)$ is the field corresponding to the internal pump mentioned earlier.

Next we use the set of equations in (4) to find the nonlinear contributions to the energy spectrum in the continuum limit. For this purpose, first we derive the expressions for densities of particles. In order to do this we use bound states $|0>$, in the sense of Bethe ansatz, derived in this model. The particle densities can be calculated by taking the expectation values of appropriate operators in the states $|0>$. We introduce the following notation:

$$I_j = <0 \mid B_j^+(x,t) \, B_j(x,t)|0> \tag{5}$$

As is indicated earlier, in this model each particle number is not constant; their rate of change is governed by the following equations

$$i\frac{\delta}{\delta t}\hat{I}_j = \left[\hat{I}_j, H\right] \tag{6}$$

where $\hat{I}_j$ is the operator corresponding to value I_j.

Introducing a new space variable, $\zeta = x - vt$, moving with constant velocity v, and using the Hartree-Fock approximation to decouple the equations for I_j's we finally arrive at the equations

$$\frac{d^2 L_1}{d\zeta^2} = \beta \left(a_{11} + a_{12}L_1 - 6L_1^2\right), \tag{7a}$$

$$\frac{d^2 L_2}{d\zeta^2} = \beta \left(a_{21} - a_{22}L_1 - 6L_2^2\right), \tag{7b}$$

and

$$\frac{d^2 L_3}{d\zeta^2} = b \left(a_{31} + a_{32}L_3 - 6L_3^2\right), \tag{7c}$$

In (7)

$$L_j = (v_j - v)I_j; \qquad \beta = \frac{g^2}{(v_1 - v)(v_2 - v)(v_3 - v)}$$

and

$$a_{11} = 2P(v_3 - v + Q - P); \; a_{12} = 2v + v_2 - 3v_3 + 8P - 4Q$$
$$a_{21} = 2PQ; \qquad\qquad\quad a_{22} = v_1 + v_3 - 2v + 4P + 4Q$$
$$a_{31} = 2Q(v_1 - v - Q + P); \; a_{32} = 2v + v_2 - 3v_1 + 8Q - 4P.$$

In these relations P and Q are two conserved quantities defined as:

$$P = L_1 + L_2 \qquad \text{and} \qquad Q = L_2 + L_3 \tag{8}$$

Solutions of (7) can be expressed in terms of Jacobi's elliptic functions. One point should be noted: our solutions must be compatible with (8) so that a periodic energy exchange among the three bosons can take place. A similar situation occurs in plasma physics also [54].

In the limiting case of infinite period, equations in (7) lead to triple solitary wave solutions, moving in unison with speed v. Various types of solutions are obtained for these equations. But in order to keep mathematical calculations simple, we treat here a special case where two boson fields have equal velocities. We choose $v_1 = v_3$ and get the following solutions:

$$L_1 = a + b \ \text{sech}^2 \mu\zeta \tag{9a}$$

$$L_2 = c + b \ \text{tanh}^2 \zeta s \tag{9b}$$

and

$$L_3 = d + b \ \text{sech}^2 \mu\zeta \tag{9c}$$

In (9) a, b, c, d and μ are constant quantities whose values are given by

$$a = 1/6 \left[v + v_2 - 2v_1 + 4P - 2Q \right],$$
$$b = 1/4(v_1 - v_2)$$
$$c = 1/12 \left[v_2 - v_1 + 4Q - 2P \right],$$
$$d = 1/6 \left[v + v_2 - 2v_1 + 4Q - 2P \right]$$

and

$$\mu^2 = \frac{(v_1 - v_2)g^2}{4(v_1 - v)(v_2 - v)(v_3 - v)}.$$

Further the velocity v and can be expressed in terms of v_1 and v_2 as follows:

$$v = \frac{5v_1 - v_2}{4} \tag{10}$$

Energy densities of the solitary waves can be obtained from $E_j = w_j I_j$. Where the w_j are the frequencies of the three waves. Using the relations $L_j = (v_j - v)I_j$ and $E_j = w_j I_j$ we can write from (8)

$$P = \frac{(v_1 - v)}{w_1} E_1 + \frac{(v_2 - v)}{w_2} E_2 \tag{11a}$$

and

$$P = \frac{(v_2 - v)}{w_2} E_2 + \frac{(v_3 - v)}{w_3} E_3 \tag{11b}$$

One can easily check that, at $d\zeta = 0$

$$E_1(0) = \frac{w_1(a + b)}{(v_1 - v)}; E_2(0) = \frac{w_2 c}{(v_2 - v)}; E_3(0) = \frac{w_3(b + d)}{(v_3 - v)} \tag{12a}$$

and at $\zeta \to \infty$

$$E_1(\infty) = \frac{w_1 a}{(v_1 - v)}; E2(\infty) = \frac{w_2 c(b + c)}{(v_2 - v)}; E_3(\infty) = \frac{w_3 a}{(v_3 - v)} \qquad (12b)$$

From these results it is not difficult to verify that the values of P and Q are unchanged. This confirms that solutions obtained by us are in conformity with relations in (8). From the results in [12] one can infer that the solitary waves are stable and they can be best carriers of energy transport in a general biological system.

6 Discussion Based on the Above Results

This demonstration may have some significance itself, but importantly it gives rise to the expectation that various physiological processes which are dissipative create a network of energy transporting beams. Further consequences do not need to be belabored: resonant absorption, point-to-point information tranfer, modulation of physiological activity at various distances to which the waves travel and further the mere speculation that a dynamic network of energy transfer may be the organizing, coordinating and cohering mechanism in living systems.

The above result leads to a co-ordinating continuum as the base on which specific biochemical interactions may cooperate to yield specificity and hierarchical control. One may ask if by the above formulation referring only to electron excitations leading to photons of various energy levels, phonons and excitons giving rise to solitary waves would not lead to scattering by the atomic nuclei. Neutron diffraction data on biological materials [22, 23] indicate that space still exists even in densely packed enzymes for focussed beam to pass through them.

In the Davydov conjecture of solitons in biology [25]-[27] energy and deformation of macromolecules provides basically the generation of solitons. In the present view for the necessary balance of forces of dispersion and non-linearity, nonlinear field interaction between atom gives the push necessary to maintain solitary waves against dissipative effects. What can such waves, or even solitons achieve? In this regard the role of selected atoms and bonds and a "loose structure" (where energy of molecular deformation is approximately of the same order as associational energy) facilitating morphological or phase change due to subtle pumping have been emphasized earlier [6].

While above may be the necessary conditions for a living state it would seem on reflection that network of pumped coherent states in an open system would be sufficient for viability of dynamic structures for physiological purposes.

By way of implications, we have to consider the following:

(i) The reason why 'high energy bonds' require, relatively, only a fraction of chemical bond energy may be that these act from an already existing background of radiation field.

(ii) Although the subsonic velocities conjectured for Davydov soliton are fast enough for biological purposes, which has assumed biological functions to be mediated in 'trunk lines' by heavily mylinated fibers conducting at around hundred meters/s, punctuated by very slow fibers in a rigorously deterministic net, the present proposal extends the communication gobally to the elasticity of the components. This is a suitable case to test specific results.

(iii) The model disregards no formulation of far from equilibrium state existing in the organism and provides, or indeed demands, specific biochemical activities to act as pumps or sinks. We may now say a word about **possible support from experimental observation.** Bioluminescence from marine organisms, fungi, unicellular organisms has been known far a very long time. Electroluminescence from isolated mitochondria due to electron transport is known. There is increasing evidence for weak but coherent photon emission from DNA in excited state (exciplex radiation). This has been reviewed [19] - [21].

That photons are trapped and stored in plants, has been shown [29]. These are certainly absorbed by retina and there is cooperative photon transfer in action centers in photosynthesis indicating photon migration in chlorophyll assemblies [30].

Photons may be involved in morphogenesis [31] and the photon spectrum may cause changes in molecules such that in a closed assembly a signal is generated. This has been used increasingly for spectrometric determination of molecules and transitions like glass transitions [32] - [34]. Photon bursts on application of AC fields in intact systems has been shown. Correspondence between carcinogenic potential of a molecular and photon interaction with trapped electron and vibronic excitations has been shown [65].

Some of the aspects of present proposal have been stated earlier, along with consideration of the role of time and temperature [9]. The words "Living State" in different settings were used earlier [39, 41], as was living matter [45] and a special state postulating minimal amounts of water bound to organelles, just sufficient to keep enzyme activity in suspended animation was suggested by the word anhydrobiosis [42]. But not attempt to a theory as a new described state essential for self-organization of phenomena of living organisms appears to have been made.

Acknowledgements

The author thanks Dr. F. A. Popp for unfailing courtesy, understanding, cooperation and support, in the preparation of this paper and Professor D. Maaß for having made this presentation possible. Thanks are due to Frau B. Hemmer for excellent secretarial assistance, and Herrn W. Weiss for exemplary cooperation and facilitation.

Appendix I

Bosons Versus Fermions

One general formulation for abolition of randomness and modification of information content has been Bose-Einstein condensation, in case of hydrogen bonds, by Fröhlich [46] leading to generation of a giant dipole. However, this cannot be regarded as the cannonical Bose Einstein condensation. In the liquification of complete shell gases like helium, argon, heat is removed from the assembly. In the Fröhlich proposition energy is pumped into a collection of bosons which are in a bath such that they constitute an open system far from equilibrium. We conjecture that they should be bosons which may be so ordered, on the following rationale:

(i) Distances between hydrogen atoms in an assembly are in the range of interaction energies which arise by non-linear interactions at interatomic distances. They are much smaller than equilibrium points in the case of Lennard-Jones potential of 1932, i.e. the 6-12 potential (see [8]) by almost half of the value, This leads to strong perturbation and excitation followed by dissipation of energy. London in 1927 using an expression in Wang's work in 1927 (see [8]) had based his theory explaining dispersion forces as simultaneous exchange of two photons generated by states due to "uncertainity generated dipoles" leading to virtual excitation. He regarded each atom as a site of multipoles to be visualised as a result of uncertainly principle. His work triumphantly explained the liquification of rare gases and heat of sublimation of hydrocarbons etc. However, one may or may not invoke this effect, but consider that the breakdown of Born-Oppenheimer approximation is fully or partially effective mechanism. For the purposes of enabling quantum chemical calculations this approximation is often made and one regards that the nuclei are stationary while the electron charge in orbitals is in fast motion. However, at the temperature of living bodies the nuclei are moving, as can be seen in neutron diffraction data [22, 23]. With respect to electron orbitals transient multipoles would occur due to differences in mobility. In this view too the interaction energies, which could be summated, would lead to a condensed phase. The photon exchange, which can be regarded as bosons, can exist, in this case too. There would be exciton-like interaction interpreted either by original London formulation and/or following the above suggestion. The phonon-like states do exist in innumerable regions of liquid crystalline phases as in membranes or nucleic acid assemblies in the living systems. The processes that give rise to the photons, in addition to London forces, are seen in fluroscence or electroluminiscence due to electron transport in respiratory chains, in phosphorescence or troboluminescence. Indeed biological processes must involve these. The case therefore exists for considering the behavior of bosons in energy and information transfer and major species of bosons would be photons, "excitons", and phonons.

(ii) Fermionic ordering has been suggested by Conrad [48]. This is worthy of consideration and further defination in the context of probability of transformation into dipoles. Ubiquitous presence of gegenions present in water at pH of 7 to 7.2 will entrap and solvate free electrons and effectively generate dipoles in character. Of course, if attraction between electrons is a necessary element then this would be helpful. Fröhlich regarded the dipolar character associated with hydrogen bonds as a bosonic species in a macromolecule. This would give rise to a giant dipole by Bose-condensation in an open system subjected to a pump. Hydrogen bonds may be due to "proton" tunneling in atoms in a macromolecule at appropriate distances between two atoms of differing electronegativities. Thus protons too (Fermionic elements) are construed to possibly generate bosons. In any case the formulation given in the present paper is pliable enough by replacing one out of three bosonic fields by a ferminonic field. This needs investigation for experimental verification of number densities.

Bullough [49] recalls Primakoff-Holstein transformation to convert Fermi oscillators of the atoms to the Bose oscillators in their collective behaviours. The Bose formulation would still remain the mechanism worthy of prior examination. The fundamental question that arises is: How does the energy propogate in biological systems? Different workers [25, 26, 27, 50] have tried to answer this question. In specific systems like alpha-helical parts of macromolecules, including protein, the energy released in the hydrolysis of adenosine triphosphate (ATP) is proposed to propagate in the form of 'solitons'. Davydov's [25]-[27] work in this direction may be recalled. He proposed that non-linear mechanisms are in operation which abrogate dispersion of energy and give rise to solitons. These solitons propagate in the biomolecules in the form of solitary waves that carry energy without loss in the system from one point to another. Davydov's solitons originate as a result of interactions between C=O streching vibrations in alpha-helices and the phonons that result due to the stretching of the hydrogen bonds in the peptide groups of these molecules. Elastic recoil from deformation is the non-linear restoring force for the soliton. We would like to mention that the peptide groups contain the HNCO atoms and as a result of stretching of the hydrogen bonds the whole peptide group is displaced from its equilibrium position giving rise to phonons.

Lastly, some features of bosons/fermions must be born in mind. If fermions show Pauli exclusion, it is not a matter of any force interaction. Bosons can accumulate in one state, not fermions. Fermions will be generated if a state is vacant. In spite of conductivity, electrons in biological systems are, so to say, in a dielectric. Finally Lifshitz stresses that if a classical oscillator process is possible in condensed matter, the quasiparticle corresponding to it is a boson [11]. These considerations seem to favor interacting bosons.

Appendix II

One of the important features is the very small amount of energy involved in elementary processes. For example, upon oxygen consumption, nerve produces heat at the rate of 0.5 calorie/g/hour. In anaerobic condition this falls to 10^{-3} calorie/g/hour. In a single shock the initial heat produced is of composite type, first a burst of heat 14×10^{-6}, then absorption of heat of 12×10^{-6} cal/g. Total body requirement of vitamin B_{12} is 1 microgram per day for an adult (70 kg). The energy of energy rich bonds is also spectacularly low, for example, in kilocal/mol, ATP (- ADP + Orthophosphate) 7.0; ATP (- AMP + Pyrophosphate) 8.6; Pyrophosphate (- 2 orthophosphates) 6.7; Creatine phosphate 10.2; Phosphoenolpyruvate 12.7; 3-Phosphoglyceryl 1-phosphate (- 3- Phosphoglycerate + orthophosphorate) 13.6; Acetyl-coenzyme A 8.2; and Aminoacyl-AMP 7.0.

It is interesting to compare this with energy for cleavage of bonds. A primary valence bond requires 50–100 kcal/mole, secondary bonds like hydrogen may require up to 5 kcal/mole. Many explanations exist for such discrepancy, including in general a concept that we are describing something different when we state a bond to be a "high-energy" bond. In case of muscle, the resting heat has been decribed by A.V. Hill to be 6.3×10^{-5} cal/g/s. With tetanic stimulation it may rise to 2×10^{-1} cal/g/s again demonstrating smallness of energies involved in life [51]. Our view in this paper here leads to a background radiation field which assists transformation induced by relatively smaller energies.

Appendix III

We already mentioned that rules for self-organization in far-from-equilibrium regimes [54]-[56], as demonstrably applicable to several systems, do need special features in the terrain if they are to be applied in intact invivo systems. These necessary features are presented by this special state, the Living State. This aspect is to be stated more explicitly.

Outstanding examples of organization of millions of molecules by chemical reactions in homogeneous media exist, like Bellousov-Zhabotinsky (B.Z.) reaction in the line of work of Bray on iodic acid-iodine oxidation couple, or oscillation of the potential in an iron rod dipped in an appropriate medium as a model of potential variation of nerve membrane. The B. Z. reaction has been extended to unstirred two or three dimensions and followed by "Oregonator" and models in stirred reactors, Briggs-Rauscher reactions and some others. These are all examples of a rudimentary "process structure" whereby a new spatio-temporal order is generated. In the living system the situation is complicated by numerous interactions whereby the input and exit of reactants is modified. There are barriers of densities like cell membranes and structured water. Oscillatory impulsions like those transmitted from a beating cardio vascular system, rotating cell organelles, pumping intracellular tubules and fibres and rate controlling factors like hormones, biomodulators thus complicate the results. Various barricades transform an individual reaction as "closed" to the required transfer of mass and momentum and specific biochemicals. The reaction therefore tends to be a closed one tending to thermodynamic equilibrium where the new orders are not possible or only so with difficulty. Dominating phase shifts at the level of phosphofructokinase, pyruvate kinase and glyceraldehyde-phosphate kinase may cause oscillations, in biochemical preparations. Similar activity may be induced by relaxed-tense conformations in allosteric enzymes or under environmental control of input-output balance of reactants. But these are in isolated stirred systems. Demonstration of oscillation in electron transport chain in mitochondria by Chance and Maitra introduces a more advanced initial structuration, but still subject to external disturbances like diffusion, pH, ionic concentrations, and influence of characteristic lengths. Perhaps the best example of overall rhythmic activity influenced by chemical synthesis-waves is control of slime mold aggregation by cAMP synthesised by *D.discoidium* cells. Here interestingly balance of adenlyic energy charge may play a role. The structural level suggested in the present plea of ours fits this quite analogously. It provides a basic structure in which background radiation field plays a crucial role in sensing and communicating. Effective final pathways demanded by a consortium of activities, result in both oscillatory and limit cycle behavior. The model also provides a structural base for hierachical control of reactions, and adiabatic elimation or suppression of some variables, leading to physiologically relevant activity in the context of persistent far-from-equilibrium state. Synchronisation, mitotic block, abortive prophase discontinuity in fusion of *Physarum polycephalum* will be facilitated in the model proposed. Morphogenetic field "gradients" of Child and proposal of co-ordinated systems by Wolpert are best supported by the present view. Computer generated plots of magnification-invariant fractal structures initially raised hopes of generation of morphology externally resembling actual natural and living forms caused by repetition of adducts of products on precursors under the same set of forces. Regrettably the external form at least is not repeated at the microscopical level under varying magnification. The mechanism may be limited, if relevant at all, to external exposed bounding surfaces, the "escaped fractal"

References

1. Mishra, R.K., Fluctuating Liquid Crystallinity: The condition for the Living State. The Fourth International Congress of Biophysics, Moscow, USSR, E-XXI a2/5, 1972.
2. Mishra, R.K., Occurrence, Fluctuations and Significance of Liquid Crystallinity in Living Systems. 20th International Liquid Crystal Conference, Kent, Ohio, Aug 22, 1972. Published in Mol. Crystals. Liquid Crystal, 1975, 29, 201-224.
3. Mishra, R.K., The Living State, The Matrix of Self-Organization, in R.K. Mishra (Ed.): 9 Physics of the Living Systems, Kluwer Academic Publishers, Dordrecht, 1990, 215-237.
4. Mishra, R.K., and Dubey, S.K., The Living State X, A system of interacting bosons, R.K. Mishra (ed.), The Living State II, . World Scientific, Singapore 1985, 492-509.
5. Mishra, R.K., Role of Charge Fluctuation Forces in Adlineation of Similar Molecules. Molecular Crystals and Liquid Crystals 10, 85-114 (Presented in 1968 in International Liquid Crystal Conference Kent, Ohio), 1970.
6. Mishra, R.K., G.C. Shukla and K. Bhownick. 1982. Lyotropic Liquid Crystallinity and its "Loose Structure", An Essential Aspect of the Living State in Applications of Physics to Medicine and Biology 1983. ICTP Trieste Conference 1982, World Scientific, Singapore, 1982, 319-336.
7. Mishra, R.K., Physical Matrix of Self-Organization in Disequilibrium and Self-Organization (C.W: Kilmister. Ed) Riedel, 1986, 185-195.
8. Mishra, R.K., Bhowmick, K., Mathur, S.C., and Mitra, S., Excitons and Bose-Einstein Condensation in Living Systems, International J.Q. Chem., 1979, 691-706.
9. Mishra, R. K. and Satish Kumar, Towards a Theory of Living State. Role of Time and Temperature, J. Appl. Mathematics and Computing (in Press), 1992.

10. Balinski, B. I. An Introduction to Embryology (4th Ed.), W. B. Saunders Co, Philadelphia, 1975, 535-563.
11. Kaganov, M. and Lifshits, I., Quasiparticles, Mir Publishers, Moscow, 1979, 35.
12. Smith, C.W., The Boundaries of the Living State in Molecular and Biological Physics of Living Systems (R.K. Mishra Ed.), Kluwer, Dordrecht, 1990, 87-100.
13. Fatt, B., Katz, B., from Eckert, R., Randall, D., Augestine, G., (Eds.), Animal Physiology, 3rd Ed., W.M. Freeman, 1988.
14. Mishra, R.K., (Ed.), The Living State, Wiley Eastern, New Delhi, 1982.
15. Mishra, R.K., (ED.), The Living State II, World Scientific, Singapore, 1985.
16. Kléman, M., The geometrical nature of disorder and its elementary excitations, J. de Physique, 1982, 43, 9, 1389-1396.
17. Eckert, R., D. Randall, G. Augustine. Animal Physiology Mechanisms and Adaptations. 3rd Ed., W. M. Freeman and Co., 1988, 151-159.
18. Ohkuma, K., Quantum three wave interaction models: Bethe- Ansatz and statistical mechanics, S. Takeno (ed.) in Dynamical Problems in Soliton Systems, Springer-Verlag, Berlin, 1985, 99-104).
19. Popp, F. A., Gurwitsch, A.A., Inava, H., Slawinski,J.,. Clento, G., van Wik, R., Chwinot, W.P., Nagl, W. Biophoton Emission, 1982, Experientia, 1981, 44, 543-600.
20. Popp, F. A., Photon Storage in Biological Systems, in: Electromagnetic Bioinformation, F. A. Popp, G. Beelcei, H. L. Koenig, W. Peschka (Eds.) Urban and Schwarzenbach, Munich, 1979, 123-149.
21. Popp, F.A., Li, K.H., and Gu, Q., Recent Advances in Biophoton Research and its Applications, World Scientific, Singapore, 1992.
22. Sakabe, N., Sakabe, K., Sasaki, K., Hydrogen atoms and hydrogen bonding in rhombohedral 2 Zn insulin crystals by X-ray analysis and 1-2 A resolution in structural studies on Molecules of Biological interest in G. Dodson, J. Glusker, D. Sayre (Eds.), Clarenden Press, Oxford, U.K., 509-526
23. Ramanadham, M., and Chidambaram, R., Amino Acids: Systematics of Molecular Structure, Conformation and Hydrogen Bonding, Adv. in Crystallography, Oxford and IBH Publ. Co. New Delhi, 1978, 81-103.
24. Margenau, H. and Kestner, N.R., Theory of Intermolecular Forces, Pergamon Press, New York 1969.
25. Davydov, A.S., Solitons in Molecular Systems: Phys. Scripta 20, 1979, 387-394.
26. Davydov, A.S., Biology and Quantum Mechanics, Pergamon, Oxford (1982).
27. Davydov, A.S., Solitons in quasi one dimensional molecular structures: Sov. Phys. Usp. 1982, 25, 899-918.
28. Ottensmeyer, F.P. and Andrew, J.W., High Resolution Microanalysis of Biological Specimen Electronic Energy Loss Spectroscopic and by Electron Spectroscopy Imagery,J. Ultrastructur Res.,1980, 72, 336-348.
29. Voegelman, T. C. and Bjrn, L.O., Plants as Light Traps, Physiol. Planrar. 68, 1986, 704 - 708.
30. Langbein, D., Theory of van der Waals attraction, Springer Tracts in Modern Physics, Heidelberg, 1974.
31. Hartmann, Haupt, W., Photomorphogenesis in Biophysics, 3rd Ed., 1983, Springer, Heidelberg. W. Hoppe, W. Lohmann, H. Markl and H. Ziegler (Eds.), 1983, 542-564.
32. Sunandan, C. S., Physical Applications of Photonic Spectroscopy, Physica, 1988, A. 2. 13.
33. Pao, Y. H., (Ed.), Optoacoustic Spectroscopy, 1977, Academic Press, Nr. 4

34. Neubacher, H., Scharrmann, A., and Lohmann, W., Applications of Photoacoustic Spectroscopy in Biophysics 3rd. Ed. W. Hope, W. Lohmann, H. Markl, H. Ziegler (Eds.), Springer, Heidelberg, 1983, 109.

35. Conrad, M. Superinformation Processing: The feasibility of proton superflow in the Living State, in Mishra R. K. (Ed.) Molecular and Biological System, 1990, 159-174.

36. Selye, H., Private communication, 1957.

37. Benninghoff, A., Form und Funktion I, II. Z. ges. Naturwiss I, 2, 1935/1936.

38. Benninghoff, A., ber Einheiten und Systembildungen im Organismus, Dtsch. med Wschr 1938.

39. Bertalanffy, L.V., Kritische Theorie der Formbildung, Nikolaus von Kues Mnchen 1938.

40. Bertalanaffy, L.V., Probleme einer dynamischen Morphologie, Biologia Generalis, 15 1941.

41. Ling, G.N., A Physical Theory of the Living State? The Association-Induction Hypothesis, Blaisdell, Watham, Mass 1962.

42. Clegg, James, S., in "Anhydrobiosis", J. Crowe and J. Clegg (Eds.), Dowden, Hutchinson and Ross, Stroudsberg, 1973, 141.

43. Mishra, R.K., Mind-Brain Relationship: a Physical Analogy, I. J. Psychiat. 73, 1965.

44. Zwicker, E. and Manley, G., in W. Lohmann, H. Markl, H. Ziegler (Eds), Biophysics 3rd. Ed., Springer, Heidelberg, 1983.

45. Renger, G., Biological Energy Conservation in W. Hoppe, W. Lohmann, M. Markl and H. Ziegler, (Eds),Biophysics 3rd. Ed., Springer, Heidelberg, 1983, 347.

46. Fröhlich, H., The Biological Effects of Microwaves and Related Questions, Electronic and Electron Physics 69, 1980, 85-152.

47. Prigogine, I., Wiame, J.M., Biologie et Thermodynamique des Phenomenes irreversibles, Experienta 2, 1946.

48. Brown, A.G., Nerve Cell and Neuron Systems, Springer, Heidelberg,1991.

49. Bullough, R.K., Puri, R.R., Hassan, S.S., Some remarks on the organization of living matter, in. R.K. Mishra (ed.), Molecular and Biological Physics of Living Systems, Kluwer, Dordrecht, 1990,1-18.

50. Scott, A.C., Biological Solitons in S. Takeno (ed.), Dynamical Problems in Soliton Systems, Springer, Berlin, 1985, 224-235.

51. Davson, H., A Textbook of General Physiology, J. & A. Churchill, London, 1959, 541, 545.

52. Bernal, J.D., Untitled comments in Disc., Faraday Soc. 1933, 29, 1082.

53. Wald, G., Life in the Second and Third Periods in M. Kasha and B. Pullman (eds), Horizons in Biochemistry, Academic Press, N.Y., 1962, 127.

54. Thacker, H.B., Exact Integrability in Quantum Field Theory of Statistical Mechanics: Rev. Mod. Phys., 1981, 53, 253-285.

55. Nicolis, G. and Prigogine, I., Self-Organization in non-equlilbrium systems. Wiley-Interscience, New York 1977.

56. Prigogine, I., From Being to Becoming. W.H. Freeman and Co., San Francisco 1980.

57. Haken, H. and Graham, R., Synergetik - Die Lehre vom Zusammenwirken, Umschau 1971, 6, 191.

58. Thom, R., A dynamical scheme for vertebrate embryology, American Math-series: Some mathematical questions in biology 4. Lectures on Mathematics in Life Sciences, 5, Providence (R.I.), USA, 1973, 3-45.

59. Kaganov, M. and Lifshits, I., Quasiparticles, Mir Publishers, Moscow, 1979, 35.
60. Schuster, H. G., Deterministic Chaos, 2nd Ed., VCH-Verlag, Weinheim, 1989.
61. Haken H., Pattern formation by dynamic systems and pattern recognition, Springer, Heidelberg, 1979.
62. Hartmann, K. M., Action Spectroscopy, in W. Hoppe, W. Lohmann, H. Markl and H. Ziegler (Eds.), Biophysics, Springer, Heidelberg,2nd Ed. 1983, 116 -130.
63. Renger, G., Photobiophysics, in Biophysics, W. Hoppe, W. Lohmann, H. Markl and H. Ziegler (Eds.), Biophysics, Springer, Heidelberg 2nd ed. 1983, 518.
64. Schuster, P. and Sigmund, K., Evolution in Biophysics, in W. Hoppe, W. Lohmann, H. Markl and H. Ziegler (Eds.), Springer, Heidelberg, 2nd Ed. 1983.
65. Mishra, R. K., Dubey, S.K. and Tyagi, R.S., Interaction between Photons, "Trapped Electrons" and Vibronic, I. J. of Quantum Chemistry, 1985, 27, 303 321.

A Model for Stimulated and Co-operative Electron Transfer in Biomolecular Systems

L. Pohlmann and H. Tributsch

1 Introduction

The paradigm of self-organization is playing an increasingly important role in attempts to reunify the sciences. Due to Prigogine's fundamental work the old paradox of emergence of highly complex order in living systems on the one hand, and the increase of entropy demanded by the second law of thermodynamics on the other hand, was resolved. Therefore one is tempted to say that modern biology must be the very natural field of self- organization theory. This is, indeed, the case in population dynamics, in the theory of emergence of life (Eigen's Hypercycle) and in some special oscillating biochemical reactions.

But, surprisingly, in other parts of biology and biochemistry, e.g. in bioenergetics or in the speciation theory (the so called synthetic theory of evolution), the ideas of self-organization today are rather neglected. In contrast, we know that all living organisms and thus all of their subunits like cells or macromolecules are thermodynamically open systems, capable of self-organized structuring of the intrinsic chemical and electrochemical processes. Therefore it is, to our understanding, a useful heuristic approach to develop alternative models which show the possibility that self-organization processes can in principle be used by nature to effectivate the capabilities of living systems (for the speciation theory see [1]).

Some of the most important electrocatalytic mechanisms such as water splitting, carbon dioxide and nitrogen fixation, sulphate and oxygen reduction are multi-electron processes, which only proceed close to the favourable thermodynamic overall potential when undesirable intermediates are avoided and electrons are transferred more or less simultaneously. These reactions are very successfully handled by biological systems at environmental temperatures, but generate significant difficulties for development in technical processes. Accordingly, the need to consider autocatalytic mechanisms as a way of making photoinduced interfacial electron transfer more efficient for solar energy conversion has been discussed by one of the authors [2].

Biological systems typically apply multicentre catalysts (e.g. cytochrome oxidase, the manganese centre of photosynthesis, or the cytochrome c_3 of sulphate-

reducing bacteria) which show significant interactions between their metal cen-
tres and probably utilise feedback mechanisms. This indicates the presence of
far from equilibrium processes, where the co-operative phenomena of irreversible
thermodynamics become possible.

The internal electronic feedback has recently been especially well demonstrated between the four heme groups of cytochrome c_3 in *Desulfovibrio vulgaris*.
Whenever an electron is exchanged, all heme groups change to a smaller or larger
extent their oxidation state and their internal redox potential [3]. This means
that in multi-electron transfer reactions the first electron to be transferred may
exert a positive feedback effect on the transfer of the second one and so on (like
chemical autocatalysis).

The aim of this paper is to prove on the basis of a simple model that far from
equilibrium co-operative electron transfer is indeed possible.

2 Modelling Strategy

Our model was inspired by an idea of Prigogine from 1965 [4] concerning a
sequence of equivalent chemical reactions in a homogeneous liquid environment
with catalytic feedback as a model for the remarkably high reaction efficiency of
certain molecular biochemical mechanisms:

$$A \xrightarrow{\;1\;} X_1 \xrightarrow{\;2\;} X_2 \xrightarrow{\;3\;} \dots \; X_n \xrightarrow{\;n+1\;} B$$

$$\uparrow\downarrow_{n+2} \tag{1}$$

$$M$$

with

$$k_1 = k_1' = \dots = k_n = k_n' = k = 1 + \alpha M$$

$$k_{n+1} = k_{n+1}' = 1 \tag{2}$$

Here, the substance M is some intermediate, which catalyses the reactions
1 to n in the reaction chain. Under non-equilibrium conditions $(A \gg B)$, but
without feedback $(\alpha = 0)$ the stationary value of the last intermediate X_n is
very small:

$$X_n = B + O(\frac{1}{n}). \tag{3}$$

However, when $\alpha \to \infty$ (strong feedback):

$$X \to A.$$

In this way certain reaction products in biochemical reactions can be, in contrast to classical expectations, produced in high concentrations due to feedback couplings. We will now try to develop this simple theoretical approach for electron transfer reactions along molecular electron transfer chains in membranes and through interfacial electron transfer complexes, which are involved in multi-step electron transfer processes. For this purpose, first we will interpret Prigogine's model as a sequence of electron transfer reactions

$$(X_i + e; X_{i+1}) \leftrightarrow (X_i; X_{i+1} + e),$$

(i.e. the X_i is interpreted as electron density at the site i) and, second, we will assume reaction rates, that are not equal for forward and backward reactions, i.e. the ratio

$$\kappa = k/k'$$

is in general not equal to unity. Therefore we obtain a set of non-linear differential equations:

$$\frac{dX_i}{dt} = \beta(kX_{i-1} - k'X_i + kX_i + k'X_{i+1}), \qquad X_0 = A, \qquad i = 1, \ldots, n-1$$

$$\frac{dX_n}{dt} = \beta(kX_{n-1} - k'X_n) - kX_n + K'B - aX_n + a'M \tag{4}$$

$$\frac{dM}{dt} = aX_n - a'M$$

with

$$k_1 = \ldots = k_n = k\beta; \qquad k'_1 = k'_n = k'\beta$$

$$k_{n+1} = k'; \qquad k'_{n+1} = k'; \qquad \kappa = \frac{k}{k'}; \qquad \kappa_a = \frac{k_{n+2}}{k'_{n+1}} \tag{5}$$

where $\beta = 1 + \alpha\kappa_a X_n$.

3 Analytic and Numerical Solutions: Stimulated Electron Transfer

In the steady state case, the corresponding set of (n+1) algebraic equations can be solved analytically:

$$X_i = \frac{\left[(\beta\kappa - \beta - \kappa)\kappa^j + \kappa^{n+1}\right]A - (1 - \kappa^j)B}{(\beta\kappa - \beta - \kappa) + \kappa^{n+1}} \quad \text{where } \beta = 1 + \alpha\kappa_a X_n \tag{6}$$

For $\kappa = 1$ the results obtained by Prigogine can be obtained for the limiting cases of vanishing and of very strong feedback. Furthermore it should be noted that this discrete model can be extended to a model with continuously changing states leading to a kind of non-linear diffusion equation with variable diffusion coefficient and with a migration term [5]. The numerical results obtained by this model are qualitatively similar to the results of the discrete one. For the sake of simplicity we will restrict our considerations here to the discrete model. For the case of "isotropic" e-transfer kinetics ($\kappa = 1$) we obtain a result similar to that of Prigogine's model: With increasing feedback parameter α the electron density at the last intermediate site also increases up to the maximal value, which cannot exceed the value of the left boundary condition A (Fig. 1). On the contrary, in the case of "anisotropic" e-transfer kinetics ($\kappa > 1$), an increase of the feedback parameter α causes an increase of the electron densities at the intermediate sites, which are not limited by the value of A. (Mathematically, this is a consequence of the nonlinearity of the equations, which appears only if $\alpha \neq 0$.) Thus, here the feedback increases the value of X_n, and therefore the resulting stationary electron flux, in a dramatic way, which we will call feedback stimulated electron transfer (Fig. 2).

4 Introducing Co-operativity. Order Parameter Equation

The model described above, shows the feature of stimulated electron transfer, but co-operative electron transfer could not be demonstrated. Co-operativity is to be understood in the sense of Haken's synergetics, i.e. when one or a few modes of a system with many degrees of freedom become unstable at a bifurcation point, then all remaining stable modes will be "enslaved" in their motion by the motion of these few "order parameters". Therefore, co-operation appears as a drastic reduction of the degrees of freedom of the complex system. Haken originally introduced his concept of synergetics for the laser, where the many independently emitting atoms become enslaved by one or a very few modes of the electromagnetic field. Consequently the atoms will emit in a very co-operative way. In our model, one can show that an autocatalytic production of the catalytically acting intermediate M, instead of the first order reaction above, leads to co-operative behaviour:

$$M + X_n \overset{\mu}{\rightleftharpoons} 2M; \qquad \frac{dM}{dt} = m\left(\mu X_n - M\right) \tag{7}$$

This model also can be solved analytically in the stationary case, which leads to three solutions, one corresponding to $M^{(1)} = 0$ (feedback is switched off):

$$X_{n,1} = \frac{\kappa^n}{\kappa^n + \kappa^{n-1} + \ldots + 1} \qquad , \tag{8}$$

the other two corresponding to $M^{(2)} = \mu X_n$ (feedback is switched on):

$$X_{n,2,3} = \frac{-\left(D \pm \sqrt{D^2 + 4\alpha\mu\kappa^n}\right)}{2\alpha\mu}, \quad D = \left(\kappa^n + \kappa^{n-1} + \ldots + 1 - \alpha\mu\kappa^n\right) \tag{9}$$

Here one solution is always negative and therefore physically meaningless. Hence there are two relevant states of the system, i.e. two positive solution branches in the bifurcation diagram with the bifurcation parameter μ. Their intersection at $\mu = 0$ is the bifurcation point. To determine the type of stability change at the bifurcation point one must carry out a stability analysis. The linear stability analysis of the model is based on the linearization of the differential equation system around a stationary state, e.g. around the state $X_{n,1}$. From the corresponding Jacobi matrix $[\partial F_i/\partial X_i]$ we obtain, following the standard procedure, the eigenvalues:

$$\sigma_1 = \mu X_{n,1}$$
$$\sigma_{i+1} = -(1+\kappa) + 2\sqrt{\kappa}\cos A_i < 0 \qquad (10)$$
$$A_i = \frac{i\pi}{n+1}, \qquad i = 1, 2, \ldots, n$$

Hence, at $\mu = 0$ only one eigenvalue, σ_1, changes its sign and all others are smaller than zero. According to the slaving principle, all $(n + 1)$ degrees of freedom reduce to only one degree of freedom. The slowest mode, which is just becoming unstable (following Haken, the order parameter, which is in our system M), is enslaving all other stable modes. Following the slaving principle (i.e. a kind of adiabatic elimination), there remains only one differential equation describing the evolution of the order parameter: After inserting the expression for the dependence of $X_n(M)$ on M into the differential equation for M,

$$\frac{dM}{dt} = M\left(\mu X_n(M) - M\right), \quad X_n(M) = \frac{\beta(M)\kappa^n}{\kappa^n + \kappa^{n-1} + \ldots + \kappa + \beta(M)},$$
$$\beta(M) = 1 + \alpha M$$

we obtain for M a non-linear differential equation of first order

$$\frac{dM}{dt} = M \cdot \left(\mu \frac{\kappa^n(1 + \alpha M)}{\kappa^n + \kappa^{n-1} + \ldots + \kappa + (1 + \alpha M)} - M\right) = F(M) \qquad (11)$$

where $X_i(\beta(t))$ are now the enslaved variables, the evolution of which mainly depends only on the evolution of $M(t)$. The shape of the right hand side function $F(M)$ is depicted in Fig. 3 for several values of the parameter μ.

Consequently the evolution of the whole system of electron transfer reactions can for $\mu > 0$ approximately be described by one variable M, i.e. a self-organized ordering process occurs at $\mu = 0$, which imposes on the electron flux in the reaction chain a "macroscopic" ordered structure. Equation (11) we will call, by analogy to Haken's laser equation, the equation of co-operative electron transfer. This analogy can be illustrated by introducing formally a potential for the motion of the variable M:

$$\frac{dM}{dt} = -\text{grad } V(M), \quad V(M) = -\int F(M)\, dM. \qquad (12)$$

The behaviour of M can then be visualized as an overdamped asymptotic motion of a ball in a one-dimensional potential field with the shape $V(M)$ [7]. The shape of the potential $V(M)$ is shown in Fig. 4 for the same parameter values as in Fig. 3.

Here one can see that for $\mu < 0$ the potential has only one stable state (minimum) at $M = 0$, which becomes neutrally stable at $\mu = 0$ and unstable with $\mu > 0$ (maximum). Simultaneously a new minimum occurs at a value of $M > 0$, which represents the new state of co-operative electron transfer. According to equation (11) in Fig. 5 is shown the time evolution of the autocatalytic order parameter M and the corresponding value of electron flux J (which is proportional to X_n). The evolution starts in the unstable state $M = 0$ and, after a small perturbation, undergoes a rapid transition to the new stable state of co-operative and stimulated electron transfer. It should also been noted that the autocatalytic production of the intermediate M could be of order higher than two. In the case of a third order reaction (a cubic non-linearity in M), after the bifurcation the system would have the choice between two stable states, separated by an unstable middle state. These stable states are equivalent to a highly conducting and a poorly conducting electron transfer pathway. Here we are dealing with a non-linear molecular electron device which could serve to build molecular computers.

5 Discussion

The present model has been used to show that stimulated and co-operative electron transfer is, in principle, possible. A molecular complex suitable for co-operative electron transfer must have suitable molecular feedback loops and autocatalytic process steps. A remarkable consequence would be the generation of a state of high electronic conductivity equivalent to a dramatic reduction of activation energies. Such a dynamic co-operative electronic state with new electronic properties would have significant consequences for several fields including material science, energy conversion and electrocatalysis as well as for the further understanding of very efficient biochemical processes. In biology the most likely system where co-operative electron transfer is to be expected is the last step of photosynthetic oxygen evolution ($S_4 \rightarrow S_0$), which occurs very close to the ideal thermodynamic potential of water decomposition [8] and which up to now cannot be reproduced artificially. An indication for the existence of non-linear positive feedback processes in photosynthesis is the occurrence of spontaneous oscillations in intact leaf tissue from spinach, induced by a sudden increase in ambient CO_2 concentration [9].

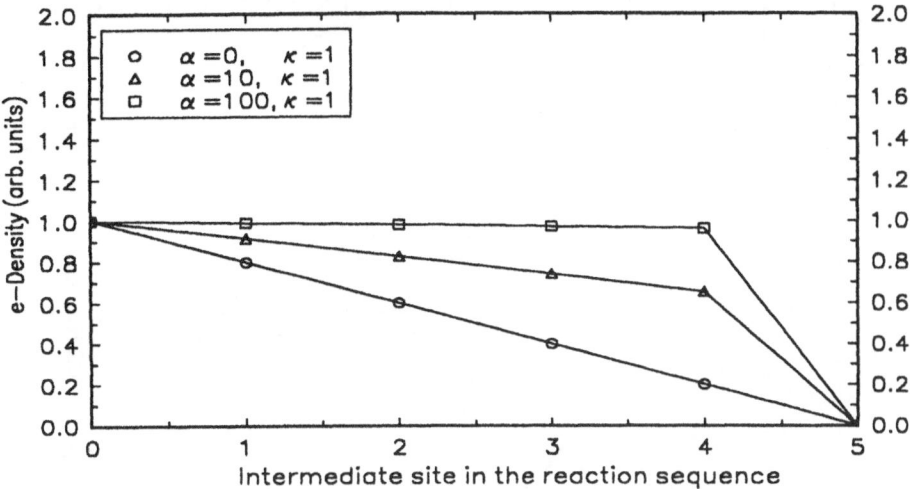

Fig. 1. Electron density along reaction chain for different values of α and isotropic electron transfer ($\kappa = 1$)

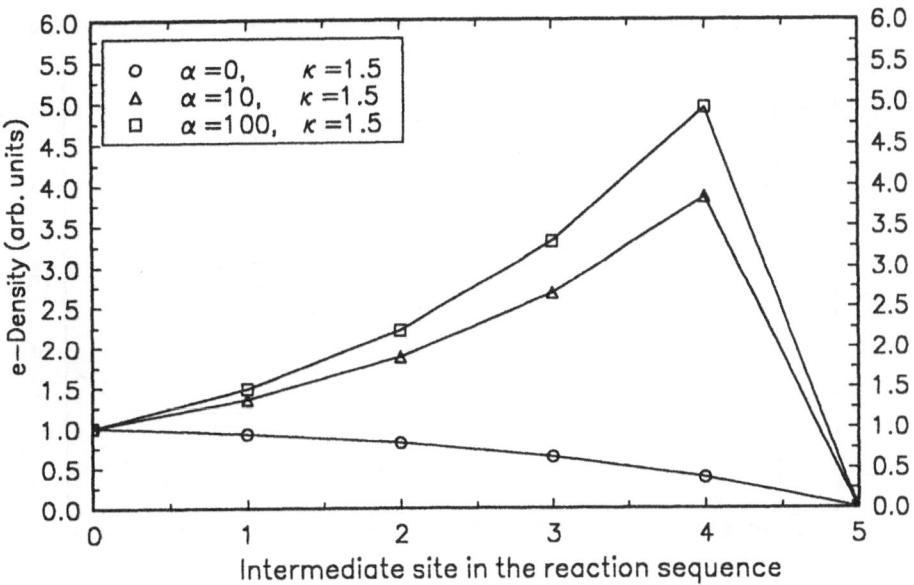

Fig. 2. Electron density along reaction chain for different values of α and anisotropic electron transfer ($\kappa > 1$)

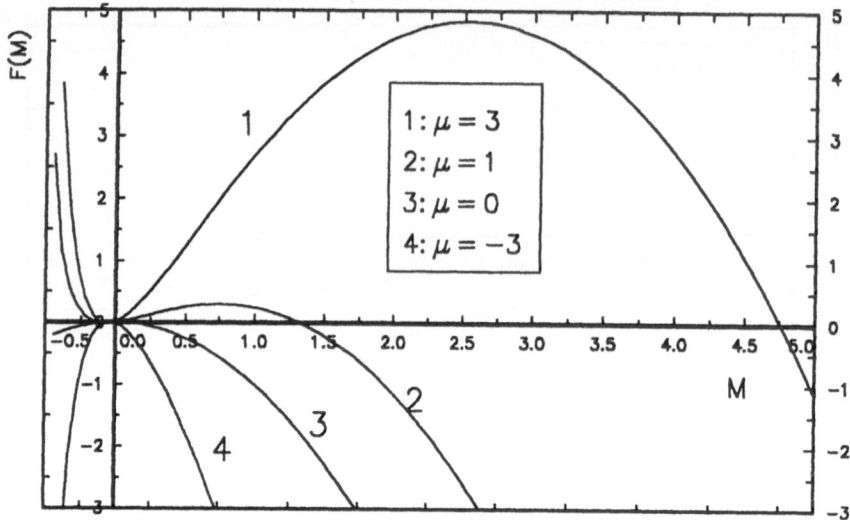

Fig. 3. Dependence of the function $F(M)$ on M for different values of parameter μ (with n=3)

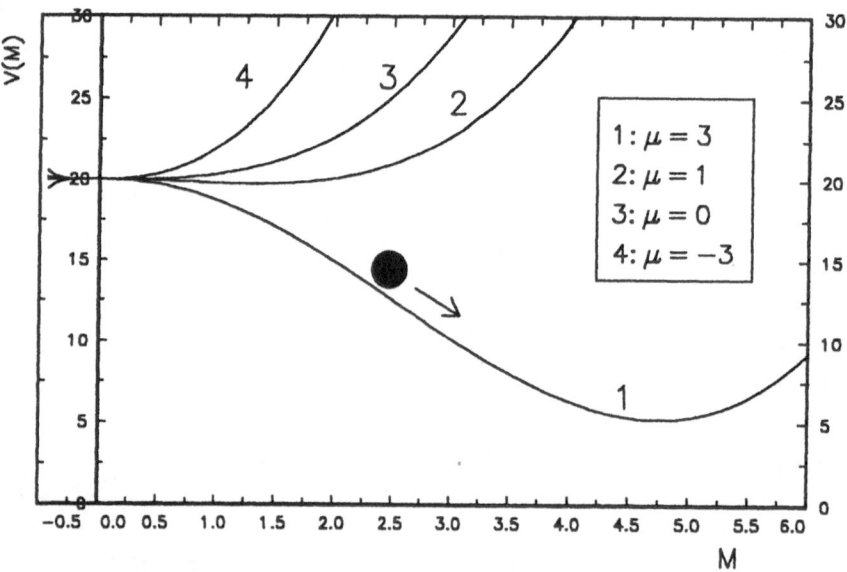

Fig. 4. Dependence of the potential function $V(M)$ on M for different values of the parameter μ (with n=3)

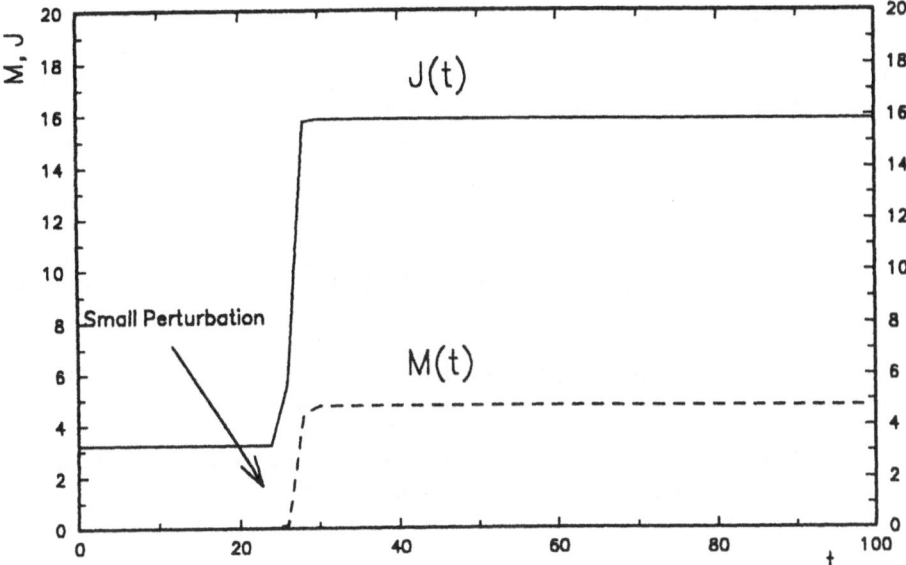

Fig. 5. Evolution of order parameter M and electron flux J ($\alpha = 10$, $\kappa = 1.2$, $\mu = 3$ and with $n = 3$)

References

1. Pohlmann, L.,Niedersen, U., Dynamisches Verzweigungsverhalten bei Wachstums-
   und Evolutionsprozessen, in: Selbstorganisation. Jahrbuch für Komplexität in den
   Natur, Sozial- und Geisteswissenschaften. Berlin: Duncker & Humblot, Band 1, S.
   63, 1990.
2. Tributsch, H., Lecture and Proceedings of the 8th International Conference on
   Photochemical Energy Conversation and Storage (IPS-8), Palermo, July 1990.
3. Akutsu, H., Fan, K., Kyogoku, Y., Niki, K., In: Charge and Field Effects in Biosystems II, Eds. Allen, M. J., Clearly, S. F., Hawkridge, F. M., New York 1989, pp.
   59-68.
4. Prigogine, I., Physica 31, 719, 1965.
5. Pohlmann, L., Tributsch, H., J. theor. Biol. 155, 443, 1992 and 156, 63, 1992.
6. Haken, H., Synergetics. An Introduction, 3rd. edn., Berlin: Springer, 1983.
7. see, e.g., Haken 1983.
8. Witt, H.T., In: The Roots of Modern Biochemistry, Kleinkauf, von Döhren and
   Jänicke (eds.), Berlin: Walter de Gruyter, pp. 713-720.
9. Peterson, R.B., Sivak, M.N., Walker, D.A., Plant Physiol. 88, 1125, 1988.

Self-Organization, Catastrophe Theory and the Problem of Segmentation

P.T. Saunders and M.W. Ho

While self-organization has been studied for some time and is now a familiar concept in science, there is as yet no general theory nor even a generally accepted definition. Much of what has been done has been within the context of dissipative structures, so much so that for many people the two ideas have become totally conflated. But the concept of self-organization is much broader; it encompasses systems described by catastrophe theory, for example. Here we draw on the experience of catastrophe theory to suggest features that a theory of self-organization should have, and we illustrate this by the problem of segmentation. We point out that self-organization provides an alternative to natural selection as an explanation of order and organization in biological systems.

1 What is self-organization?

What exactly do we mean by self-organization? Words like order and organization are recognized as a source of confusion in science, and much has been written on how they should be defined. Yet the term self-organization seems to have crept into the literature without much discussion.

One of the distinctions that is commonly drawn between order and organization is that organization is a process, whereas order is a pattern. Where organization produces order, it is a *dynamic* order, as in Bénard convection cells, or the concentric circles or spirals of the Beloussov-Zhabotinsky reaction. Even the apparently stable patterns of animal coat colours or segmentation have arisen from dynamic processes.

Another distinction is that organization is thought of as being *for* something. We say that a crystal has order but that a watch is organized, because the parts of the latter have been put together in a certain way for a purpose, viz. to keep time. In contrast, *self*-organization creates order without conscious design and so the concept of function hardly arises. No one asks what the red spot on Jupiter is organized for. Biologists, of course, automatically seek explanations in terms of function, but as a theory of self-organization develops it should counter this tendency.

We shall use the term self-organization to refer to all situations in which some form of order appears spontaneously in a system with no direct external control.

On Self-Organization
© Springer-Verlag Berlin Heidelberg 1994

The crucial point – the implication of the prefix "self" – is that the order that interests us arises *within the system and at the level we are considering*. It can be found neither in the external influences nor at lower levels of the system. As a result, to derive a general theory of self-organization, i.e. to derive results that are applicable to many different systems, we have to work at the level of the phenomena.

In most cases, and here catastrophe theory provides a good example, many different lower level mechanisms lead to the same kind of order at a higher level. This makes the order a phenomenon to be studied in its own right, as are the elementary catastrophes. It also creates a problem for conventional modelling, because if many different mechanisms can produce the same sort of order, that we observe a particular order is not good evidence for any one mechanism. We can derive Snell's Law of refraction by supposing that photons try to minimize the travel time between two points, but that does not prove that is what photons do; we may be able to fit the behaviour of dung beetles and other organisms to game theoretic models but again it remains to be seen how much that tells us about what is really going on. (See Saunders, 1989).

The development of a theory of self-organization will have two main strands:
a) to discover what sorts of order are likely to occur by some process of self-organization, to classify these and to discover their properties.
b) to develop a methodology for applying this knowledge to the study of real problems.

Considerable progress has been made on both of these in catastrophe theory, which supports our contention that it is properly considered as part of a theory of self-organization.

2 Catastrophe theory – a theory of self-organization

The aim of catastrophe theory (Thom, 1972; for a simpler introduction see Saunders, 1980) was to solve the general problem of form and the succession of form, to "construct an abstract, purely geometrical theory of morphogenesis, *independent of the substrate of forms and the nature of the forces that create them.*" (p.8) Elementary catastrophe theory, which is what most people think of as catastrophe theory, was intended to be only the first stage of the programme. It deals with the "... local accidents of morphogenesis, which we will call *elementary catastrophes,* whereas the global macroscopic appearance, the form in the usual sense of the word, is the result of the accumulation of many of these local accidents."

Elementary catastrophe theory is typically concerned with a system whose behaviour is usually smooth but which occasionally exhibits discontinuities, either in time or in space. We suppose that the system can be described by a large but finite number of state variables, and a number of control variables which may be controllable parameters or even just the independent variables in a set of differential equations.

We then suppose that the system can be represented adequately by any one of a large class of mathematical models. This includes systems that are at local equilibria of (possibly large) systems of ordinary differential equations. (Note that catastrophe theory is not, as is sometimes claimed, restricted to gradient dynamics; it is sufficient that there be a Lyapunov function.) It also includes systems that are governed by the sorts of partial differential equations that are customarily used in physics and biology.

For such systems, catastrophe theory provides a remarkably strong result. The number of qualitatively different patterns of discontinuities that can occur (locally) depends not on the number of state variables, which can be very large, but on the number of control variables, which is generally small. If there are four, the number of patterns is 7, if there are five it increases to 10. If there are six or more the situation is more complicated but still tractable: the case of eight, for example, has been extensively analyzed and there are applications in the literature.

The condition that the number of control variables must be small is far less restrictive than it may sound. What matters is not the total number of variables that might influence the system in any way, but only the number of variables or combinations of variables that can have qualitatively different effects on the discontinuities: in effect the number of independent essential eigenvariables. You can imagine that if a system is exhibiting discontinuities of one kind or another and if six or more variables are influencing these and all in quite different ways, then it is going to be very hard to make sense of what is going on by any means.

The clearest example of the patterns of discontinuities to which catastrophe theory applies is in caustics (Berry and Upstill, 1980), the bright lines that you see when light is focussed onto a surface. Caustics are determined by the characteristics of the wave equation, and catastrophe theory tells us that no matter whether the focussing is by reflection or refraction or a combination of the two, and no matter what the shapes of the surfaces and lenses involved, all caustics must have shapes that are qualitatively like those found in the catastrophes with three or fewer control variables. Since there are only five such catastrophes, the range of possibilities is very limited.

What makes catastrophe theory useful is that it is possible to classify the elementary catastrophes, that these describe a wide range of systems, and that for reasonable numbers of control variables the number of distinct classes is small. The elementary catastrophes themselves thus become the entities with which we can work, and we do not need to know in any detail what is going on at the lower levels to produce them.

This is what we should be aiming for in a theory of self-organization. Note, however, that even in the case of elementary catastrophe theory the classification is not entirely successful. It is based on the concept of structural stability, and Smale (1966) has shown that in more than two dimensions the structurally stable systems are not dense, which means that we cannot always choose a structurally stable model. Zeeman (1989) has suggested a concept of stochastic stability to cope with this problem, but the full implications of this have not yet been worked out. In any case, a theory of self-organization will almost certainly not be a complete theory, and the methodology that develops will have to recognize this.

3 Catastrophe theory – a methodology

Catastrophe theory also has much to offer to the development of a methodology for self-organization, both in the ways it is applied and in the questions that it has raised.

A common first reaction to catastrophe theory is that it sounds too good to be true. You put in practically no theory and very little data about the system, and you get out results. Surely, one feels, this must be a violation of some first law of applied mathematics, that you cannot get more information out of an anlaysis than you put in.

In fact this is not the case. A complete description of the behaviour of a system is a great deal of information. A qualitative description of the pattern of discontinuites is a very much smaller amount. It may be the most interesting or useful information, but that is a different matter altogether.

In statistics, one statistic is said to be more efficient than another if it can give as good an estimate of the same parameter using fewer data. In the same way we can say that catastrophe theory is more efficient than conventional methods at dealing with patterns of discontinuities. This is because it deals with them directly, typically telling us little if anything about the rest of the behaviour of the system.

A general theory of self-organization should be efficient in the same sense. Once we have reason to believe that a system is of a certain class, or at least are prepared to make the hypothesis that it is, then it will follow immediately that we may expect the system to have the properties that are typical of that class.

The word "typical" is very important. In more conventional modelling we generally have a particular mechanism in mind. From this we derive firm, often quantitative, predictions about what should be observed. Thus we can use Newton's laws to predict where Mars will be on a given date some thousand years hence. We know that there is some form of *ceteris paribus* assumption involved, because our prediction will be wrong if in the meantime the Earth is destroyed or some massive body comes too close to the Solar System, but no one sees this as a problem for classical mechanics.

Most applications of catastrophe theory are different, in that the *ceteris paribus* assumption is a much more important part of the model, and more explicit as well. We almost always have to include some form of simplicity postulate, generally along the lines that the mechanism that underlies the phenomena is no more complicated than it has to be to produce what we observe. Even when we start our analysis without detailed observations, we assume that what we observe is the result of a single process and hence an elementary catastrophe.

Contrary to what some have claimed, we can make predictions in this way. They will usually be qualitative rather than quantitative, though this is not necessarily a disadvantage because in complex systems it is often the qualitative results that we really want. Just as with the conventional approach, we can use them to gain information about the system that we might otherwise not have found. The waves predicted by Zeeman (1974) in embryonic development were later observed experimentally (cf. Zeeman, 1978), but without the prediction it is unlikely anyone would have looked for them.

An important difference from conventional modelling is that our results generally depend not on a specific mechanism but on a simplicity postulate of one kind or another, and it is therefore usually this simplicity postulate which is either confirmed or proved false. If the predictions are correct, we infer that there is nothing unusual happening, even if the apparent complexity of the phenomena might lead us to believe that there were. The similarity between the shapes that appear when the Beloussov-Zhabotinsky reaction is carried out on a Petri dish and those observed during the aggregation of *Dictyostelium discoideum*, suggests that there should be a relatively simple explanation of the latter. This is the same lesson that is to be learned from the similarity between the form of a hydra and that of a drop of fusel oil that has been allowed to fall into water (D'Arcy Thompson, 1917).

By the same token, if the predictions are wrong, then since the assumption of simplicity is one of the few hypotheses we invoked, that is the one that is likely to be wrong. In other words, one obvious inference is that the situation is more complicated than we thought. For example, the pattern of caustics on the bottom of a sunlit swimming pool may look ordinary enough. When we notice that there can be three lines meeting at a point, which is not what "typically" happens, this leads us to investigate more closely, and to discover that each line is in fact a pair of lines.

It may also be that one of the other hypotheses is wrong instead, and since there are so few of them it is relatively easy to discover which it is. For example, Bazin and Saunders (1979) analyzed sudden jumps that had been observed in the specific growth rate of the amoeba *Dictyostelium discoideum*. When they assumed that the amoebae were responding to the prey density, they found that the pattern of jumps was not that predicted by catastrophe theory. (See Saunders, 1980, for details.)

Since it seemed unlikely that the amoebae would employ a complicated mechanism to govern their growth, Bazin and Saunders tried the alternative hypothesis that the amoebae were responding to the prey:predator ratio. On this hypothesis the pattern of jumps is as predicted by catastrophe theory. This means that a mechanism based on the ratio could be less complicated than any based on prey density, and hence the inference is that it is the ratio to which the amoebae respond.

This result led Bazin and Saunders to ask how the amoebae could measure this ratio. Since it was known they are sensitive to folic acid and that the prey secrete it, they predicted that the amoebae must degrade it. The observation (Pan and Wurster, 1979) that the amoebae do in fact degrade it is strong evidence that the inference from catastrophe theory was correct. An important aspect of this example is that it demonstrates how work in self-organization is likely to involve close interaction between theory and the real situation; it is not like theoretical physics where we can often abstract problems and deal with them almost as though they were pure mathematics.

The next section, in which we discuss self-organization using as illustration the problem of segmentation maintains the connnection with catastrophe theory

because almost everything that has so far been accomplished within that theory
has involved elementary catastrophes. In his book, however, Thom also described
what he called generalized catastrophes. These, he wrote, arise when a stable
regime which originally governs an entire region, breaks down, and gives rise
to a number of local regimes. This is essentially what happens when we alter a
parameter in a Turing system and thereby produce pattern out of an originally
homogeneous tissue. So we are still within the scope of what Thom envisaged,
even if this part of his programme has so far remained incomplete.

4 Segmentation

Segmentation, or metamerism, the subdivision of the body into repeated parts,
occurs in the vast majority of living organisms. It is seen in all multicellular
plants and even in such unicellular algae as *Acetabularia*. Among metazoans,
segmentation is common to arthropods, vertebrates and annelids, and, in a somewhat different form, molluscs, coelenterates, echinoderms and most other phyla.
Even the strange Cambrian animals of the Burgess shale, which provided at least
20 new phyla for systematists (cf. Gould, 1990) are commonplace in at least one
respect: they are all segmented. It is thus difficult to imagine that segmentation
per se is encoded in specific genes.

Yet this is precisely what biologists tend to assume. The currently most
widely accepted model of segmentation in *Drosophila* (Meinhardt, 1988) is firmly
tied to genetic mechanisms, so much so that it is difficult to see that the model
does anything more than redescribe the data, and it is neither generic nor robust.

The fundamental problem is that no matter how great the powers of control
and pattern specification we may suppose genes to possess, they cannot *generate*
patterns. The most they can do is to respond to and stabilize a pattern generated
by something else (see Ho, 1984). There is nothing inherent in protein synthesis
or in the turning on and off of genes that makes patterns; on the contrary, it is
the spatial distribution of the gene products themselves that must be explained.
The problem becomes obvious when we remind ourselves that Liesegang rings,
analogous to segmental patterns, can be produced in a test-tube system that has
no genes at all.

So what we have here is a generic property of self-organization. It is generic
because not only is it widespread, it arises in many different systems and through
different mechanisms.

If there were a general theory of self-organization, we would be able to predict
that segmentation should be common, to specify the class of dynamics which
should give rise to it, and to list many of its typical properties. As there is not,
and as one of our aims is precisely to work towards a general theory, we have to
start at the other end. We look first at one particular model which we take, at
least provisionally, to be a typical member of this class, and see how it behaves.
We then consider how general our results may be.

5 The Turing model

A generic mechanism that can account for metamerism is Turing's (1952) reaction-diffusion system. This been used to model segmentation as well as other features such as animal coat patterns (e.g. Gierer and Meinhardt, 1972; Murray, 1981; Meinhardt, 1982). But while Turing mechanisms are generic in the sense that they do not depend strongly on a particular choice of chemical reactions, in their original form they lack robustness. The chemical wavelengths are highly sensitive to the size of the domain in which they occur and to other parameters as well. Embryos of different sizes would be expected to have different numbers of segments, which in general they do not. (See Saunders and Ho, 1992, for further discussion.)

We suggest that an effective way of producing a required number of structures is to proceed by successive bifurcations. The idea of successive bifurcations itself is not new, but has been discussed by a number of previous authors. Here we show why we expect it to occur frequently, especially, though not only, where a fixed number of segments must be produced.

Successive bifurcation works because it greatly reduces the accuracy required in subdivision at each stage. The principle is actually the same one that we would automatically use if we had to divide a baguette into eight roughly equal pieces, which shows how generic it is. For a simple reaction-diffusion model (Turing, 1952; for details see, e.g. Murray, 1989), the basic equations are,

$$u_t = \Gamma f(u, v) + u_{xx},$$
$$v_t = \Gamma g(u, v) + D v_{xx}, \tag{1}$$

where u and v are the morphogen concentrations, Γ is a rate constant, $f(u, v)$ and $g(u, v)$ are functions which are assumed known, and D is the diffusivity of v, that of u being taken equal to unity. Subscripts denote differentiation.

Many of the properties of the system can be discovered by considering the linearized equations, the solutions of which are of the form

$$\sum C_n \exp\{\alpha_n (n\pi/p)^2 t\} \cos(n\pi x/p).$$

Here n is an integer representing the mode, the α_n are complex numbers that depend on n and the parameters of the system, p is the length of the domain and the C_n are real constants.

For large t, the only modes that contribute to the solution are those for which $Re(\alpha_n) > 0$. In many cases, the fastest growing mode, i.e. that for which $Re(\alpha_n)$ is the greatest, determines the final pattern. In such cases, integration of the full non-linear system confirms that this will be essentially proportional to $\cos(N\pi x/p)$ where N is the index of the dominant mode. The domain is then divided into $N + 1$ sub-domains.

Now for given parameters of the system, the maximum and minimum wave numbers that correspond to unstable (i.e. growing) modes and the wave number of the dominant mode are all proportional to the length of the domain, apart from the small discrepancy due to the fact that the domain increases continuously whereas wave numbers are integers. Treating N as continuous for simplicity, and writing $p = \beta N$, where β is a constant, if precisely $N + 1$ sub-domains are to be formed, p must satisfy

$$\beta(N - \tfrac{1}{2}) < p < \beta(N + \tfrac{1}{2}).$$

Hence the permitted range for p is approximately $\pm\beta/2$, i.e. the fractional variation permitted is approximately $\pm 1/2N$.

For small N, this is only an order of magnitude calculation because the discrepancy caused by the discreteness of wave numbers is relatively large. Nevertheless, it is clear that to specify 16 sub-domains accurately the length of the domain must be within about 3% of its nominal value, whereas to specify two sub-domains requires far less precision.

Note that while $N = 1$ is the best case, the other low modes are also relatively robust. Hence while we expect bifurcation to be the most common, reliable division into three, four or even five is not so unlikely as into 16.

If segmentation occurs in one fell swoop, an embryo that is as little as 5% too long or too short will develop the wrong number of segments. In contrast, one that is as much as 20% too long or too short will still divide correctly into two, if that is what it is supposed to do. Assuming the two halves to be of approximately equal length, as the model predicts they should be, each of them, being again within about 20% of its correct length, will bifurcate properly when the process reeaches that stage. Even if the division is not equal, there is still considerable leeway before one of the sub-domains is so much larger or smaller than it should be that it will not bifurcate properly.

It has also been found (cf. Murray, 1989, p.405) that the interactions caused by the non-linearities are more complex for the higher modes than for the lower ones. We would expect this to lead to further lack of robustness, especially in the light of the observation of Hunding et al (1990) that while taking the two neglected dimensions of the embryo into account does not affect the pattern of the system, it does adversely affect its stability. This is especially relevant because the precise shape, as well as the length, will vary among different individuals. Thus, a pattern of seven bands, corresponding to the transcript pattern of the pair-rule genes, can only be reliably generated by successive bifurcation in the supercomputer simulation of Hunding et al (1990).

For purposes of exposition, we have spoken so far only of variations in the length of the embryo. In fact, of course, this is not the only parameter of the equations that can vary, either between individuals or in time in a single embryo. In this context it is important to distinguish between physical variables on the one hand, and what we may call the essential control variables of the process on the other. In mathematical modelling we usually work with as few parameters as are necessary to produce the full range of behaviour of the system. This makes it much easier to see what is happening, but it can also lead us to forget that in any actual system there may be a number of different variables which produce similar effects (Saunders, 1993a; for an example see Zuckerkandl and Villet, 1988).

Thus even though in the model the period doubling is brought about by an increase in one parameter, it is not necessary to postulate a single substance increasing in concentration throughout the process. Even in the simple model (1), we can increase the length, decrease the diffusion rate, or both; the effect on the pattern will be the same. What is more, the changes in different parameters do not have to be coordinated precisely; even the order in which they occur may not matter. This gives additional robustness to the process. It also means that the same result may be achieved in different organisms by different mechanisms, i.e. that successive bifurcation and the robustness it confers are generic properties which can be considered in their own right as fundamental processes in development (cf. Saunders, 1988).

The most obvious way to bring about successive bifurcations is for structures to start to form at each stage in the process. These then act as boundaries for the next stage. A model for this has been proposed by LeGuyader and Hyver (1990), though it seems unlikely in any case that nothing should even start to happen until the complete prepattern has been set up. It is, however, not necessary that any structures form. As Lacalli et al (1988) comment, pattern subdivision is a common feature of reaction-diffusion models. Both Lacalli (1981) and Arcuri and Murray (1986) found that as the parameters change, the peaks typically broaden and split into two. In effect, the existing pattern elements act almost as boundaries for the next stage. Of course if real boundaries have started to form, the effect will be enhanced, thus conferring additional robustness on the process. Lacalli et al considered that a pattern could be stabilised if minima were fixed by preexisting patterns, but it would seem simpler simply to reinforce those that the mechanism itself produces.

Another reason for expecting successive bifurcations has to do with the way the process starts. A typical reaction-diffusion scheme involves two substances which are supposed to be initially at their equilibrium values. For a pattern to form, the parameters of the system must be such as to make it unstable against perturbations, but this would make it hard to set up. We would expect that until the morphogens are at their appropriate initial concentrations the parameters will be in the stable regime, and then they will alter so as to destabilize the system. This might happen for example when fertilization takes place.

There are essentially two ways of entering the unstable regime. One can induce the high mode directly, or one can move up through lower modes. The former has the severe problem of regulation we have just described. Arcuri and Murray (1986) investigated the latter case, allowing a parameter which depends on the size of the domain and a diffusion constant to increase in time. They found what they describe as a distinct tendency towards frequency doubling, i.e. successive bifurcations. Hence this can occur even without the early formation of actual structures or primordia to fix sub-domain boundaries (cf. LeGuyader and Hyver, 1990).

This result also helps us to understand how segmentation can proceed reliably even though the embryo may be growing during the process. Increasing the size of the domain is just one way of increasing the key parameter. It can also be done by increasing the rates of formation of morphogens. In the former case we get embryos of the short germ type, which starts with little more than a head and a tail and the intermediate segments are added during growth, and in the latter, embryos of the long germ type, in which segments are formed by subdividing the whole body without growth. Thus the difference between long and short germ development may be the result of only a small transformation.

Finally, it is unlikely that the important parameter will cross its threshold value at exactly the same time at every location. It is more likely that there will be a "primary" wave of increase (Zeeman, 1977) which passes through a region to destabilize it. This means that even if the size of the domain remains constant, that of the portion that is unstable will not. Hence it is predicted that a boundary will form as a secondary wave, and that it will move after it is formed, since the effective length from the standpoint of the reaction diffusion system will be changing. Nevertheless, we would expect both that successive bifurcations should occur and also that by this means a reliable number of structures will be produced.

There is evidence to suggest that anterior and posterior signalling processes are involved in segmentation in insects, and that the segmental boundaries move along the body after they are formed (Sander, 1984). Now the primary wave of destabilization is followed by a secondary wave as the region stabilizes in its new configuration (i.e. the bifurcation boundaries reflecting segmental boundaries). Saunders and Ho (1985) showed that if there is a gradient in some substance along the embryo, the secondary wave can be in the opposite direction to the primary wave. On the successive bifurcations model, therefore, there is no need to postulate two separate, coordinated waves or signalling processes in opposite directions. One source suffices and we get the second wave and the coordination for free.

6 Other models

While we have used the reaction-diffusion model to demonstrate the principle, the result applies to other models as well. For example, the vibrational modes of a membrane are governed by the same equation

$$\nabla^2 w + k^2 w = 0$$

as the one that determines the eigenmodes of the linearized reaction-diffusion equations. This led Xu *et al* (1983) to investigate (by holographic interferometry) the vibration patterns on a thin plate, and they found that these do indeed resemble the patterns found by numerical solution of the reaction diffusion equations (Murray, 1981). It is particularly interesting that Xu *et al* then realized that the plates they used were too stiff to be considered as membranes and that the eigenmodes were therefore governed by the biharmonic equation

$$\nabla^4 w - \alpha^2 w = 0$$

They point out, however, that if the plate is simply supported, the eigenmodes of this equation are the same as those of the one above.

Another approach to pattern formation in embryology is the mechano-chemical model (Oster, Murray and Harris, 1983). This is more complicated to analyze than the Turing model and has more parameters, which may make it a less likely candidate for a mechanism where reliability is important. As a model of segmentation, however, it does appear to behave much like the reaction-diffusion model, at least for some parameter values. The crucial point is that the analysis holds for those cases in which the fastest growing modes dominate the final pattern, and this includes the majority of situations, at least in one dimension. This means that our analysis probably holds for all segmentation mechanisms of the type involving self- organization rather than detailed external control or genetic specification.

7 Empirical evidence for successive bifurcations

Detailed evidence in support of successive bifurcation and low mode multifurcation is given elsewhere (Saunders and Ho, 1992), where it is shown to apply not only to all insects, but possibly also to vertebrates, to soft-bodied precambrian fossil biota, and to extant and extinct echinoderms.

Here, we mention only the evidence from ether treatment of *Drosophila* embryos (Ho, 1990). In the model, each splitting of a region into two serves as the starting point for the next. Hence if one bifurcation fails to occur within a subregion, none of the subsequent ones can occur. The action of ether consists in suppressing the formation of bifurcation boundaries that happen to be forming at the time of exposure. The result is the formation of corresponding gaps in the segmental pattern. This allows us to construct a taxonomic map of all possible transformations and their neighbourhood relationships. We find that only 676 transformations can be generated by the process, whereas if each segment could appear or not independently of the others the total would be 2^{16}, i.e. more than 60 000. All the observed segmentation abnormalities fall within the taxonomic map. Furthermore, a rational taxonomy or classification system can be devised on that basis (Ho, 1990), which is itself a strong evidence in favour of the model. The successive bifurcations model in *Drosophila* is also consistent with many other observations in which early development is perturbed by centrifugation, ligation and other means (see Saunders and Ho, 1992).

8 Conclusion

The experience of catastrophe theory gives us an idea of what a general theory of self-organization can be and what some of the problems are. The work on segmentation shows many of the same features. Both Turing's original demonstration that segmentation can occur through a simple mechanism and the result described above that successive bifurcation implies robustness apply to more than just the reaction-diffusion system. They clearly hold in any system for which the linear approximation leads to the harmonic equation, providing that there is no significant interaction between modes. They will also work in systems which lead to the biharmonic equation, or indeed any equation with the same eigenvalues as the harmonic. They may also work in other systems as well, but this remains to be seen; in any case the class to which they are applicable is already large enough to be useful.

An attractive feature of the hypothesis of successive bifurcations is that it does not require a separate control mechanism. As we have seen, it can be a matter of nothing more than the order in which certain parameters are varied. Hence we do not have to appeal to natural selection to explain how the regulation arose. And indeed, it is by no means obvious what advantage there is in having a fixed number of segments unless some feature of the organism requires it, yet such a feature would be unlikely to evolve in an organism that did not already have reliable segmentation.

The development of a theory of self-organization will have profound consequences in biology. Biologists have tended to assume that where organisms exhibit order and organization, this must be the result of design, either consciously by the Creator or else through Natural Selection. There has, however, been a minority who dissented from this view. These workers, including D'Arcy Thompson, Turing and Thom, have for the most part concentrated on one aspect of self-organization, the generation of form and pattern. Now the approach is being extended to other organismic properties as well. We have discussed one kind of regulation here; another example is the maintenance of the environment on the whole Earth (Lovelock, 1979, 1988).

It is significant that much of the opposition to Lovelock's claim that the Earth is self-regulating has been on theoretical, rather than empirical grounds. "No Darwinist" writes Maynard Smith (1988), "could accept the 'Gaia' hypothesis, according to which the earth is analogous to a living organism, because the earth is not an entity with multiplication, variation and heredity." Yet Watson and Lovelock (1983) have shown in their "Daisyworld" model how regulation can arise in a self-organizing system simply through the interaction of positive and negative feedbacks. What is more, regulation that arises naturally is likely to be more robust (Saunders, 1993b). The more we learn about self- organization, the more we will be able to reverse the argument of the neo-Darwinists: even if an entity has multiplication, variation and heredity, this does not imply that all its organismic properties, including regulation, arose through natural selection.

References

1. Arcuri, P. & Murray, J.D. (1986). Pattern sensitivity to boundary and initial conditions in reaction-diffusion models. *J. math. Biol. 24*, 141-165.
2. Bazin, M.J. & Saunders, P.T. (1979). Determination of critical variables in a microbial predator prey system by catastrophe theory. *Nature 275* (1978) 52-54.
3. Berry, M.V. & Upstill, C. (1980). Catastrophe optics: morphologies of caustics and their diffraction patterns. In *Progress in Optics 18* (E. Wolfe, ed.). North Holland, Amsterdam, pp. 256-346.
4. Gierer, A. & Meinhardt, H. (1972). A theory of biological pattern formation. *Kybernetik 12*, 30-39.
5. Ho, M.W. (1984). Heredity and environment in development and evolution. In *Beyond Neo-Darwinism: An Introduction to the New Evolutionary Paradigm* (M.W. Ho & P.T. Saunders, eds). Academic Press, London, pp. 267-288.

6. Ho, M.W. (1990). An exercise in rational taxonomy. *J. theor. Biol. 147*, 43-57.

7. Hunding, A., Kauffman, S.A. and Goodwin, B.C. (1990). *Drosophila* segmentation: Supercomputer simulation of prepattern hierarchy. *J. theor. Biol. 145*, 369-384.

8. Lacalli, T.C. (1981). Dissipative structures and morphogenetic pattern in unicelluar algae. *Phil. Trans. R. Soc. Lond. B294*, 547-588.

9. Lacalli, T.C., Wilkinson, D.A. and Harrison, L.G. (1988). *Development 104*, 105-113.

10. Le Guyader, H. & Hyver, C. (1990). Modelling of the duplication of cortical units along a kinety of Paramecium using reaction-diffusion equations. *J. theor. Biol. 143*, 233- 250.

11. Lovelock, J.E. (1979). *Gaia: A New Look at Life on Earth*. Oxford, Oxford University Press.

12. Lovelock, J.E. (1988). *The Ages of Gaia*. Oxford, Oxford University Press.

13. Maynard Smith, J. (1988). Evolutionary progress and levels of selection. In *Evolutionary Progress* (M.H. Nitecki, ed.) Chicago, University of Chicago Press, pp. 219-230

14. Meinhardt, H. (1982). *Models of Biological Pattern Formation*. London, Academic Press.

15. Meinhardt, H. (1988). Models for maternally supplied positional information and the activation of segmentation genes in *Drosophila* embryogenesis. Development 104 (Suppl) 95-110.

16. Murray, J.D. (1981). A pre-pattern mechanism for animal coat markings. *J. theor. Biol. 88*, 161-199.

17. Murray, J.D. (1989). *Mathematical Biology*, Springer-Verlag, Berlin.

18. Oster, G.F., Murray, J.D. and Harris, A.K. (1983). *J. Embryol. exp. Morph. 78*, 83.

19. Pan, P. and Wurster, B. (1979). Inactivation of the chemoattractant folic acid by cellular slime molds and identification of the reaction product. *J. Bacteriol. 136*, 955-959.

20. Sander, K. (1984). Embryonic pattern formation in insects: basic concepts and their experimental foundations. In *Pattern Formation* (G. Malacinski and S. Bryant, eds.), Macmillan, New York.

21. Saunders, P.T. (1980). *An Introduction to Catastrophe Theory*. Cambridge University Press, Cambridge.

22. Saunders, P.T. (1989). Mathematics, Structuralism and the Formal Cause in Biology. In *Dynamic Structures in Biology* (B.C. Goodwin, G.C. Webster & A. Sibatani, eds). Edinburgh University Press, Edinburgh, pp 107-120.

23. Saunders, P.T. (1993a). The Organism as a Dynamical System. In *Thinking about Biology* SFI Studies in the Sciences of Complexity, Lecture Notes Vol. III (F. Varela & W. Stein, eds). Addison Wesley, New York.

24. Saunders, P.T. (1993b). Evolution without natural selection: Further implications of the Daisyworld parable. *J. theor. Biol.* (in press).

25. Saunders, P.T. and Ho, M.W. (1985). Primary and Secondary Waves in Prepattern Formation. *J. Theor. Biol., 114* (1985) 491-504.

26. Saunders, P.T. and Ho, M.W. (1992). Successive bifurcations and the stability of segmentation. (to be submitted)

27. Smale, S. (1966). Structurally stable systems are not dense. *Amer. J. Math. 88*, 491-496.

28. Thom, R. (1972). *Stabilité Structurelle et Morphogénèse*. Benjamin, Reading.

29. Thompson, D'A.W. (1917). *On Growth and Form.* Cambridge University Press, Cambridge.
30. Turing, A.M. (1952). The chemical basis of morphogenesis. *Phil. Trans. R. Soc. Lond. B237*, 37-72.
31. Watson, A.J. & Lovelock, J.E. (1983). Biological homeostasis of the global environment: the parable of Daisyworld. *Tellus (1983), 35B*, 284-289.
32. Xu, Y., Vest, C. & Murray, J.D. (1983). Holographic interferometry used to demonstrate a theory of pattern formation in animal coats. *Appl. Optics 22*, 3479-3483.
33. Zeeman, E.C. (1974). Primary and secondary waves in developmental biology. In *Some Mathematical Questions in Biology VIII: Lectures in Mathematics in the Life Sciences,* Vol. 7, (S.A. Levin, ed.). Providence, American Mathematical Society, pp. 69-161.
34. Zeeman, E.C. (1978). A dialogue between a mathematician and a biologist. *Biosci. Commun. 4*, 225-240.
35. Zeeman, E.C. (1989). A new concept of stability. In *Theoretical Biology: Epigenetic and Evolutionary Order from Complex Systems* (B.C. Goodwin & P.T. Saunders, eds) Edinburgh, Edinburgh University Press, pp. 8-15.
36. Zuckerkandl, E. & Villet, R. (1988). Concentration affinity equivalence in gene regulation: Convergence of genetic and environmental effects. *Proc. Natl. Acad. Sci. USA 85*, 4784- 4788.

Self-Organization, Entropy and Order

P. T. Landsberg

1 Introduction to Chaos and Self-Organisation

Biology offers simple examples of self-organisation, and we use it to illustrate the key concept of <u>bifurcation</u> or branching. The tree shown in Fig. 1 develops branches on a yearly basis. Each branch can grow a new one (drawn to the right and numbered) after it has existed for two years. On this basis the number of branches counted at the end of each year form a series named after Leonardo, son of Bonaccio (1170-1250):

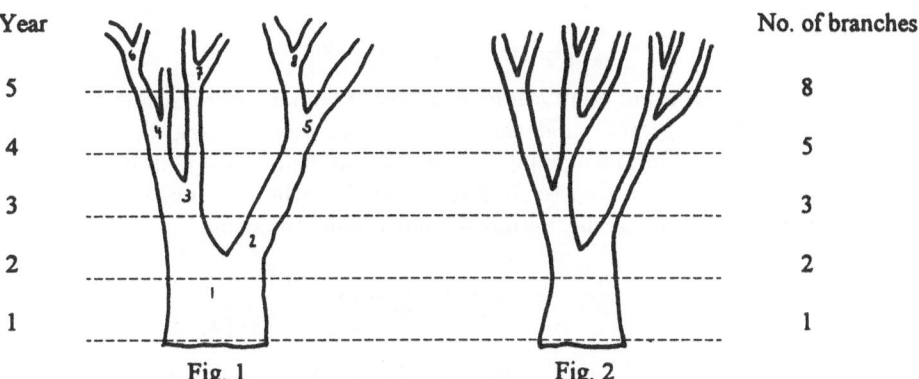

Fig. 1. A tree whose number of branches as a function of time yield the Fibonacci series.

Fig. 2. Fig. 1 rearranged.

Springer Series in Synergetics, Vol. 61   **On Self-Organization**
Eds.: R.K. Mishra, D. Maaß, E. Zwierlein   © Springer-Verlag Berlin Heidelberg 1994

$$F_1 = 1, F_2 = 2, F_3 = 3, F_4 = 5, F_5 = 8, \ldots \tag{1}$$

This Fibonacci series has interesting properties (though they are not essential to self-organisation). By growing new branches to the right it is easy to see what is happening. By rearranging the branches somewhat, one obtains a more convincing tree (Fig. 2, [1]).

The occurrence of this series in biological phenomena is well known. The subject involved is known as phyllotaxis and includes the study of the arrangement of leaves on stems of plants and related matters. The Fibonacci numbers often occur. In fact the number of petals of common flowers are the Fibonacci numbers

Iris 3, primrose 5, ragwort 13, daisy 34, michaelmas daisy 55 and 89.

The Fibonacci spirals can be seen in sunflower heads and pine cones, and this pattern has recently been reproduced in physics laboratory experiments as well as in numerical simulations [2]. The series is connected with the golden section, which also occurs in physics; for example, in connection with quasi crystals and with an infinite set of resistors [3].

Bifurcations are part of many theoretical models, the simplest being taken from population studies. Thus the equation of normal reproduction is normally taken in continuous time, but has a counterpart for discrete times $t_k = k\tau$ where τ is a given time interval and k is a positive integer:

$$\dot{n} = rn \rightarrow n(t) = n(0) \ \exp(rt), \tag{2}$$

or

$$n_{k+1} = rn_k \rightarrow n_{k+1} = r^k n_1 \tag{3}$$

This leads to a Malthusian population explosion (Thomas Robert Malthus, 1760-1834). This explosive growth holds for some social phenomena, e.g. the growth in the number of scientific journals [4], Fig. 3. The Malthusian population regime neglects limits on growth due to food shortages, deaths, etc. On taking these into account, one comes to the Verhulst-Pearl (1838) equation:

$$\dot{n} = rn \left[1 - \frac{n}{n^*} \right] \rightarrow n(t) = \frac{n(0)}{\frac{n(0)}{n^*} + \left[1 - \frac{n(0)}{n^*} \right] e^{-rt}} \tag{4}$$

There are now two steady-state solutions: To the Malthusian, but irrelevant, solution $n(t) = 0$ has been added

$$n(t) = n^*.$$

In fact for $n^* = \infty$ one goes back to the Malthus equation.

By writing $n/n^* \equiv m$, and then changing notation back to n, one obtains the usual form of the so-called logistic equation and its discrete version:

$$\dot{n} = rn(1 - n) \tag{5}$$

$$n_{k+1} = rn_k(1 - n_k) \tag{6}$$

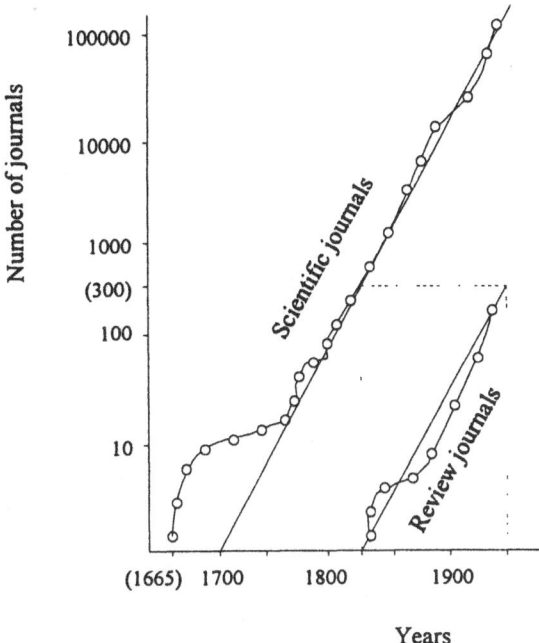

Fig. 3. The number of scientific journals over the years [4].

As is explained in many places, n_k reaches stable values $n^* = (r-1)/r$ for $1 < r < 3$, periodic values for $3 < r < 3.5$ and one finds chaos for $r \sim 3.58$. The well-known construction of solutions to the logistic equation is illustrated in Fig. 4. (The stable values are solutions of (6): $n^* = rn^*(1-n^*)$.) The bifurcation diagram (Fig. 5) shows the two possible long-term values of n^* for $r \sim 3.2$, between which there are oscillations. As each occurs with probability 1/2, one can obtain an entropy of $k \ln 2$. In the chaotic regime the maximum entropy is approached, except for the drop in entropy in the windows of periodicity. Here N is the number of subdivisions used for the unit interval, and it has the value 40 for the entropy graph shown [5]. Thus the entropy concept is here not essential - it merely summarises one aspect of a probability distribution.

2 Phase Transitions

Our interest is in <u>transitions</u>. So far we have considered bifurcations in the somewhat academic setting of population dynamics. Physics provides related examples, which are less academic, since they can be can be subjected to more rigorous experimental investigation. There are two main classes of these so-called phase transitions:(i) equilibrium ones and (ii) non-equilibrium ones. We shall start with class (i), which is more elementary and more widely familiar.

The common feature of the transitions of both classes is that there is one parameter (Φ): the volume v in a van der Waals gas and the magnetisation M

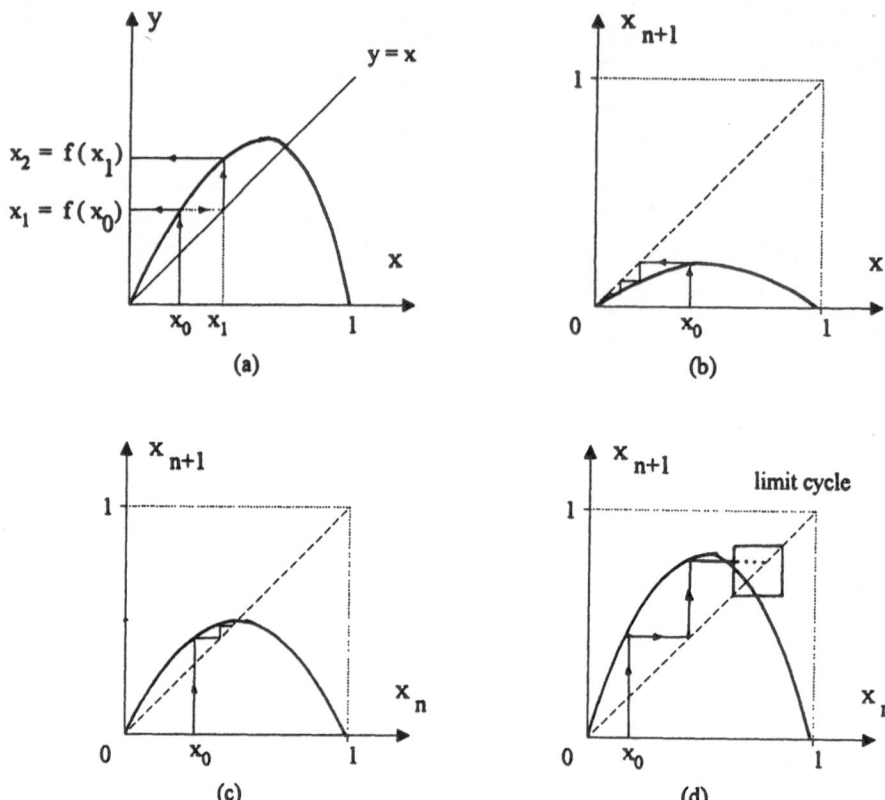

Fig. 4.

(a) Procedure for obtaining $x_{n+1} = f(x)$ from x_n for $n = 0$ and 1. Also shown are $x_{n+1} = r x_n (1 - x_n)$ (the logistic equation) for the first few n-values for $r < 1$ in Fig.(b), $r = 2$ in Fig.(c) and $r = 3*2$ in Fig.(d).

in a Curie-Weiss ferromagnet (Figs. 6, 7). Further, the equation governing the transition is cubic in Φ (> 0) with $a_3 < 0$ (see Table 1). It is given by:

$$\Lambda'(\Phi) \equiv a_0 + a_1\Phi + a_2\Phi^2 + a_3\Phi^3 = 0 \qquad (7)$$

We write this function as Λ' because one could imagine a "free-energy"

$$\Lambda(\Phi) = a_0\Phi + \frac{1}{2}a_1\phi^2 + \frac{1}{3}a_2\Phi^3 + \frac{1}{4}a_3\Phi^4 \ \ldots$$

which gives rise to (7) on looking for an extreme with respect to Φ. Table 1 gives four distinct interpretations of these equations. In interpretations 3 and 4 $\Lambda'(\Phi)$ is actually $d\Phi/dt$ and one sees that for this case Φ_1 and Φ_3 are stable solutions which are separated by the unstable solution Φ_2. Figs. 6 and 7 show that in both the van der Waals case and the ferromagnetic case the temperature is a kind of control parameter.

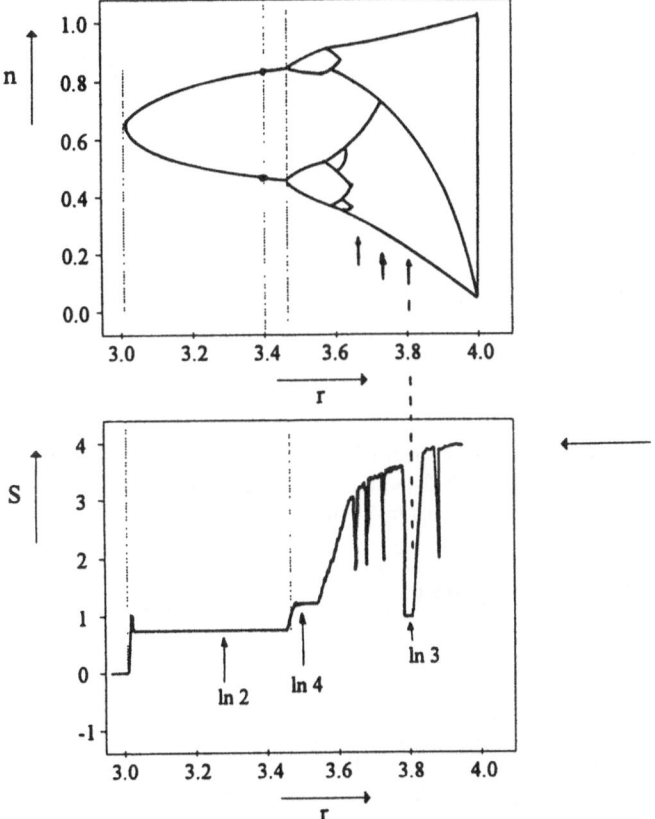

Fig. 5. Bifurcation and entropy diagrams for the logistic equation $n_{n+1}^* = r n_n^* (1 - n_n^*)$. For $r = 3.4$ the system oscillates between $n^* = 0.48$ and 0.83, indicated by $\downarrow$. Windows of periodicity are marked by $\uparrow$. For 40 cells in the range $[0,1]$ the maximum entropy is given by $\frac{1}{k} S_{max} = \ell n 40 = 3.69$, indicated by $\leftarrow$.

For the van der Waals gas (Fig. 6)

$$\left(p + \frac{a}{\sigma^2} \right) (v - b) = AT \tag{8}$$

The critical point C corresponds to $\nu = \tau = 1$, where

$$\Phi = \nu \equiv \frac{v_c}{v} \equiv \frac{3b}{v} \text{ and } \tau \equiv \frac{T}{T_c} \equiv \frac{27 AbT}{8a}.$$

The diagram shows v (rather than Φ) as a function of p .

For the ferromagnet-paramagnet transition we show $\Phi = M$ against T. This represents the order parameter as a function of the control parameter (See Table 1). The arrows in Figs. 6, 7 represent a first-order phase transition. In the case of the ferromagnet the transition at C is a second order transition, made smoother

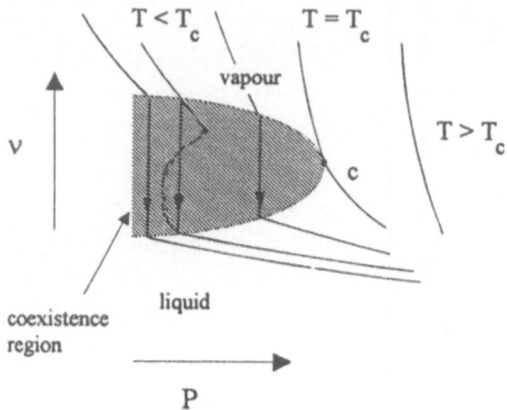

Fig. 6. Curves of constant temperature and constant pressure for a van der Waals gas. C denotes the critical point. The van der Waals equation gives typically the wavy curve shown. The arrows represent the Maxwell construct which links the co-existing phases.

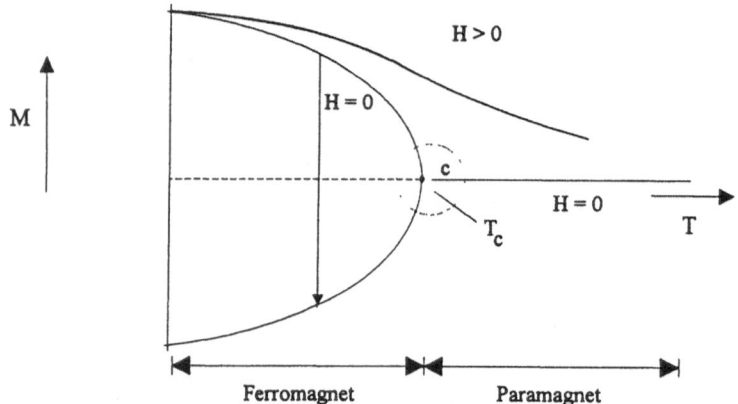

Fig. 7. Curves of constant magnetic field for the ferromagnet ↔ paramagnet transition. C denotes the critical point.

if there is an external magnetic field ($H \neq 0$). Of course the equation (for $H = 0$)

$$bM^3 = (T_c - T)\, a\, M \quad (a, b, \text{ are constants}) \tag{9}$$

always has $M = 0$ as a solution. But for $T < T_c$ it is no longer stable, the stable solutions being

$$M = \pm\left[\frac{a}{b}(T_0 - T)\right]^{\frac{1}{2}}.$$

Here the bifurcation is seen quite clearly: $M = 0$ and $M \neq 0$ are both solutions, but $M = 0$ survives only as the unstable one for $T > T_c$. The symmetry of the $M = 0$ state is broken for $T < T_c$, bringing about a greater degree of "order".

Table 1. Cubic equations for phase transitions

1 and 2 represent equilibrium phase transitions
3 and 4 represent non-equilibrium phase transitions
Φ and a_1 are abstract order and control parameters respectively
$\Lambda'(\Phi) = a_0 + a_1(\Phi) + a_2\Phi^2 + a_3\Phi^3(+\ldots) = 0$

	Φ	a_0	a_1	a_2	a_3
1. V.d.Waals gas	ν	π	η	3	-1
2. Ferromagnetism	M	H	$-(T-T_c)a$	0	$-b$
3. Chem. reaction	x	$r_1 b$	$-f_1 a$	$f_2 b$	$-r_2 q$
4. S C Model (a)	n	Y^s	$-Y_1 - \ell$	B^s	0

Explanation of 3.:

$$A + X \underset{r_2}{\overset{f_1}{\rightleftharpoons}} B \qquad\qquad P + 2X \underset{r_2}{\overset{f_1}{\rightleftharpoons}} 3X + Q$$

$$(\dot{x} = r_1 b - f_1 a x) \qquad\qquad (\dot{x} = f_2 p x^2 - r_2 q x^3)$$

The non-linearity of the simple equations given guarantees the possibility of phase transitions between the different solutions. It is, however, only by extending the argument to open, non-equilibrium systems that the possibilities of self-organisation are realised. This can be illustrated by a semiconductor under external influences such as incident radiation or electric fields. These promote electrons from traps or from the valence band to the conduction band. If n is the electron concentration in the conduction band, one has a birth – death situation or, in our case, a generation – recombination regime:

$$\dot{n} = g(n) - r(n). \tag{10}$$

Discretising as in (3) and (6),

$$n_{k+1} = j\psi(n_k), \tag{11}$$

where $\qquad \psi(n_k) \equiv n_k + \tau[g(n_k) - r(n_k)] \tag{12}$

This suffers from the trouble that all transitions occur at the precise times $t_k = k\tau$. If they are allowed to be spread throughout the intervals τ one finds

$$\psi(n_k) = n_k + \frac{\tau g(n_k)}{[1 - \tau g(n_k)]^2} - \frac{\tau r(n_k)}{[1 - \tau r(n_k)]^2} \tag{13}$$

The situation has been shown to give rise to the possibility of chaos [6], which is indeed found in a number of semiconductor experiments [7].

Our interest is, however, in phase transitions and one then needs to give specific models which furnish the functions $g(n)$ and $r(n)$. Simple chemical reactions, as in example 3 of Table 1 [8], give one example. The model (a) (Fig.

8) is perhaps the simplest non-equilibrium phase transition in a semiconductor [9]. It presumes: one generation process $Y_1 n$ by impact ionisation, two loss terms $B^s n^2$ by electron-hole recombination and one other loss term ℓn, and an intrinsic semiconductors (in which the concentration p of holes is equal to n). A band-band excitation process Y^s is not essential. Like the magnetic field H in the ferromagnetic case, it tends to smooth out the transition. The main effect of increasing the electric field is to increase Y_1 which is therefore the control parameter. We have

$$\dot{n} = (Y_1 - \ell)n - B^s n^2. \tag{14}$$

The steady-state solution $n_1 = 0$ is stable for $Y_1 < \ell$, and for $Y_1 > \ell$ the solution

$$n_2 = (Y_1 - \ell)/B^s$$

is stable and increases with Y_1 (Fig. 8).

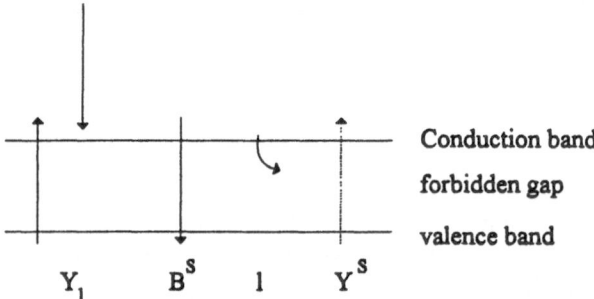

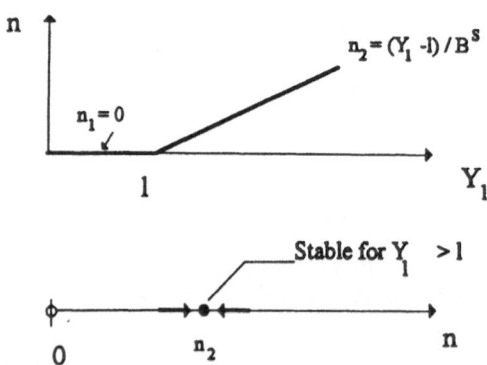

Fig. 8. Semiconductor model (a), pumped by the impact ionisation Y_1 , and showing a second order phase transition at $Y_1 = \ell$.

The model is equivalent to assuming chemical reactions:

$$e + h \xrightarrow{B^s} A, \qquad e \xrightarrow{Y_1} 2\,e + h, \qquad e \xrightarrow{\ell} C \tag{15}$$

where e and h refer to electron and hole and A and C are "chemicals" whose concentrations are kept constant. If we regard sth order autocatalysis to be represented by

$$A + sn = B + (s+1)n, \qquad (16)$$

we see that semiconductor model (a) involves first order autocatalysis. The validity of mass action laws in chemical and semiconductor kinetics has here been assumed throughout.

Thus impact-ionisation in the kinetics for semiconductors takes the place of autocatalysis in chemistry.

3 Example of Self-Organisation: Non-Equilibrium Phase Transitions, Hysteresis and Switching in Semiconductors

The semiconductor models have been elaborated beyond model (a) in the fifteen years since the subject was initiated. But we will confine ourselves here to introducing only one more complicated model in which one has (i) <u>two</u> populations, viz electrons and holes of concentrations n and p which need not be equal, and (ii) two impact ionisations (X_1 and X_4), both from traps, which replace Y_1 of model (a) (Fig. 9).

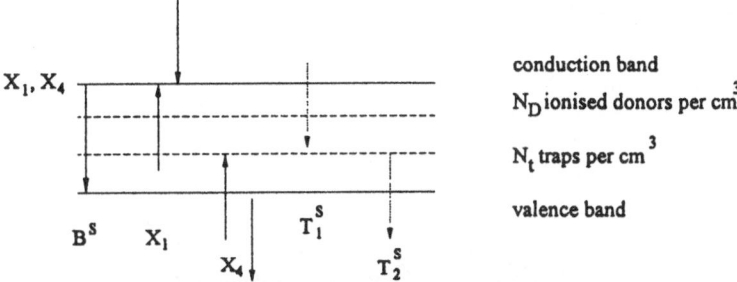

Fig. 9. A semiconductor model pumped by impact ionisations X_1 and X_4 . For $n \neq p$ it has four types of steady state solutions.

In process Y_1 an energetic conduction band electron collides with a valence band electron, and, by losing kinetic energy, it promotes it to the conduction band. This promotion can occur also for a trapped electron (process X_1). Similarly an energetic valence band hole can be filled by a valence band electron, thus promoting another valence band electron into a trap. This creates an extra hole by another autocatalytic process (X_4). The equations are

$$\dot{n} = (X_1 n_t - B^s p)n, \qquad \dot{p} = (X_4 p_t - B_s n)p, \qquad (17)$$

where

$$n_t = N_D - n + p \ , \ n_t + p_t = N_D.$$

This presumes the presence of N_D fully ionised donors which serve merely to supply some electrons. The concentration of trapped electrons has been denoted by n_t , while $p_t \equiv N_D - n_t$ is the concentration of trapped holes.

Calculation shows [10] that the four steady state solutions have overlapping regions of stability of Fig. 10 using the control parameter plane (X_1, X_4). The solutions are

$$(i) \quad n_1 = p_1 = 0, \qquad (ii) \quad n_2 = 0, p_2 > 0$$
$$(iii) \quad n_3 > 0, p_3 = 0 \qquad (iv) \quad n_4, p_4 > 0 \tag{18}$$

In fact, for the record, but not essential here,

$$p_2 = N_t - N_D, \qquad n_3 = N_D$$

$$n_4 = -\frac{B^s(N_t - N_D)X_4 - X_1 X_4 N_t}{B^s(X_1 + X_4 - B^s)}$$

$$p_4 = -\frac{B^s N_D X_1 - X_1 X_4 N_t}{B^s(X_1 + X_4 - B^s)}$$

As the electric field E is raised, X_1 and X_4 increase and the system moves typically on the trajectory $t(E)$. The "threshold" field E_{th} characterises a jump as solution (ii) goes over into solution (iii). On returning, the jump (iii) $\rightarrow$ (ii) occurs at the lower "holding" field E_h . It is thus the overlap of the regions of stability which is responsible for the hysteresis. A different field dependence of X_1 and X_4 would yield a different trajectory, typically $t'(E)$, in Fig. 10.

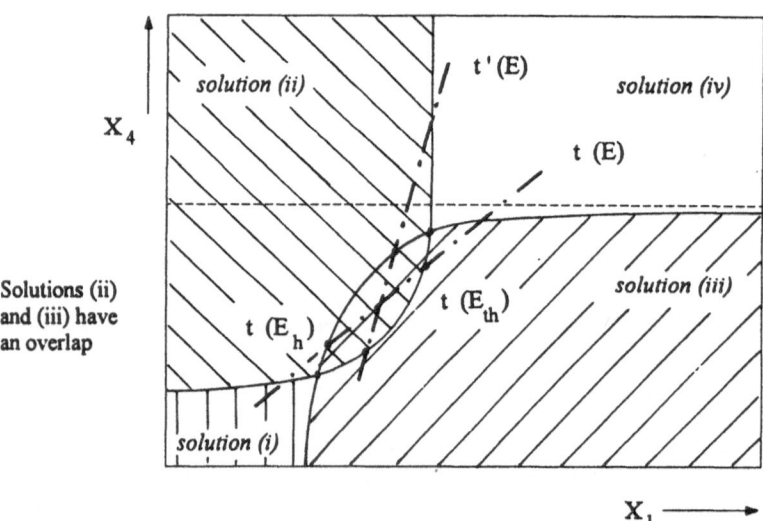

Fig. 10. The (X_1, X_4) plane of model (17) showing the stability regions of the four steady-state solutions.

One notes that this apparently highly simplified model gives nevertheless interesting features of switching phenomena, and a literature is developing on this (eg [11, 12]) and related topics (eg [13]).

So far we have discussed the non-equilibrium steady states of our model. We next turn to the dynamics, illustrated in Fig. 12, which can be obtained by plotting the integral curves of (17) in the (n, p)-space when the control parameters B^s, X_1, X_4, N_t, N_D have fixed values. Indeed, in Fig. 11 the effect of T_1^s and T_2^s is also included. This diagram corresponds to a typical point in the overlap region of Fig. 10. The main feature is that the dot-dash line (the "separatrix") divides the phase space into regions attracted to solution (iii) and (ii) respectively. The diagram shows that a given state (n, p) will develop in time by moving along the typical flow lines shown, ending up in the stable steady state (ii) or (iii). Of course the flow lines will be different if a point outside the overlap region is envisaged. For these details the original paper should be consulted [7, 10].

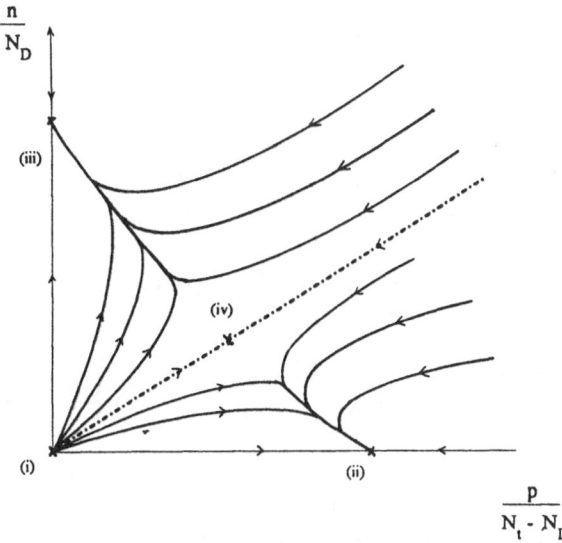

Fig. 11. (n,p) transient values from the overlap region of Fig. 10 behave as shown with lapse of time, reaching eventually steady-state solution (ii) or (iii). For this diagram $B^S = 10^{-10} \ cm^{-3}s^{-1}$, $T_1^S = 10^{-12} \ cm^{-3}s^{-1}$, $T_2^S = 10^{-14} \ cm^{-3}s^{-1}$, $x_1 = 3 \times 10^{-11} cm^{-3}s^{-1}$, $x_4 = 6 \times 10^{-11} cm^{-3}s^{-1}$, $N_t = 1.5 \ N_D$.

One learns from these semiconductor studies, and other similar ones cited in [13], that external forces acting on open systems can move a system to non-equilibrium states which have a surprisingly rich reservoir of structure. This is precisely what one means by self-organisation. A force is applied to a system and some unexpected structures emerge. If many of these situations occur in temporal sequence and involve larger and more complex systems then the results can be highly organised. One could imagine, by a vast extrapolation, the emergence of more complex and even biological systems via stationary and possibly some unstable states in some such manner.

4 Entropy and Order

We have used the entropy concept only as a statistical measure so far (in Fig. 5). But it is related in a more basic way to our discussion because of the popular perception:

$$\text{self-organisation } \longrightarrow \text{ greater order } \longrightarrow \text{ lower entropy(S).} \qquad (19)$$

This relationship will be examined, but we now consider a closed rather than an open system.

Note first that if ρ is the density operator, the fine-grained entropy of quantum statistical mechanics is for a closed system at time t the same as at time $t = 0$:

$$S_t = kTr(\rho_t ln\rho_t) = kTr\left(\sum_{i=0} a_i\rho_t^i\right) = k\sum_i a_iT_r\left(\rho_t^i\right) = k\sum_i a_iTr\left[(U^{-1}\rho_0 U)^i\right]$$

$$= k\sum_i a_iTr\left(\rho_0^i\right) = kTr(\rho_0 ln\rho_0) = S_0. \qquad (20)$$

A power series has here been written for $\rho \ln \rho$; the time development is known to be due to a unitary transformation, using $U \equiv U_t$, and $Tr(U^{-1}AU) = TrA$ for any operator A. We now have the elegant and highly suggestive logical implications that, given that we work in Hilbert space,

$$\left.\begin{array}{r}\text{Fine-grained entropy } \Rightarrow \dot{S} = 0 \\ \text{Hence} \qquad \dot{S} > 0 \Rightarrow \text{ entropy has to be coarse-grained}\end{array}\right\} \quad (21)$$

If we take $\dot{S} > 0$ for an adiabatically isolated system (which includes the possibility of the system being completely isolated) from thermodynamics, then coarse-graining is seen to be necessary for statistical mechanics.

A typical way of coarse-graining is to lump sets of g_i states together and associate with each state not the original probability but the averaged probability. Then one can indeed derive an H-theorem, i.e. for our purposes show that $\dot{S} \geq 0$ for a closed system (see eg [14], p.146). This partial uniformisation of the probabilities leads to

$$S_{coarse} \geq S_{fine} \qquad (N \text{ states in both cases}). \qquad (22)$$

When the averaging process extends over all states, the limiting case for the left-hand side, then $S' = k \ln N$ - the maximum value the entropy can have for given N . Another, less usual, way of coarse-graining is to refuse to recognise certain states. Suppose for example that the probability of finding system A in state A_j is p_j $(j = 1, 2, \ldots n)$. Another part B of the system can then be in states $B_\ell(\ell = 1, 2, \ldots \ldots m)$ with probabilities $q_{j\ell}$ $(\sum_{\ell=1}^m q_{j\ell} = 1)$. Then the entropy of the part B of the combined system (given the part A is in state j) is

$$S_j(B) = -k \sum_{\ell=1}^{m} q_{j\ell} \ln q_{j\ell} \tag{23}$$

The average of all $S_j(B)$ is then what we may call the <u>conditional entropy of B,
given A</u>:

$$S_A(B) = -k \sum_{j=1}^{n} p_j \sum_{\ell=1}^{m} q_{j\ell} \ln q_{j\ell} \tag{24}$$

The Shannon inequality refers to the entropy $S(AB)$ of systems A and B regarded as a joint system and is

$$S(AB) = S(A) + S_A(B) \leq S(A) + S(B) \tag{25}$$

It takes into account all $N = nm$ states of the system. Eq. (25) states that the conditional entropy of B, knowing the state of A and averaged over all states of A, is less than the entropy of scheme B in the absence of such knowledge. One can now regard $S(A)$ as a coarse-grained entropy, in the sense that the states B_ℓ are not recognised

$$S_{coarse} \equiv S(A) \ (n \text{ states}) \leq S_{fine} = S(AB) \ (nm \text{ states}) \tag{26}$$

The smallest value of the left hand side is obtained when $n = 1$ so that the system has to be in the one coarse state with probability 1, making $S_{coarse} = 0$. Results (22) and (26) can be combined:

$$\left. \begin{array}{l} k \ln nm \ \geq \ S_{coarse}(AB)(nm \text{ states}) \ \geq \ S_{fine}(AB) \ (nm \text{ states}) \\ \qquad\qquad \leq \ S_{coarse}(A)(n \text{ states}) \ \geq \ 0 \end{array} \right\} \tag{27}$$

The inequality (27) will now be utilised.
  It is often said that

> the statistical entropy is decreased as one
> gathers more information about a system. $\qquad\qquad$ (28)

In order to examine (28) take

$$S = -\sum_{i=1}^{N(t)} p_i \ln p_i \ \leq \ k \ \ln N(t) \ \equiv \ S_{max} \tag{29}$$

as the statistical entropy definition, where p_i is the probability of finding the system in state i at a certain time t and $N(t)$ is the number of available states. Then one can say:

A. Information given at <u>constant</u> $N(t)$ (the usual case).
  a. If only the value of $N(t)$ is given, then $S = S_{max}$
  b. Add information which makes at least one p_i unequal to $1/N(t)$. Then at least one other p_i is also unequal to $1/N(t)$ and the entropy is decreased, in accordance with (28).

c. Add information that the system is in a particular definite state. Then $S = 0$. This is also in agreement with (28).

B. Variation in N(t) (not normally considered).

a. If the "information" is that $N(t)$ is greater than thought previously, $M(t) > N(t)$ say, but no other information, then the entropy is increased from $k \ln N(t)$ to $k \ln M(t)$, in contradiction with (28).

b. If a description of the system is made less coarse (i.e. finer) by recognising more states, then the entropy is increased by (26), in contradiction with (28).

Following earlier ideas ([14], p.366; [15]), an intensive variable "disorder" will be defined in a technical sense by

$$D = S/k \ln N(t). \tag{30}$$

It is a number which lies between 0 and unity in virtue of equation (29). The "order" Ω can therefore be defined by the condition

$$\Omega + D = 1. \tag{31}$$

A simple application of this idea will now be made in order to elucidate the definitions (30) and (31).

Suppose a system has two equally probably states (the "initial" situation),

$$S_i = k \ln 2. \tag{32}$$

Improved apparatus resolves the upper state into r states, equally spaced about the upper energy level, so that the total energy is unchanged. Assume the probability of $1/2$ to be now equally shared among these r states (the "final" situation), then

$$S_f = k \ln(2r^{\frac{1}{2}}). \tag{33}$$

Fig. 12 gives the resulting values of S_f and of

$$\Omega_f = 1 - \ln(2r^{\frac{1}{2}})/\ln(1 + r) \tag{34}$$

and they are both seen to <u>increase</u> with r, in agreement with what we had already anticipated. Thus for an expanding or growing system, as studied in cosmology, biology, etc., this is at least one effect resulting from a rapidly increasing number of observable states.

Suppose for example that the number of observable states increases "super-exponentially" according to (s, d are positive constants)

$$\ln N(t) = [\ln N(0)] \exp[(s + d)t] \tag{35}$$

Suppose also that the entropy grows exponentially

$$S[N(t)] = S[N(0)] \exp(st) \tag{36}$$

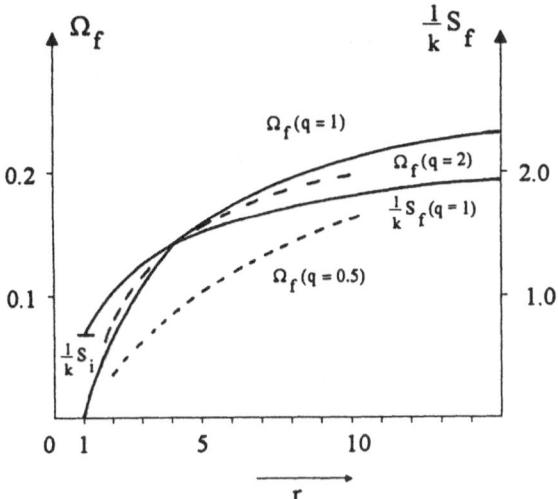

Fig. 12. One entropy and three "order" curves for a two-level system upon resolving the upper level into r levels according to equations (33) and (34). Also shown are some "order" (Ω_f) curves for the Tsallis entropy.

then (30) gives a decreasing disorder even though entropy goes up:

$$D[N(t)] = D[N(0)] \exp(-dt) \tag{37}$$

For other examples, e.g. for estimating the "order" of the universe by this method, see eg [16].

For a radiation-dominated universe one finds for the so-called scaling parameter of cosmology a time-dependence $R(t) = Bt^{\frac{1}{2}}$, where B is a constant, while the entropy S remains constant in the expansion.

Assuming $N(t) \propto R(t)^3$ this also yields a decreasing disorder

$$D = \frac{S}{k \ln B^3 t^{3/2}}, \tag{38}$$

and hence increasing order as the universe expands.

This completes our remarks on self-organisation and order, leaving for later sections a discussion of alternatives to entropy and a deeper study of it. What, then, is self-organisation? People write a great deal about it, but a hint at a definition is less easy to find. Let me, from what we have seen, say that it occurs in macroscopic non-equilibrium systems which exhibit phase transitions and can be described in part by order parameters. This is in fair agreement with Haken's remark in section 3 of [17]. More loosely, however, it indicates a spontaneous increase in the structural complication of a system. That it is a meaningful term has been shown here particularly by the semiconductor examples. It furnishes (mainly via mathematical formulations) many interdisciplinary links to population dynamics, biology, psychology, meteorology etc as often noted in discussions of synergetics.

Let us add a footnote on alternative entropies.

The usual statistical entropy expression (20) has here been adopted. Other functions have in fact been proposed. Thus let us reconsider the problem which led in Fig. 12 to a "q-entropy" [18] (see also e.g. [19]) due to C Tsallis:

$$S_q \equiv k \left(1 - \sum_{j=1}^{m} p_j^q \right) /(q-1) \qquad (39)$$

Here q is some real number. In this case

$$\begin{aligned} S_i &= -k \left[1 - 2 \left(\tfrac{1}{2} \right)^q \right] /(q-1) \, , \quad D_i = 1 \\ S_f &= -k \left[1 - \left(\tfrac{1}{2} \right)^q - r \left(\tfrac{1}{2r} \right)^q \right] /(q-1), \end{aligned} \qquad (40)$$

$$D_f = \frac{1 - \left(\tfrac{1}{2} \right)^q - r \left(\tfrac{1}{2r} \right)^q}{1 - [1/(r+1)]^{q-1}} \qquad (41)$$

This yields the "order" Ω_f for general q . Curves for $q = \tfrac{1}{2}$ and $q = 2$ are shown in Fig. 12 and show that this generalisation needs a theory to tie down the value of q. (For $q = 1$ one recovers the normal entropy). The entropies for $q = \tfrac{1}{2}$, 2 are not shown, but they turn out to be considerably in excess of the entropies for $q = 1$. There exist of course information measures which we have not used here. We refer instead to the surveys [20], [21] (though [21] does not refer to [18] or [20]).

Some authors have introduced an "entropy decrease"

$$\delta S \equiv k \ln N - S \qquad (= \Omega \, k \, \ln N)$$

of a steady non-equilibrium state; it measures also the "order" [22], [23]. Although almost equivalent to Ω, this measure of order is extensive rather than intensive, and one may therefore prefer to use Ω. The "S-theorem", discussed in the above references, states that δS increases as one departs more from thermodynamic equilibrium.

Among the most notable alternative entropies is that defined by A. Rényi

$$\tilde{S}_q \equiv \frac{k}{1-q} \, \ln \left(\sum_{j=1}^{n} p_j^q \right) \left(= \frac{k}{1-q} \, \ln \left[\frac{1}{k}(q-1)S_q \right] \right) .$$

5 Biological Applications

The possibility of increase of both order and entropy (Fig. 12 and [15]) has been utilised to suggest that biological evolution can be in some way an example of the operation of the second law of thermodynamics [24]. Bearing in mind that Darwinian evolution is still subject to much discussion, this view must be expected to be subject to controversy (see, e.g. [25]-[27]). In fact the systematic applications of entropy and information theoretical concepts to biology goes back at least to Gatlin [28] who gave graphs, and estimated the ranges of information theoretical quantities, for a variety of biological objects.

To examine biological applications, one might consider a DNA sequence in order to draw on an alphabet of, say, 4 letters. We shall make it N letter in order to follow the earlier notation. Then $S(A)$ involves only the states A_j of probabilities $p_j (j = 1, 2, ...)$, and not the interactions q with a neighbouring chain B. In that sense $S(A)$ is the entropy of a Markov chain of first order. Gatlin defines a divergence D of $S(A)$ from equal probabilities. (Her notation is compared with the present notation in Table 2.)

A measure for the interaction, or interdependence, of A and B is very important. It is the so-called mutual information

$$S(A:B) \equiv S(A) + S(B) - S(AB) = S(A) - S_B(A) = S(B) - S_B(B) \quad (42)$$

There is no interaction if $S_A(B) = S(B)$ or $S_B(A) = S(A)$. This suggests introducing divergences from maximum entropy and from independence:

$$\left. \begin{array}{l} D_1 \equiv k \ln N - S(A) \\ D_2 \equiv S(A) - S_B(A) \end{array} \right\} \quad (43)$$

Therefore (Fig. 13)

$$D_1 + D_2 = k \ln N - S_B(A) = R \ln N \quad (44)$$

where the "redundancy" R is defined by

$$R \equiv k - S_B(A)/\ln N \quad (45)$$

Table 2. Lila L. Gatlin's notation [28]

Here	N	$S(A)$	$S(AB)$	$S_B(A)$
Gatlin	a	H_1	H_2^D	H_M

For experimental plots of such quantities to investigate DNA codings we refer to the literature, e.g. [25], [26], [28]. Among other biological applications of the entropy concept, we show in Fig. 14 the entropy production per effective radiating surface area in a human lifespan which is of order 0.15 joule $m^{-2} sec^{-1} K^{-1}$ [29]. It represents a measure of activity and is related to the energy annihilation rate. A clear understanding of the principles involved would put much botanical and biological information into a good theoretical framework.

Let us fix ideas by some simple examples. The entropy (divided by k) of an abstract language which uses 1 space and 26 equally likely letters is given by

$$\ln 27 = 3.29 \text{ or } \log_2 27 = \frac{\ln 27}{\ln 2} = 4.75 \text{bits}$$

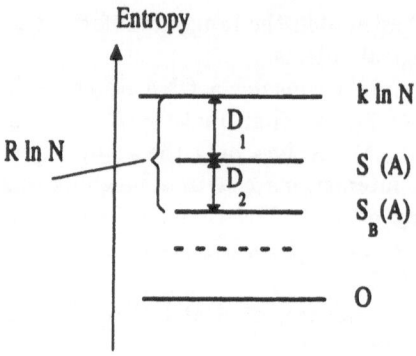

Fig. 13. Entropy scales showing that the entropy is lowered upon taking into account more interactions.

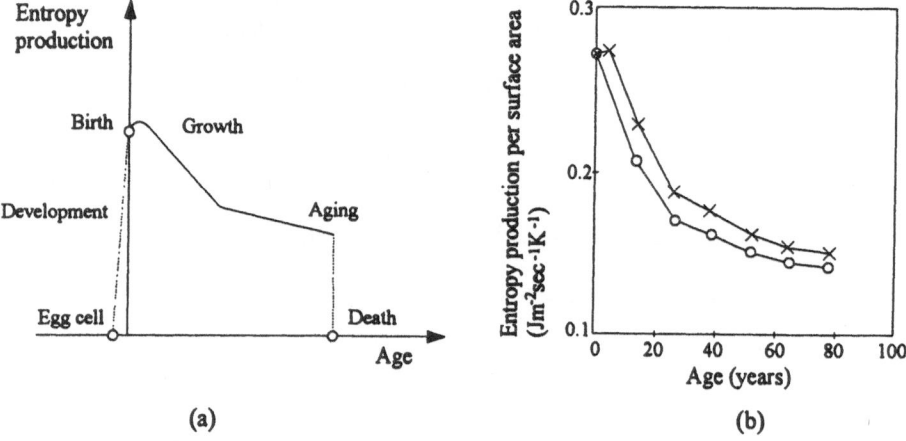

(a)   (b)

Fig. 14.

(a) Schematic profile of the entropy production per surface area in a human life [29].
(b) Entropy production per effective radiating surface area for Japanese males (x) and Japanese females (O) as a function of age [29].

per letter. Spelling rules and unequal probabilities reduce this to about 2 bits per letter for the English language. This yields a redundancy of

$$1 - \frac{2}{4.75} = 0.58$$

If this were not so, any sequence of letters would constitute a word and would make sense. In fact, by virtue of redundancy, one finds misprints and can guess omitted letters. The so-called channel capacity of the optic nerve which carries, say 200 words/min or about 2000 bits/min is of the order of 33 bits/sec. For a pianist with 300 operations per min. per finger, when each finger is equally likely to be put on one of three notes, the channel capacity is

$$300 \times 10 \ \log_2 3 = 4800 \ bits/min \ = \ 80 \ bits/sec$$

One may also recall the extensive work on the entropy of languages done in the 1950s by W. Fucks and others (cited in [30]).

6 Entropy or Variance?

First choice: entropy

"We look for connecting links which lead with mathematical precision from thermodynamic principles to characteristically quantum mechanical notions" I said on p.373 of my 1961 book under the heading "Thermodynamics as a precursor of quantum mechanics" [30]. Such unexpected precursors can be found. It was shown that for a finite entropy at low temperatures the energy of the particles of a system must not go to zero continuously, but thermodynamic arguments show that there must indeed be a lowest energy level á la quantum mechanics. The proof is general, but a little complicated. For the present purposes it suffices to note that the expected continuous spectrum formula for the entropy, based on (20) and (29), is

$$\frac{1}{k}S = -\int p(x)\ln p(x) \ dx \qquad \text{(with } \int p(x) \ dx = 1\text{)}. \qquad (46)$$

For the ideal monotonic gas (with a continuous density of states proportional to $\sqrt{E - E_0}$ starting at some particle energy E_0) it gives the Sackur-Tetrode result for N particles in volume v:

$$\frac{S}{kN} = \ln\left(\frac{N_0}{N}e^{5/2}\right), \ N_0 \equiv \left(\frac{2\pi mkT}{h^2}\right)^{3/2} v \qquad (47)$$

Indeed for $N_0 \rightarrow 0, S$ diverges, showing thermodynamically that for low energies the continuous spectrum is a wrong model. A rather obvious fact, but hardly ever pointed out.

Let us look at this in a different way. Let us consider a triangular probability distribution $p(x) = \lambda x$ (λ is a constant) (Fig. 15), whose entropy (46) is readily shown to be

$$\frac{1}{k}S(x) = \frac{1}{2}(1 - \ln 2\lambda) \qquad (48)$$

If $\lambda_a = \frac{2}{9}, \lambda_b = 8$ as in Fig. 15,

$$\frac{1}{k}(S_a - S_b) = \frac{1}{2}\ln(\lambda_b/\lambda_a) = \ln 6$$

i.e. $$\frac{1}{k}S_b = \frac{1}{k}S_a + \ln\frac{1}{6} \qquad (49)$$

The transformation $p(x) = \lambda_a x$ (case a) $\rightarrow q(y) = \lambda_b y$ (case b) can produce the negative entropy values shown. Generalisations of (49) involve Jacobians.

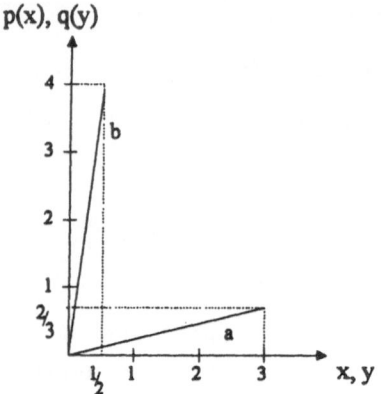

Fig. 15. Two normalised triangular probability distributions.

The narrower distribution of case b gives us more information and has a lower entropy. But we did not want formula (46) to overdo it and produce a negative entropy (- 0.885k as against +0.905 k)! Of course in case b, $p(x)$ exceeds unity for 3/4 of the range of x, so one would expect a negative entropy on mathematical grounds.

One could say: (46) or (47) were never <u>meant</u> to hold at low temperatures [31]. But then: down to what temperatures <u>can</u> it be trusted? To-day the issue is not urgent as we know that quantum mechanics must be employed.

A more general explanation is that the change to continuous variables x has to be made by the replacement

$$p_i \rightarrow p(x_i) \; \Delta \quad (\sum_i p(x_i)\Delta = 1),$$

so that

$$\frac{1}{k}S \rightarrow \frac{1}{k}S(p) = -\sum_i p(x_i) \; \Delta \; \ln p(x_i) - k \; \ln\Delta \tag{50}$$

This corresponds to dividing the parameter space into equal cells Δ such that the probability of an event is $p(x_i)$ in the ith cell. The last term is independent of the probability distribution and can be ignored if different probability distributions for the same Δ are compared.

For small Δ one then obtains (46). Transformations $p(x) \rightarrow q(y)$ with $S\{p(x)\} = S\{q(y)\}$ occur only if they are volume-conserving, i.e. have a Jacobian of unity. This is not the case in the example of equation (49).

Second choice: information gain or relative entropy

A closely related function, using two normalised probability distributions $p(x)$ and $p_0(x)$ with $p_0(x) \neq 0$ for all x, is

$$\frac{1}{k}K\{p(x), p-0(x)\} \equiv \int p(x) \ln \left[p(x)/p_0(x)\right] \; dx \tag{51}$$

It is always non-negative by virtue of the inequality

$$\ln z \geq 1 - 1/z,$$

which implies that

$$\frac{1}{k} K\{p(x), p_0(x)\} \geq \int p(x) \left[1 - \frac{p_0(x)}{p(x)}\right] dx = 0. \qquad (52)$$

It vanishes only if $p(x) = p_0(x)$ for all x. The discrete analogue is

$$\sum p_i \ln (p_i/p_{0i})$$

These functions, introduced by S. Kullback in 1951, are called information gain, surprisal, relative entropy, contrast, Kullback information, etc. If the reference distribution is chosen to be $p_0(x) = 1/n$ for all x, then K is seen to be simply related to S:

$$\frac{1}{k} K\{p(x), p_0(x) = \frac{1}{n}\} = -\frac{1}{k} S(p) + \ln n$$

But K is not affected by the Jacobians upon a transformation of variables. For a recent general reference see [32].

If we know $p_0(x)$ and additional knowledge leads us to $p(x)$ then a <u>candidate</u> for the information gain involved is

$$-\frac{1}{k} [S(p) - S(p_0)] = \sum_i \{p_i \ln p_i - p_{0i} \ln p_{0i}\} \qquad (53)$$

However, (53) is not minimised by $p(x) = p_0(x)$ whereas $K(p, p_0)$ is minimised by $p(x) = p_0(x)$ and is therefore preferable.

<u>Third choice: variance</u>

Among the other measures of disorder, or lack of certainty, or spread of a probability distribution, is of course the variance. For some quantity like the energy, E_i say, it is

$$\sigma_E^2 \equiv \langle (E - \langle E \rangle)^2 \rangle.$$

σ is also called the standard deviation. What is the relation between the dimensionless entropy $\frac{1}{k} S(p)$ and σ? One answer is this: Given σ, the normalised probability distribution of largest entropy is the one-dimensional normal distribution. We can therefore write

$$\frac{1}{2} \ln(2\pi e \sigma^2) \geq \frac{1}{k} S(p) \qquad (54)$$

On the left-hand side we have $(1/k)$ times the entropy of a <u>normal</u> distribution and on the right-hand side we have the entropy of a general one-dimensional distribution with the given variance.

Let us denote by $p_0(x)$ the one-dimensional normalised distribution of maximum entropy, subject to a given moment

$$M_r \equiv \left\{\int_{-\infty}^{\infty} |x - a|^r p(x) \, dx\right\}^{1/r} \qquad (55)$$

about a mean a. Its entropy is denoted by S_0. The normal distribution with the same r^{th} moment be denoted by S_1. Since $M_2 = \sigma$, a plot of $(S_0 - S_1)/k$ as a function of the order, r, of the given moment has to yield positive values with a zero at $r = 2$, as shown in Fig. 16, taken from a little known paper [33]. In fact, if (55) is given, rather than M_2 , the pretty result (54) generalises to ([30], p.249, [33])

$$\ln \left\{ 2(er)^{1/r} \Gamma \left(1 + \frac{1}{r} \right) M_r \right\} \geq \frac{1}{k} S(p). \tag{56}$$

On the left we have again $(1/k)$ times the entropy of the one-dimensional distribution of maximum entropy subject to the rth moment being given. This entropy increases with the measure M_r of the "spread" of the distribution, as one would expect. A list of entropies of probability distributions also exists [34].

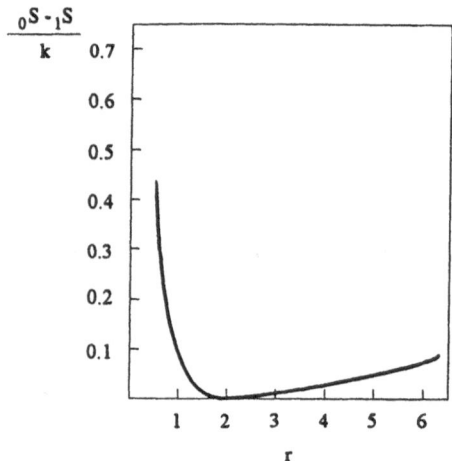

Fig. 16. The difference between the entropy $(_0S)$ of a one-dimensional distribution of maximum entropy subject to the rth moment being given and the entropy of the normal distribution $(_1S)$ having the same r^{th} moment, plotted as a function of r.

Uncertainty relation

Entropy, variance and the measures of spread are usually related and can be used as alternatives. In fact, the entropy – variance alternative is of considerable interest in connection with the uncertainty relation of quantum mechanics (Table 3). The standard uncertainty relation (Table 3b), in spite of its beauty, has been criticised because (i) the right-hand side can be zero, making the inequality useless, for example if j is an eigenstate of A or B; (ii) more generally, the right-hand side depends on the state ψ, whereas it would be better if it depended only on the properties of A and B and (iii) the use of the standard deviation as a measure of spread. This is not a suitable measure for some of the situations which arise in quantum mechanics [36]. One can go further and show that even classically the standard deviation type of uncertainty can for the position of a free

Table 3. Uncertainty relations

a. Heisenberg microscope experiment of 1927. By inspection of experiments, yielding "lacks of precision" δp, δq of conjugate momentum and coordinate, $\delta p \delta q \geq h$.

b. Robertson's standard theoretical uncertainty relation of 1929. For operators A, B representing observable

$$\sigma_A \; \sigma_B \geq \frac{1}{4} | < i \, (AB - BA) >_\psi |$$

where $< >_\psi$ indicates an expectation value for the state represented by the wave function ψ.

c. Maassen and Uffink's relation of 1988 [35] Let $|a_j>, |b_k>$ be the complete sets of normalised eigenvectors of A and B respectively, let $c \equiv \max_{j,k} | < a_j | b_k > |$, and let

$$M_r(| < b_j | \psi > |^2) \equiv \{ \sum_j | < b_j | \psi > |^{2(r+1)} \}^{1/r}.$$

for any state $|\psi>$. Then

$$M_r(| < b_j | \psi > |^2) M_s(| < a_k | \psi > |^2) \leq c^2$$

and, in particular for $r = s = 0$,

$$\frac{1}{k} \left[S\{| < a_k | \psi > |^2\} + S\{| < b_j | \psi > |^2\} \right] \geq \ell n \; c^{-2}$$

particle in a one-dimensional box be subject to a significant uncertainty principle [37, 38]. These drawbacks have led e.g. in [39, 40], to the use of entropies in alternative uncertainty relations. Table 3 gives a recent satisfactory and elegant result along these lines. It involves on the right-hand side only properties of the operators involved.

7 Outlook

We have made a transition from continuous to discrete variables [eqns(2),(3)] and conversely [eqn.50]. The ultimate in discreteness is the cellular automaton, as it is discrete in space and in time. For example, given a "local" rule (i.e. a rule about interaction of a unit with its neighbours) and an initial configuration of occupied (labelled by 1) and empty (labelled by 0) d-dimensional lattice units, the system will evolve in a sequence of "moves". For $d = 2$ it could be draughts (or possibly chess) (The world draughts champion, Dr Marion Tinsley, is currently - August 1992 - playing a world championship match against "Chinook" which is a computer program. He managed to retain his title.)

There need not be an underlying reversibility in the moves so that the second law of thermodynamics need not apply and one may expect the development of order or of self-replication, as has been shown in simple models [41]. Unsurprisingly, one therefore finds models for the origin of life based on theses ideas [42].

In the Game of Life (due to J. H. Conway) one has a rectangular array of squares which are "live" or "dead". A live square survives from time t to t+1 only if it has 2 or 3 live neighbours. A dead square at t will become live at $t+1$ only if it has 3 live neighbours.

These precise rules were smeared in [43] by introducing a probability with a "temperature" parameter T such that Conway's rules were recaptured for $T = 0$. With a reasonable entropy definition (which would take us too far to discuss here) the expected decrease of entropy with time was found. The comparative ease of achieving, on paper at least, self-organisation by using discrete moves on a lattice, raises the question of the physical realisation of such processes. The methods have certainly been used to simulate the growth of crystals, phase transitions, the behaviour of biological populations, etc. The vocabulary that has evolved is due to the fact that many results of physics can be expressed as sequences (possibly very long ones) of zeros and ones, as for the binary arithmetic in a computer. The vocabulary is therefore primarily appropriate for mathematics. Here we give rough explanations of a few terms:

a Algorithm (in logic or mathematics) – A general method for the solution of all problems of a given class. Given data is transformed into output data using a finite number of rules.

b Compressibility (logical, algorithmic) – The property that given data can be expressed more succinctly by the provision of a (physical) theory or a code. A string of symbols 0101010101 is compressible to $(01)^5$; 3.1415926... is compressible to (circumference/diameter) of a circle. The existence of science means that the world is algorithmically compressible.

c A number is random if the smallest algorithm capable of specifying it for a computer (this means here a universal computer, e.g a Turing machine) requires about the same number of bits of information as the number itself. As a consequence any minimal program is random.

d Algorithmic entropy $H(s)$ (or information content or complexity or randomness) of a binary sequence is the number of binary units (bits) in the minimal program which can be used to compute it on a universal computer.

e Logical depth of a sequence is the time required to generate it by an effectively incompressible computer program. The processes giving rise to the objects which are logically deep are likely to have been slow and intricate.

At first sight algorithmic entropy is an inappropriate term for d., and algorithmic complexity might be better, as the connection with the statistical mechanical entropy is not clear. Also it is usually hard or impossible to calculate $H(s)$ for a given situation. However, it is clearly large for a sequence produced by the tosses of an unbiased coin, and small for the first thousand digits of π (rewritten in binary notation). However, at least this is true: for a macrostate specified by the probability distribution $p(x)$

$$\frac{1}{k\ln 2}S(p) < \sum_x p(x)H(x) \leq \frac{1}{k\ln 2}S(p) + H(p) + O(1) \qquad (57)$$

where $\log_2 p = \ln p / \ln 2$ has been used [44]-[46]. This gives the required link between S and the average of H .

The origin of life which was touched on above, cannot adequately be discussed using only classical physics. This leads to the question of how quantum mechanics can be made to be compatible with the production of self-replication. This question goes back to the 1960s [47]-[50] and is still being discussed [51]. The problem is that it is hard to obtain exact replication, since the number of quantum mechanical equations to be satisfied greatly outstrips the number of unknowns. There are various statistical ways of weakening this requirement and so, if life is given, one can save quantum mechanics. Going into greater detail such as DNA sequences, reproduction has been attributed to the interlocking autocatalytic units which are linked into some kind of hypercycle [52]. This mechanism which has been much discussed elsewhere, seems to go some way towards explaining autopoiesis (the ability of biological systems to renew themselves), a term introduced in 1973 by Chilean biologists H. Maturana and F. Varela (see [53] for a general survey.)

The thermodynamics of evolution, [54]-[57] being often semi-philosophical, but not always (see for example [58]), is unfortunately outside the present scope. But it is clear that life is maintained and proceeds to greater degrees of organisations, by virtue of an energy flux which passes through the living system. It arrives as a low entropy flux and departs as a high entropy flux. In this sense the living system is like a thermodynamic engine with the work output per cycle (W) reinterpreted as the energy useful for the living system. The entropy inequality

$$\frac{Q_h}{T_h} \leq \frac{Q_c}{T_c} = \frac{Q_h - W}{T_c}$$

yields

$$\frac{W}{Q_h} \leq 1 - \frac{T_c}{T_h} \tag{58}$$

Here c and h refer to the cold and hot reservoirs and Q_h is the heat supplied at temperature T_h in each cycle. The export of entropy keeps the system going.

We have said nothing about gravity. Gravity is unique because it is <u>always</u> attractive. Hence density fluctuations in a cosmic cloud of gas can give rise to the clumping of matter and, with pre-existing angular momentum, solar systems and rings of Saturn can arise. This is a great ordering tendency, but the entropy of gravitating systems is still in the research stage. Although we know that the black holes have such large entropies that, if the laws of physics can be extrapolated indefinitely, the universe might eventually exist with all matter collapsed into a black hole. Using entropy comparisons with that final state the universe may be regarded as currently rather orderly [59]!

References

1. A. Rényi, *A Diary on Information Theory* (Chichester: Wiley, 1984), p.87.

2. J.H.M. Thornley, Phyllotaxis I. A mechanistic model, Ann. Bot. **39**, 491 (1975). F. Rothen and A-J. Koch, Phyllotaxis, or the properties of spiral lattices. I Shape invariance under compression, J. Phys. France, **50**, 633 (1989). S. Douady and Y. Couder, Phyllotaxis as a physical self-organized growth process, Phys. Rev. Lett. **68**, 2098 (1992).

3. R. Nelson, Quasicrystals. Sci. Am. **255**, 32 (August 1986). T.P. Srinivasan, Fibonacci sequence, golden ratio and a network of resistors, Am. J. Phys. **60**, 461 (1992).

4. V.I. Arnol'd, *Ordinary Differential Equations* Berlin: Springer 1992, p.22). Reproduced from V.V. Nalimov and A.M. Mul'chenko, *Scientometry* (Moscow: Nauka, 1969).

5. G.L. Baker and J.P. Gollup, *Chaotic Dynamics*, Cambridge: University Press 1990, p.88.

6. P.T. Landsberg, E. Schöll and P. Shukla, A simple model for the origin of chaos in semiconductors, Physica **D30**, 235 (1988).

7. E. Schöll, *Nonequilibrium Phase Transitions in Semiconductors* (Berlin: Springer, 1987).

8. F. Schlögl, Chemical reaction models for non-equilibrium phase transitions, Z. Phys. **253**, 147 (1972).

9. P.T. Landsberg and A. Pimpale, Recombination - induced non-equilibrium phase transitions in semiconductors, J. Phys. **C9**, 1243 (1976).

10. D.J. Robbins, P.T. Landsberg and E. Schöll, Threshold switching as a non-equilibrium phase transition, Phys. Stat. Sol. (a) **65**, 353 (1981).

11. K. Nicolic, Steady-states in a cooled p-Ge photoconductor via the Landsberg-Schöll-Shukla model, Solid-State Electronics **35**, 671 (1992).

12. W.I. Khan, Generation-recombination induced non-linear characteristics of solid-state bipolar devices, J. App. Phys. **65**, 4891 (1989).

13. E. Schöll, Instabilities in semiconductors including chaotic phenomena, Physica Scripta **T29**, 152 (1989).

14. P.T. Landsberg, *Thermodynamics and Statistical Mechanics*, (Oxford: University Press 1978). Pages 146-7 have been clarified in the 1990 Dover reprint of this book in response to comments by Dr. Kenneth Denbigh (1983).

15. P.T. Landsberg, Can entropy and "order" increase together?, Physics Lett. **102A**, 171 (1984).

16. P.T. Landsberg, Thermodynamics and black holes in *Black Hole Physics* (Ed. V.de Sabbata and Z. Zhang) (Dordrecht: Kluwer, 1992).

17. H. Haken, Application of the maximum information entropy principle to selforganizing systems, Z. Phys. **B61**, 335 (1985).

18. C. Tsallis, Possible generalization of Boltzmann Gibbs statistics. J. Stat. Phys. **52**, 479 (1988).

19. A.M. Mariz, On the irreversible nature of the Tsallis and Rényi entropies, Physics Lett. **A165**, 409 (1992).

20. A.J. Stam, Some inequalities satisfied by the quantities of information of Fisher and Shannon, Information and Control **2**, 101 (1959).

21. I.J. Taneja, On generalised information measures and their applications, Adv. in Electronics and Electron Physics **76**, 327 (1989).

22. Y.L. Klimontovich, S-Theorem, Z Phys. **B66**, 125 (1987).

23. W. Ebeling, H. Engel and H. Erzel, *Selbstorganisation in der Zeit* (Berlin: Akademic Verlag, 1990).

24. D.R. Brooks and E.O. Wiley, *Evolution as Entropy: towards a Unified Theory of Biology* (2nd ed., Chicago University Press, 1988).

25. S. Banerjee, P.R. Sibbald and J. Maze, Quantifying the dynamics of order and organization in biological systems, J. Theoret. Biol. **143**, 91 (1990).

26. A. Hariri, B. Weber and J. Olmsted III, On the validity of Shannon-information calculations for molecular biological sequences, J. Theoret. Biol. **147**, 235 (1990).

27. J. Collier, Entropy in evolution, Biology and Philosophy **1**, 5 (1986).

28. L.L. Gatlin, *Information Theory and the Living System* (New York: Columbia University Press, 1972).

29. I. Aoki, Entropy principle for human development, growth and aging, J. Theoret. Biol. **150**, 215 (1991).

30. P.T. Landsberg, *Thermodynamics with Quantum Statistical Illustrations* (New York: Interscience, 1961).

31. E. Schrödinger, *Statistical Thermodynamics* (Cambridge University Press, 1948), p.59.

32. F. Schlögl, *Probability and Heat* (Braunschweig: Vieweg, 1989).

33. B.S. Westcott and P.T. Landsberg, Entropy and nth-law rectification, Int. J. Electr. **37**, 219 (1974).

34. A.C.G. Verdugo Lazo and P.V. Rathie, On the entropy of continuous probability distributions, IEEE Trans. on Information Theory IT **24**, 120 (1978).

35. H. Maassen and J.B.M. Uffink Generalized entropic uncertainty relations, Phys. Rev. Lett. **60**, 1103 (1988).

36. J.Hilgevoord and J. Uffink, The mathematical expression of the uncertainty principle in *Microphysical Reality and Quantum Formalism*, Int. Conf. held in Urbino, 1985 (Dordrecht: Reidel, 1988), p.91. Edited by F. Selleri, A. van der Merwe and G. Tarozzi.

37. J. Peslak Jr., Comparison of classical and quantum mechanical uncertainties, Am. J. Phys. **47**, 39 (1979).

38. P.T. Landsberg, Uncertainty and Measurement in *Quantum Theory without Reduction*, Int. Conf. held in Rome in 1989 (Bristol: Adam Hilger, 1990), p.161. Edited by M. Cini and J.-M. Lvy-Leblond.

39. I. Bialynicki-Birula and J. Mycielski, Unvertainty relations for information entropy in wave mechanics, Commun. Math. Phys. **44**, 129 (1975).

40. D. Deutsch, Uncertainty in Quantum measurements, Phys. Rev. Lett. **50**, 631 (1983).

41. C.G. Langton, Self-reproduction in cellular automata, Physica **D10**, 135 (1984).

42. P. Tamayo and H. Hartmann, Cellular automata, reaction-diffusion systems and the origin of life in *Artificial Life* (Reading, MA: Addison-Wesley, 1988) p.105. Edited by C. Langton.

43. L.S. Schulman and P.E. Seiden, Statistical mechanics of a dynamical system based on Conway's Game of Life, J. Stat. Phys. **19**, 293 (1978).

44. C.H. Bennett, The thermodynamics of computation - a review, Int. J. Theoret. Physics **21**, 905 (1982).

45. K. Lindgren, Microscopic and macroscopic entropy, Phys. Rev. **A38**, 4794 (1988).

46. W.H. Zurek (Ed.) *Complexity, Entropy and the Physics of Information* (Reading, MA: Addison-Wesley, 1990).

47. E.P. Wigner, The probability of the existence of a self-reproducing unit, in *The Logic of Personal Knowledge*. Essays presented to Michael Polangi (London: Routledge and Kegan Paul, 1961), p.231.

48. P.T. Landsberg, Does quantum mechanics exclude life?, Nature **203**, 928 (1964).

49. L. Bass, Biological replication by quantum mechanical interactions, Foundations of Physics **7**, 221 (1977).

50. P.T. Landsberg, Two general problems in quantum biology, Int. J. of Quantum Chemistry: Quantum Biology Symposium **11**, 55 (1984).

51. J.C. Baetz, Is life improbable? Foundations of physics **18**, 91 (1989).

52. M. Eigen, The hypercycle. A principle of natural self organisation, Naturwiss. **64**, 541 (1977), **65**, 7 (1978), **65**, 341 (1978).

53. E. Jantsch, *The Self-Organising Universe* (Oxford: Pergamon Press, 1980).

54. I. Progogine, G. Nicolis and A. Babloyantz, The thermodynamics of evolution, Physics Today, November 1973, p.23.

55. S. Black, On the thermodynamics of evolution, Perspectives in Biology and Medicine, Spring 1978, p.349.

56. J.S. Wicken, A thermodynamic theory of evolution, J. Theoret. Biology **87**, 9 (1980).

57. A. Peacocke, Thermodynamics and Life, Zygon **19**, 395 (1984).

58. E.M. Chudnovsky, Thermodynamics of natural selection, J. Stat. Phys. **41** 877 (1985).

59. P.T. Landsberg, The physical concept of time in the 20th century in *Physics in the Making*, Eds. A. Sarlemijn and M.J. Sparnaay (Amsterdam: Elsevier, 1989), p.131.

Self-Organization, Valuation and Optimization

W. Ebeling

Abstract

Following the valoric interpretation of entropy by Clausius, self-organisation is connected with increasing value of the energy inside the system. Starting from this physical process, the essence and the dynamics of other non-physical valuation processes is discussed. The metaphor of optimization strategies is transferred to problem solving.

1 Introduction

Valuation and optimization are concepts which are intimately connected with self-organization. Further there exist several successful approaches to transfer natural strategies to problem solving [1]-[9]. In spite of the importance of understanding the relation between these concepts, there are still many open questions. We will start here from a discussion of entropy from the valoric point of view proceeding then with other (nonphysical) forms of valuation.

Discussion of the role of entropy and value in physics dates from the very beginnings of the theory of macroscopic matter in the last century. The relationship between them engaged such pioneers as Mayer, Clausius, Helmholtz, Boltzmann and Ostwald, and later Szilard, Delbrück, von Neumann, Schrödinger and Turing. For example Robert Mayer thought about the value of energy and found that the flow of light from the sun to the earth is the main source of high-value energy, which makes life on earth possible. The role of entropy for life processes was first studied by Boltzmann, Ostwald and Delbrück and was reconsidered in Schrödinger's famous book "What is life?", and a few years later in Prigogine's early work, including his paper with Wiame. In spite of their long history these concepts are still far from being clear and are still the subject of lively discussion [10]. Schrödinger stated in clear terms that self-organization and life require that the system exports entropy. Schrödinger expresses this as the system's requiring negentropy; one may also say that the system must be fed high-value energy. In this way self-organization is always connected with an increase of the value of energy inside the body. The role of entropy export has been worked out in detail by Prigogine and his school. A special role in the discussion was played by the valoric interpretation of the thermodynamic entropy given by Clausius [10]. The valoric concept of entropy was worked out later by Helmholtz and Ostwald. These researchers considered (neg)entropy as a measure of the work value of the energy contained in the system. Maximal entropy (thermodynamic equilibrium) means that the energy has zero work value and low entropy means that the energy has a relatively high work value. The role of entropy lowering in self-organization was discussed by Klimontovich and others [10, 11]. At the condition of fixed energy, the entropy lowering increases with increasing distance from equilibrium due to self-organization. Helmholtz introduced free energy F as that part of the energy which is able to do work and showed that it can be defined over the entropy S and the temperature T by:

$$E = F + TS \tag{1}$$

The free energy decreases in thermal processes relaxing to equilibrium and it increases in processes of self-organization.

Ostwald once remarked, that he had been unable to understand the entropy concept before he discovered the valoric interpretation. He said that the best way to understand the meaning of entropy is to consider it as a measure of the value of energy to do work. Due to strong opposition to this concept, expressed especially by Kirchhoff and Planck, the valoric interpretation was nearly forgotten, except by a few authors, among them especially Schöpf. As a matter of fact, however, the valoric interpretation of entropy was for Clausius the key point. What many physicists do not know is that the entropy concept taught today at universities is much nearer to the reinterpretation given by Kirchhoff and Planck than to Clausius' original one. Here we take up the original interpretation in terms of a value concept in connection with some more recent developments [10, 11]. As we will see, the valoric interpretation of entropy (which in physics is just an alternative formulation) yields an excellent starting point for considering the role of values in general [11, 12].

Besides the valoric interpretation of entropy, modern science employs other value concepts such as information value, selection value in biology and exchange value in economy. All these value concepts have several features in common [1, 11]:

(i) Values assigned to elements (subsystems) of a system incorporate the system as a whole; they cannot be understood by a merely viewing the subsystem apart its whole environment. In other words, the whole is more than the sum of its parts.

(ii) The values are central for the structure and the dynamics of the entire system; they determine the relations between the elements and their dynamical modes, as well as the dynamics of the system. Competition and selection between elements according to their values are typical features of the dynamics.

(iii) The dynamics of systems with valuation is irreversible; it is intrinsically connected with certain extremum principles for the evolution of the values over time. These extremum principles may be very complex and, only in the simplest cases, can they be expressed by scalar functions and by total differentials. In general they are of the type of the Glansdorff-Prigogine inequality [5, 7].

As we have shown above, the simplest form of valuation is connected with the entropy subject to the second law. This fundamental law of nature expresses the natural tendency in isolated systems to a devaluation of energy. In order to increase the value of energy in a body, pumping is necessary. In general self-organization is connected with an increase in the value of the energy in the system [7], but self-organization may also be connected with an increase in other types of values. In some sense, evolution is connected with optimization in spite of the fact that there is no general optimization criterion for evolution. Even the Glansdorf-Prigogine criterion, which is rather general, has no universal character. Valuation and the tendency to optimize values play a special role in competition and selection processes. Value is connected with fitness in the sense of Darwin. Competition is always based on some kind of valuation. Apparently the concept of values was first introduced into science by Adam Smith in the 18th century in an economic context. The fundamental ideas of Adam Smith were worked out later by Ricardo, Marx, Schumpeter and many other economists. In another social context the idea of valuation was used at the turn of the 18th century by Malthus. Parallel to this development in the socio-economical sciences a value concept was developed in the biological sciences by Darwin and Wallace. Wright developed the idea of fitness landscape (value landscape) which was then elaborated by many authors [5]-[7]. As all these concepts are very abstract and qualitative great difficulties arose in the field of mathematical modelling [5]-[7]. Our point of view is that values (beside negentropy) are abstract, non-physical properties of subsystems (species) or dynamical modes in certain systemic contexts. Values express the essence of biological, ecological, economical or social properties and relations with respect to the dynamics of a given system. From the point of view of modelling and simulation values are given 'a priori', i.e. they must be considered as axiomatic elements of the dynamic models. Due to the fundamental character of values, it is very difficult to avoid a tautologies in their definition. However, the same difficulty was already discussed by Newton with respect to mass and force and by Poincaré with respect to energy. In his lectures on thermodynamics (1893) Poincaré says: In every special instance it is clear to us what energy is and we can give at least a provisional definition of it; it is impossible, however, to give a general definition of it. If one wants to express the law in its full generality, applying it to the universe as a whole, one sees it dissolve before one's eyes, so to speak, leaving only the words: There is something that remains constant.

It is quite possible that after a critical discussion of value concepts the only possible statement one can make is: There exists something in evolutionary processes that has the tendency to increase. However as the example of energy shows, this still does not mean that the concept is useless.

2 Dynamical Aspects of Valuation Processes

In the simplest case of valuation processes the value of a subsystem (species) with respect to a competition process is a real number. In other words each element (species) $1, 2, \ldots, i, \ldots, s$ is associated with a number

$$V_1, V_2, \ldots, V_2, \ldots, V_s; (V_i \in \Re) \tag{2}$$

Since real numbers form an ordered set, the subsystems are ordered with respect to their (scalar) values. In such systems competition and valuation may be induced by the process of growth of subsystems having high values and decay of subsystem having low values (or vice versa). In many cases the growth of "good" subsystems is subjected to certain limitations.

Valuation of this type is of special importance for the self-organization of life. This was first pointed out in the fundamental papers of Manfred Eigen on the origin of life. A standard case already studied in 1930 by Fisher and since 1971 by Eigen is competition by average [1], [5]-[7]. Here all subsystems better than the average taken over the total system

$$\langle V \rangle = \sum_{i=1}^{s} V_i N_i / N \tag{3}$$

grow and all others, being worse, decay. This leads to a monotonic increase in the average value over time

$$(d/dt)\langle V \rangle \geq 0 \tag{4}$$

In statistical thermodynamics the values are given by the distribution itself, e.g.,

$$V = -\log(N_i/N). \tag{5}$$

Here valuation favours equal distribution and the averages correspond to entropies. According to Boltzmanns concept, entropy maximizing is subject to the condition of constant energy and the elements are defined by a partition of the phase space of the molecules. The second law of thermodynamics expresses a general tendency to disorder (equipartition) corresponding to a devaluation of energy.

Many competition situations in biology, ecology and economy are connected with a struggle for common raw material or food [6]-[11]. Here, the result of the competition can still be predicted on the basis of a set of real numbers (scalar values). In more general situations the values are not well-defined numbers, but merely a property of the dynamic system [1, 11].

Information processing is also closely connected with valuation. This aspect was worked out in the fundamental papers by Manfred Eigen and his coworkers on the self-organization of macromolecules [6]. Another key point is the origin of information [11]. We know that the struggle for existence of all living beings is intimately connected with their ability to process information. Our approach to this question is based on an investigation of the entropic aspects of life and information processing. In this respect we shall analyze in particular the potential value of entropy for information processing. The existence of all living beings is based on entropy export, but it is also intimately connected with information processing. A living system may be characterized as a naturally ordered and information processing macroscopic system [12]. The basic thermodynamic characteristics of living systems are openess, entropy export and super-critical distance from thermodynamical equilibrium. The central aspect of the evolution of life is the evolution of information processing, which we consider as an especially high form of self-organization [11, 12].

Let us discuss now some aspects of the dynamics of valuation processes. As we have shown, the simplest of valuation processes is the entropy change; the characterization of entropy as a work value of energy has been well established in physics [10]. A special formulation of the second law says that the work value of energy in isolated systems cannot increase spontaneously. In other words the value of energy in isolated systems is a Lyapunov function

$$\delta S(E, X, t) \geq 0, \quad \partial_t \, \delta S(E, X, t) \leq 0 \qquad (6)$$

In order to increase the (work) value of energy in a system one has to export entropy, which is in fact pumping with high-valued energy. The dynamics connected with changes in the entropy was worked out by Onsager about 60 years ago and is considered now as the basis of the thermodynamics of irreversible procecces. A special case, isothermal processes will be discussed in the next section. For this type of processes the free energy is decreasing in relaxation to equilibrium and increasing in self-organization processes. Isoenergetic self-organization processes are characterized by

$$\partial \, \delta S(E, X, t) > 0 \text{ at } E = const \qquad (6a)$$

There exist however other (non-physical) irreversible processes which are not driven by entropy changes but by other forms of valuation. Such processes appear first in the evolution of biological macromolecules [11]. The origin of life on earth is intimately connected with the creation of non-physical valuation processes. Let us briefly sketch the typical dynamical processes appearing in this connection in evolutionary processes. Let us first label the qualitatively different elements of evolution (molecules, species etc.). We assume that the different elements form a countable set (the set of species) and we denote them by their index of counting $i = 1,2,3, \ldots$. The counting may already be a nontrivial problem. In most cases, however, the subsystem participating in the evolutionary game may be characterized by strings. A string may always be associated with an integer, its Gödel number. Let us assume that the objects (species) which are behind the strings (or behind the index numbers) are subject to certain evolutionary dynamics. In a general sense we consider the strings i\$ etc. as the "genotypes" of the objects participating in our evolutionary game. All possible strings (words) may be considered as elements of an abstract, metric space, the genotype space G. We assume that each genotype is connected with a set of properties forming the "phenotype"; in this way we also introduce a phenotype space Q. The phenotype is valuated during evolution. Mathematically this means that any element is associated with a set of real numbers

$$(V_i^{(1)}, V_i^{(2)}, \ldots) \subset V. \tag{7}$$

Valuation is connected with the homomorphism

$$G \longrightarrow Q \longrightarrow V.$$

Here V is a real vector space. These ideas are closely connected with the concept of an evolutionary landscape, which is one of the basic concepts in the modern theory of evolution [5]-[7]. Our next assumption is that for each object i either an occupation number N_i (stochastic picture) or a concentration (or fraction) x_i is defined. Here N_i denotes the number of representatives of the objects of kind i in the system and x_i a corresponding fraction or concentration. The simplest model of an evolutionary dynamics is the Fisher-Eigen model which is based on the assumption that the competing objects $i = 1, 2, \ldots, s$ have different reproduction rates E_i . These scalar quantities play the role of the values. The dynamics of the fractions is given by the differential equations [1], [5]-[7]

$$\dot{x} = (E_i - k_0(t))\, x_i\, (t) \tag{8}$$

where $E_i = A_i - D_i$ ist the difference between reproduction and death rates and $k_0(t) = < E >$ is a general dilution flux. The resulting dynamical equation

$$\dot{x} = (E_i - \langle E \rangle)x_i \tag{9}$$

shows that the subsystem with values better than the "social" average $\langle E \rangle$ will succeed in the competition and the others will fail. Finally only the subsystem with the largest rate E_m will survive

$$E_m > E_i \ ; \ i = 1, 2, \ldots, s \ ; \ i \neq m \tag{10}$$

The Fisher-Eigen model is only the simplest of all models of competition. There exist more realistic models [1, 6, 7]. A generalization of eq. 8 takes into account that the reproduction rates as well as the death rate (and possibly also the other values) depend on the age of the individuals belonging to the subsystem i :

$$A_i = A_i(\tau), D_i = D_i(\tau) \tag{11}$$

In this case the theory yields, under the condition of constant overall number, an eigenvalue problem. The winner of the competition is the subsystem with the maximal eigenvalue. In most cases early reproduction is of advantage in the competition [7]. In contrast with our earlier case, valuation is now concerned with functions of time, the aging functions.

Aging is only part of a general strategy of evolution which we call Haeckel strategy [10]. The essence of this strategy is the following: In the early days of life on earth, living systems consisted of one simple cell only, which had – at least in principle – an infinite lifetime. Like little machines these cells were complete at birth, they were able to consume free energy, to move in space, to react to external factors and to produce offspring. With increasing complexity the cell organisms developed a life-cycle consisting of several periods such as youth, a period of growth and learning, a period of self-reproduction and then death. Aging and development is typical especially for multi-cellular organisms. This was a great achievement in evolution which made possible the formation of complex structures by individual development, learning and teaching [7]. The processes which lead to the formation of a new animal or plant from cells derived from one or more parents (eggs or seeds) are the subject now of developmental biology. This field of science was pioneered by Haeckel who detected close relations between ontogeny and phylogeny. Haeckel's biogenetic law postulates that, to some extent, ontogeny recapitulates phylogeny. Modelling Haeckel strategies leads to quite complicated integro-differential equations [7]. In this context valuation appears as a functional problem.

As we have shown above, self-reproduction in connection with valuation, competition and selection is an essential part of evolutionary processes. For purely physical reasons (natural fluctuations, external random influences) any self-reproduction process is subject to errors. The copy may not be exact, but only similar to the original. Error copies introduce new elements into the system, which are then subject to valuation, competition and selection. This opens new paths for the evolutionary process and introduces stochastics into it. Since any error reproduction is a stochastic choice between many different possibilities, the process has a branching character. With each self-reproduction process, one of many branches of the paths into the future may be realized; therefore the actual chain of evolution is like a bifurcating network. The number of possibilities increases with time. In order to describe the complex dynamics of those processes, a stochastic language is required. The stochastic theory of evolutionary processes was pioneered by Eigen and elaborated out subsequently by several authors [1], [5]-[7]. For our model we assume that subsystems of different types are present in the system. The number of types may be very large or even infinite. Each self-replication or death process changes only the number of a single type of subsystem

$$N_i \longrightarrow N_i + 1 \qquad \text{or} \qquad N_i \longrightarrow N_i - 1$$

An exchange process, on the other hand, is accompanied by the change of two occupation numbers:

$$N_i \longrightarrow N_i + 1 \qquad \text{and} \qquad N_j \longrightarrow N_j - 1$$

Stochastic descriptions are very important for the modelling of evolution since the description of the initial phases of innovative instabilities is possible only on the basis of stochastic models. This is because the innovation leading to a new subsystem n is always a zero-to-one transition

$$N_n = 0 \longrightarrow N_n = 1.$$

Such a birth process is strongly influenced by stochastic effects. The occupation numbers N_i are functions of time. In contrast to the smooth variation of $x_i(t)$, the dynamics of $N_i(t)$ is a discrete, stochastic hopping process. Averaging the stochastic process over an ensemble of systems, the deterministic equations follow for $N_i \rightarrow \infty$. It should be stressed, however, that the deterministic equations remain true only on average. In contrast to the deterministic case, it is possible in the stochastic realm, for instance, to pass over the barrier separating two different stable equilibria [6]. In other words, valuation can turn into a very complex stochastic phenomenon. The 'goals ' of the evolutionary game are reached only with certain probability.

3 Optimization Strategies from Nature

In the course of natural evolution several basic strategies for reaching certain goals have developed [2]-[9]. As we have shown above, there exist thermodynamic, reproductional and developemental strategies. The translation of these metaphors to optimization problems seems to be promising [2]-[9]. In order to explain these ideas, let us briefly summarize the basic strategies of evolution from the optimization point of view.

3.1 Thermodynamic Strategies

Before life appeared on earth the most important goal of nature was to optimize certain thermodynamic functions. We will demonstrate the idea of thermodynamic strategies by the following model. Let us consider a numbered set of states $i = 1, 2, \ldots, n$, each of them characterized by a scalar U_i (the potential energy) and a probability of occupation at time t denoted by $p_i(t)$. A dynamics which finds the minimum among the set of scalars (the minimum energy) is given by:

$$\partial_t \, p_i(t) = \sum [A_{ij} p_j(t) - A_{ji} p_i(t)] \tag{12}$$

$$A = \text{const} \begin{cases} 1 & \text{if} & U_i < U_j \\ \exp\left[\beta(U_j - U_i)\right] & \text{otherwise.} \end{cases} \tag{13}$$

One can show easily that the distribution converges to

$$p_i(t) \longrightarrow p_i^0 = Z^{-1}\exp[-\beta U_i] \tag{14}$$

where β is the reciprocal temperature and Z the sum over states. Since the minimum of the potentials has the highest probability, the Boltzmann process solves the task of finding the minimum in a set of scalars

$$U_m \leq U_i, \qquad i = 1, 2, \ldots, m - 1, m + 1, \ldots, n$$

One can further show, that during the search process the function

$$K(t) = \sum_1 p(t) \log[p_i(t)/p_i^o] = \beta[F(t) - F_0] \tag{15}$$

in monotonically decreasing. Here $F(t)$ has the meaning of a "free energy" of the system with the equilibrium value F_0 . For an effective search it is useful to increase the β-parameter in the course of search. In this way we have given a model of the strategy of "simulated annealing".

3.2 Strategy of Replicating Populations

Let us consider now, following sect. 2, a population of subsystems with the density $x_i(t)$ and the replication rate E_i . In addition we include a symmetrical mutation matrix A_{ij} assuming the dynamics:

$$\partial_t x_i(t) = [E_i - < E >]x_i(t)$$
$$+ \sum [A_{ij} x_j(t) - A_{ji} x_i(t)] \qquad (16)$$

The total population N is an invariant of the system. In the special case when mutations are very seldom, the target state is that with the maximal value of the replication rates [5, 6]:

$$x_m^0 = N \qquad \text{if} \qquad E_m > E_i. \qquad (17)$$

During the search process the average of the rates is monotonically increasing

$$\partial_t < E > \geq 0 , \; < E > \longrightarrow E_m \qquad (18)$$

There exist much more complicated strategies of replicating populations, among them are the sexual reproduction and (as a special form) the strategies based on the "genetic algorithm" [3, 4].

3.3 Strategy of Developing Populations

In the populations with simple replication discussed above, the members of the ensuing generations are complete (fully developed) immediately after birth. In nature this simple strategy is used only by primitive organisms. With increasing complexity most of the biological species developed special life cycles consisting of several periods as, for example, birth, youth, period of learning, self-reproduction, taking care of the next generation, and death. This strategy of individual development (including morphogenesis and learning) appears to be an extremely successful strategy for solving difficult tasks such as survival in a complex and changing environment. In spite of the evidence that individual development is a potential successful strategy [2, 9] the theoretical background for this type of optimization strategies is very poor. On the basis of earlier investigations of populations with age structure [7, 8] we will now treat the model of an exactly solvable optimization problem of this type.

Developing individuals of type i are characterized by an internal time which starts to run immediately after the birth ($\tau = 0$). The population densities and the birth and death rates are depending on the age: $x_i(t, \tau), b_i(\tau), d_i(\tau)$. We assume that the dynamics of the population densities is described by the equations:

$$\partial_t x_i = [-\partial_\tau - d_i(\tau) - E(t, \tau)]x_i(t, \tau) \qquad (19)$$

$$x_i(t, 0) = \int_o^\infty d\tau [b_i(\tau)x_i(t, \tau) + \sum b_{ij}(\tau)x_j(t, \tau)] \qquad (20)$$

Here $x_i(t, 0)$ is the density of the newborns and further $b_i(\tau)$, $d_i(\tau)$ and $b_{ij}(\tau)$ are the age-dependent birth, death and mutation rates. We have introduced an age-dependent dilution rate $E(t, \tau)$ which is derived from the condition of constant overall number of individuals

$$N = \sum \int d\tau x_i(t, \tau) = \text{const} \tag{21}$$

Equation (19) describes the basic dynamics of populations including aging. The birth processes are modelled here by a source term (the derivative with respect to τ). According to eq. (20) the exact or error reproduction of an individual leads to a new individual (a newborn) with the age $\tau = 0$. By means of the condition (21) competition is introduced into the game, since any newborn individual must push out another elder individual. By integration of eqs (19-21) and using the definitions

$$x_i(t) = \int d\tau x_i(t, \tau) \tag{22}$$

$$E_i(t) = \int d\tau [b_i(\tau) - d_i(\tau)] x_i(t, \tau)/N \tag{23}$$

we get the Fisher-Eigen equations except that the rates are now time-dependent. In this way we have shown that eqs. (19-21) constitute a direct generalization of the Fisher-Eigen dynamics. The time-dependence of the rates confronts us, however, with some difficulties. In order to proceed led us assume that the limits for infinite time are well defined

$$E_i(t) \longrightarrow E_i \qquad \text{for} \qquad t \longrightarrow \infty. \tag{24}$$

For simplicity we shall further assume that the death rates are not species-dependent. Since the result of the competition is decided only in the infinite-time limit, we may conclude, as for the simple Fisher-Eigen case, that the winner is that species with the largest rate

$$E_n > E_i \qquad (i = 1, 2,).$$

The corresponding stationary solution of eqs. (19-20) is

$$x_i^0(\tau) = x_i^0(0) g(\tau) \exp[-E_i \tau] \tag{25}$$

$$g(\tau) = \exp\left[-\int_0^\infty d(s) ds\right]$$

By introducing this solution into eq. (20) we finally obtain the functional equation

$$\int_0^\infty d\tau \, b(\tau) \, g(\tau) \exp[-E_i \tau] = 1 \tag{26}$$

This is an implicit equation for the determination of the limiting rates. We have to remember that these rates are not given a priori, but only the age-dependent rates. After finding all the limiting rates from eq. (26) we know by ordering, who is the winner of the competition. This species is replicating in a stable way with the age distribution

$$x_m^0(\tau) \;=\; x_m^0(0)g(\tau)\exp[-E_m\tau] \tag{27}$$

All the other species disappear in the final state except for a small contribution due to mutations. In this way we have shown that in simple model cases valuation and optimization can mean simple scalar ordering, eigenvalue ordering as well functional ordering.

4 Discussion

In the first part of this paper we have given a general discussion of several aspects of valuation and optimization in self-organization and evolution. After showing that the second law allows an interpretation by valuation, we discussed other models, such as the competition by average. In sect. 3 we tried to transfer these general metaphors to the solution of optimization problems.

As we have shown first, the elementary strategies 1) and 2) solve the task of finding the maximum (or minimum) in a set of real numbers

$$\alpha_1, \alpha_2, \qquad \ldots, \qquad \alpha_n.$$

Several successful evolutionary strategies for this class of problems based on the ideas of simulated annealing or on genetic algorithms were developed [3, 4]. As we discussed already in sections 2 and 3, nature also uses much more developed strategies; animals in particular use a whole repertoire of rather complex problem-solving methods. We have studied here the simplest case of a strategy based on individual development. As we have shown, developmental strategies allow the solution of more complex problems, which are in general of functional character. The example treated here in an explicit way solves the problem of finding a function out of a set of real-valued integrable functions defined on the positive axis $x \geq 0$

$$b_1(x), b_2(x), \ldots, b_n(x)$$

The target of the search is that function $b_m(x)$ which has the maximal increment $\lambda_m > \lambda_i$. The increments λ_i are defined by the functional relation

$$\int_0^\infty dx \; b_i(x) \; g(x) \; \exp[-\lambda_i \; x] \;=\; 1 \tag{28}$$

Here the $g(x)$ is a rather arbitrary integrable function which is related to the death rates by

$$d(x) \;=\; -d\,[\ln\,g(x)]\,/\,dx \tag{29}$$

In other words, functional optimization problems of the type (28) may be solved by simulating a competition of developing species. A stochastic game which simulates the processes described above may (in the simplest case) be carried out as follows. We model on a computer an ensemble of N searchers each associated with an individual age and an individual birth rate [2, 9]. At stochastic times we take out an arbitrary pair of searchers and compare the values of their birth rates. The searcher with the poorer value of the rates is replaced by a specimen of the better one (but with age 0). This introduces replication and competition into the game. In order to include mutations, we may take out at stochastic times anyone of the searchers and replace it by another one (associating again zero time to it). The principle is, that any replacement of an individual is considered as a birth process, what means that the internal time (the age) starts at zero. This simple scheme may be modified in several ways. As we have demonstrated elsewhere by simulations [2, 9] search strategies including development are quite successful in problem solving; also for the standard problem of finding extrema. In conclusion we may say that valuation and optimization are an emergent part of the paradigm of self-organization. Furthermore, we have shown that the application of physically or biologically oriented search strategies leads to effective search methods and also to an extension of the class of optimization problems which may be treated.

References

1. W. Ebeling: Elements of a Synergetics of Evolutionary Processes. In: Evolution of Dynamical Structures in Complex Systems, (R. Friedrich, A. Wunderlin, eds.). Springer-Verlag, Berlin, Heidelberg 1992
2. T. Boseniuk, W. Ebeling: Boltzmann-, Darwin-, and Haeckel-Strategies in Optimization Problems. In [3].
3. H.-P.Schwefel, R. Männer: Parallel Problem Solving from Nature, Springer, Berlin 1991
4. H.-P. Schwefel: Numerische Modellierung von Computermodellen mittels der Evolutionsstrategie, Birkhäuser, Basel 1977
5. W. Ebeling, A. Engel, R. Feistel: Physik der Evolutionsprozesse, Akademie-Verlag, Berlin 1990
6. P. Schuster: Structure and Dynamics of Replication-Mutation Systems, Physica Scripta 35 (1987) 402
7. R. Feistel, W. Ebeling: Evolution of Complex systems, Kluwer, Dordrecht 1989
8. W. Ebeling, A. Engel, V.G. Mazenko: Modeling of Selection Processes with Age-dependent Birth and Death Rates. BioSystems 19 (1986) 213
9. W. Ebeling: Applications of Evolutionary Strategies. Syst. Anal. Model. Simul. 7 (1990) 3
10. W. Ebeling: On the Relation Between Various Entropy Concepts and the Valoric Interpretation, Physica A 182 (1992) 108
11. W. Ebeling, R. Feistel: Theory of Self-Organization and Evolution, the Role of Entropy, Value and Information, J. Nonequil. Thermodyn., 17(1992)303
12. W. Ebeling, M.V. Volkenstein: Entropy and the Evolution of Biological Information. Physica A 163 (1990) 398

Symbolic Dynamics
and the Description of Complexity

Bai-lin HAO

Symbolic dynamics provides a general framework to describe complexity of dynamical behaviour. After a discussion of the state of the field special emphasis will be given to the role of the transfer matrix (the Stefan matrix) both in deriving the grammar from known symbolic dynamics and in extracting the rules from experimental data. The block structure of the Stefan matrix may serve as another indicator of the complexity of the associated dynamics.

1 Introduction

It is a commonplace nowadays that many physical systems are capable of undergoing sharp transitions to more organized states when control parameters are tuned away from the trivial, near equilibrium, or linear, regimes. Intuitively speaking, these organized states are characterized by lower symmetry, lower entropy, more information, and a higher degree of complexity. By the way, it is the change of symmetry, explicit or hidden, that makes the transition sharp, because symmetry is a property that must either be present or absent; it cannot be accumulated gradually. However, when one comes to the stage of characterizing complexity, we see that this is a far from trivial task.

First of all, complexity is a notion which has been used in so many different contexts and which can hardly be defined in general. On 10th September, 1992, there were 302 books in the U. S. Library of Congress that have the word "complexity" in their title[3]. The most frequent usages include algorithmic complexity (AC), biological complexity, computational complexity (CC), developmental complexity, ecological complexity, economical complexity, evolutional complexity, grammatical complexity (GC), language complexity, etc.

Second, simplicity and reductionism has been a guideline in science. Many scientists believe that the fundamental laws of Nature must be simple. Although they encounter "complex" phenomena everyday and everywhere, they still try hard to reduce them to something simpler. Indeed, science knows a number of ways by which simple things can get more complex. For, example:

1. Projection onto lower-dimensional space may make things look more complex or, put the other way around, adding new dimensions, sometimes in the

---

[3] We thank Dr. Ming-zhou Ding, Florida Atlantic University, for checking the number.

Springer Series in Synergetics, Vol. 61   **On Self-Organization**
Eds.: R.K. Mishra, D. Maaß, E. Zwierlein   © Springer-Verlag Berlin Heidelberg 1994

parameter space, sometimes in the configuration space, may simplify the description. Some nonlinear problems may be embedded into higher dimensions as linear ones, some non-Markovian processes may be made Markovian by adding more stochastic variables; even a discretized version of a continuous model may turn out to be more complex.

2. Repeated use of simple rules may lead to more complex behaviour. Everyone knows that iteration of a quadratic polynomial may yield chaos, iteration of "complex" (yet another context of the word) maps may produce the beautiful patterns of the Julia and Mandelbrot sets. Simple nearest neighbour rules of cellular automata may simulate the complexity of universal computers, as was conjectured by S. Wolfram.

Furthermore, complexity appears in our description of Nature simply because we are not clever enough. For instance, the use of a wrong system of reference may bring about unnecessary complications, as was the case with the Ptolemy system compared to Copernicus. The modern notation of the Maxwell equations for the electromagnetic fields, e.g., using the notion of exterior differential d, codifferential δ and differential 2-form Θ:

$$
\begin{aligned}
d\Theta &= 0, \\
\delta\Theta &= J,
\end{aligned}
\tag{1}
$$

look much "simpler" than the original ones in Maxwell's paper in *Philosophical Transaction* or in his *Treatise*. In order to view things simply, you must stand high.

We see that the problem of complexity and its description are much like the problem of beauty and the appreciation of the beautiful. One needs a definite framework into which to set the problem and an objective way to estimate the complexity. Historically, the quantitative description of information has experienced similar problems. In a sense, it is a correct attitude to start with simple situations, where complex behaviour is generated by a comprehensible mechanism. If we cannot deal with these situations, there is little hope of coping with "real" complexities.

Comparison of various classes of orbits in one-dimensional mappings provides a first challenge toward this goal. Here is the arena where symbolic dynamics and its natural relation to formal language theory may play an instructive role. This development comes concurrently with the need to characterize various chaotic attractors at a more fundamental, "microscopic", level.

In recent years there has been a new upsurge of interest in quantifying and measuring complexity. Many meetings have been devoted to this problem, see, e.g., [1]–[6]. In particular, the complexity of chaotic attractors and trajectories have been the subject of many discussions. A number of new definitions of complexity have been suggested. We have in mind the use of grammatical complexity to characterize the complexity of cellular automata by Wolfram [7], the extensive study of grammatical complexity of symbolic sequences in one-dimensional maps and in cellular automata by Grassberger and his definition of "set complexity", "true measure complexity", "effective measure complexity" [8], the construction of logical trees and calculation of "grammatical complexity" in a narrower sense in [9, 10], and the "hierachical approach" in [11], etc. An essential advance is the understanding that some conventional measures of complexity, e.g., entropy, are in fact measures of randomness, not complexity. Although periodicity and pure randomness are two extremes on the scale of randomness, they are close to each other on the scale of complexity. Complexity is always associated with a certain degree of organization and lies somewhere between simple periodicity and pure randomness on the scale of randomness.

Nevertheless, we would like to note that the state of the field is far from satisfactory. For example, some of the definitions mentioned above assign infinite complexity to the orbits at the accumulation point of the period-doubling sequence of one-dimensional maps [8, 12], while we know that the orbits there are only quasiperiodic and the symbolic dynamics is quite simple. The definition in [10] gives finite complexity to the limit of so-called Fibonacci sequence, i.e., the sequence of orbits, whose periods grow as Fibonacci numbers. This limit is also a quasiperiodic orbit and the symbolic dynamics is even more simple. We will return to this issue later.

To provide a remedy for these shortcomings, some authors introduced hierachical definitions which yield non-zero complexity only for systems in higher than one dimensions [12], hence put aside the whole problem of describing complexity of orbits in one-dimensional maps. Anyway, one-dimensional dynamics may be a result of projection from higher dimensions and it may be even more complex.

Furthermore, while there are convincing arguments [8] that there must be infinitely many orbits in one-dimensional maps whose complexity should go beyond "regular language", the lowest step in the Chomsky ladder of language complexity [17], no explicit way of constructing or approaching these orbits has been indicated so far. On the other hand, some seemingly simple languages such as $R^n L^n$ or $R^n M^n L^n$ are more complex than regular (the former being non-regular, and the latter non-context-free), but they are not admissible in any known symbolic dynamics for one-dimensional maps.

We will not go into details of any of the definitions mentioned above, nor analyze their pros and cons. Our main aim is to stress the usefulness of transfer matrices as a bridge between the dynamics and the underlying grammar. Since the most workable definitions of complexity are related to symbolic dynamics and our discussion of grammatical complexity will also be restricted to the context of symbolic dynamics, a few words on symbolic dynamics may be in order.

2 Symbolic Dynamics

Symbolic dynamics is a coarse-grained description of dynamics. Instead of tracing the trajectories in the phase space in full detail, one divides ("coarse-grains") the phase space into a number of regions and labels each region with a letter from some alphabet. The total number may be finite or infinite; only the finite case can be studied in any great detail. The time evolution of a trajectory shows up as a series of letters; a numerical trajectory is replaced by a symbolic sequence. The correspondence may be many-to-one, thus opening the possibility for classification of trajectories. For theorem-proving, the exact way of partitioning does not matter in most cases. However, when one divides the phase space according to the "physics" or "geometry" of the dynamics, many detailed rules may be derived in the case of one- or two-dimensional mappings.

In a sense, symbolic dynamics is nothing but what experimental-physicists do every day. Using an analog-digital converter of, say, 12 bits precision, there are no more than 4096 different readings, i.e., symbols, from the instrument, yet one intends to draw reliable conclusions on invariant, robust properties of the system.

Symbolic dynamics as a chapter in abstract mathematics has a long history from the work of Morse in the 1920s; see [18, 19] for the development until the late 1970s. Applied aspects of symbolic dynamics have been developed mainly since the 1973 paper of Metropolis et al. [20] through the work of many mathematicians and physicists, see, e.g., [13, 15, 23]. Our group at the Institute of Theoretical Physics, Academia Sinica, Beijing, has made some contributions to symbolic dynamics of one-dimensional [21, 22] and two-dimensional [26]–[29] mappings, and in their application to systems, described by ordinary differential equations [21].

The essence of symbolic dynamics is rather simple. I will take a simple example to demonstrate some of the most important ingredients.

In the simplest case of unimodal maps, one divides the phase space (the interval) into two regions: one to the Left of the Central or Critical point, labeled by L, another to the Right of C, labeled by R. These three letters have a natural order

$$L < C < R, \qquad (2)$$

which is the basis for ordering all possible symbolic sequences. The ordering rule is simple. Suppose two symbolic sequences

$$\Sigma_1 = \Sigma^* \sigma \cdots$$

and

$$\Sigma_2 = \Sigma^* \tau \cdots$$

have a common leading part Σ^*, and the next letters σ and τ are different. Since they are different, they must have been ordered according to the natural order (2). Then the order of Σ_1 and Σ_2 are the same as σ and τ if the common leading part Σ^* contains an even number of the letter R; or the order be reversed if there is an odd number of R in Σ^*. This rule holds for any one-dimensional map, not only for unimodal ones, provided the R counting in Σ^* extends to counting all letters which represent a decreasing branch of the mapping function.

Although the letter C occurs only at one point, the sequence starting with C, or, to be more precise, starting with the first iterate of C, i.e., $f(C)$, where $f(x)$ is the mapping function, plays an important role. It has acquired a special name "kneading sequence" [13]. A map is best parametrized by its kneading sequence (or sequences, when there is more than one critical point). Given the kneading sequences, everything about the symbolic dynamics is determined.

Take, for example, a period 5 kneading sequence $(RLLRC)^\infty$ in the unimodal map. This is the symbolic sequence of the superstable period 5 orbit which starts from the critical point C of the map. It is a good convention to denote a symbolic sequence by the first number that starts the iteration. Since iterations of the map correspond to consecutive shifts of the symbolic sequence, we have the following alternation of numbers-sequences:

$$\begin{aligned}
x_0 &\equiv C = CRLRRC\cdots, \\
x_1 &= RLRRC\cdots, \\
x_2 &= LRRC\cdots, \\
x_3 &= RRC\cdots, \\
x_4 &= RC\cdots,
\end{aligned} \tag{3}$$

then it repeats. Using the ordering rule of symbolic sequences, it is easy to check that these sequences, and consequently the corresponding points, are ordered as follows:

$$x_2 < x_0 < x_3 < x_4 < x_1. \tag{4}$$

Besides the ordering rule, another important issue is the admissibility condition. Obviously, not every arbitrarily chosen symbolic sequence can correspond to a realizable trajectory in a given dynamics. One needs some criterion to check the admissibility. Referring to, e.g., [22]–[23], for a detailed formulation of the admissibility condition, we note only that in case of unimodal maps the condition reduces to shift-maximality of the symbolic sequences.

3 Coarse-Grained Chaos

A good example of how symbolic dynamics embodies the idea that complexity lies somewhere in between simple periodicity and pure randomness is the notion of coarse-grained chaos, based on a generalized composition rule [25].

We start from a superstable fixed point, which corresponds to the kneading sequence C^∞. By disturbing C a little, it goes either to R or to L, according to the natural order (2). Omitting the infinite power in the notations, we have a symbolic fixed point window:

$$(L, C, R). \tag{5}$$

Applying repeatedly the substitutions

$$\begin{aligned}
R &\to RL, \\
C &\to RC, \\
L &\to RR,
\end{aligned} \tag{6}$$

to the fixed point window (5), we get the symbolic representation of the whole period-doubling cascade.

The substitutions (6) hint at a generalization: find the conditions that the following substitutions

$$R \to \rho,$$
$$L \to \lambda,$$
(7)

ρ and λ being strings made of R and L, applied to any admissible symbolic sequence, would yield another admissible sequence. These conditions were found in [25] and they happened to be a natural generalization of the well-known $*-$composition in symbolic dynamics [15], hence the name generalized composition rule.

Now take the much-studied case of chaotic maps, namely, the map

$$x_{n+1} = 1 - 2x_n^2.$$
(8)

Its kneading sequence is RL^∞. It has many nice properties, for example:

1. There is a continuous distribution $\rho(x)$ for the orbital points x_i for almost all choices of initial points;
2. Each initial point leads to a different symbolic sequence; there are as many different symbolic sequences as real numbers in the interval $(-1, 1)$ (This is the symbolic statement of sensitive dependence on initial conditions);
3. There exist homoclinic orbits;
4. It is a crisis point;
5. It is a band-ending point, beyond which a chaotic band no longer exists;
6. It is a surjective map;

and so on.

Applying the generalized composition rule (7) to the kneading sequence RL^∞, we get infinitely many kneading sequences of the form $\rho\lambda^\infty$ for infinitely many different choices of ρ and λ. Almost everything said about the map with kneading sequence RL^∞ may be carried over to maps with kneading sequences $\rho\lambda^\infty$. The RL^∞ map is the most random map with maximal topological entropy. The $\rho\lambda^\infty$ maps have lower entropy, but if viewed with lower resolution, i.e., taking each of the strings ρ and λ for a single letter, they are as random as the RL^∞ map. However, if we look at them with higher resolution, we see patterns of organization, embodied in the repeated structures of ρ and λ. This is what we call coarse-grained chaos.

Now we have been prepared enough to introduce the transfer matrices.

4 Stefan Matrix

The main message I would like to convey [14] is that the transfer matrix, or Stefan matrix, as it is often called in this case [15], may be very useful both in deriving the grammatical rules from a given symbolic dynamics and in inferring the unknown rules from experimental data. Moreover, the Stefan matrix has a direct relation to the transfer functions used in constructing the non-deterministic finite automaton (NDFA), as well as the corresponding deterministic finite automaton (DFA), which accepts the language. The block structure of the matrix may serve as another indicator of complexity for the automaton and the language. In fact, the topological entropy calculation makes use only of the largest eigenvalue of the matrix, the construction of the automaton utilizes more information in the matrix. I will devote the rest of this paper to the explanation of what has been said.

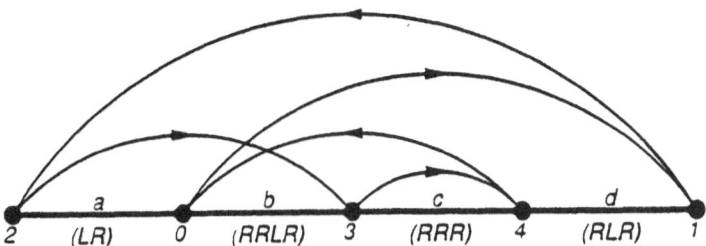

Fig. 1. Construction of the Stefan matrix for the period 5 kneading sequence $(RLRRC)^\infty$.

Although we are more concerned with the characterization of infinite aperiodic sequences, we continue with the example of the period 5 kneading sequence $(RLRRC)^\infty$ for simplicity of presentation. In fact, we will try to construct everything using this example, then the generalizations needed for infinite sequences will become clearer.

First of all, the ordered numbers in (4) are shown in Fig. 1. Also shown in the figure is the way the orbit visits these points. The latter divide the interval into four segments, denoted by a, b, c, and d. If one shifts the initial point of iteration from the superstable orbital points to other points in the interval, then continuity considerations alone tell us that the four segments will transform into one another in the following way:

$$
\begin{aligned}
a &\to c + d, \\
b &\to d, \\
c &\to b + c, \\
d &\to a.
\end{aligned}
\tag{9}
$$

Written in matrix form, they define the Stefan matrix

$$
S = \begin{vmatrix} 0\,0\,1\,1 \\ 0\,0\,0\,1 \\ 0\,1\,1\,0 \\ 1\,0\,0\,0 \end{vmatrix}.
\tag{10}
$$

The trace of S^n in the $n \to \infty$ limit, hence the largest eigenvalue of S, gives the number of different periodic orbits of length n. In fact, the logarithm of the largest eigenvalue, in our example $\lambda = 1.51288$, yields the topological entropy [16], which, as we have said, is rather a measure of randomness, not complexity. However, we will see that other approaches of relating complexity to the underlying grammar of the symbolic sequences just make more use of the same Stefan matrix, going beyond its largest eigenvalue.

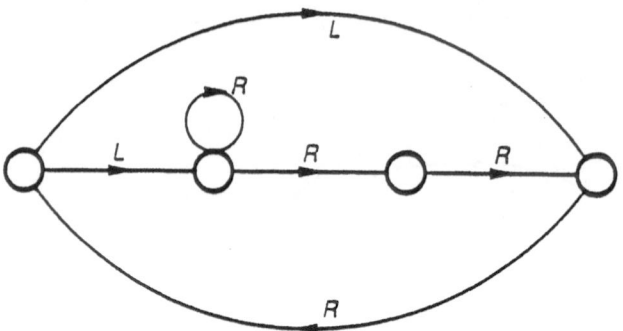

Fig. 2. Non-deterministic finite automaton constructed from the Stefan matrix of the period 5 kneading sequence $(RLRRC)^\infty$.

At each of the numbered points in Fig. 1 we have one of the symbolic sequences given in (3). They contain the letter C. By shifting from C to the left or to the right, we get two symbolic sequences, located on the two sides of the numbered points according to the ordering rule. By comparing the sequences at the two ends of each segment, we get the first few symbols for any sequences starting from that segment. These are the words (LR), $(RRLR)$, (RRR), and (RLR), written under a, b, c, and d, respectively, in Fig. 1. From the meaning of the transfer matrix we deduce that these 4 words appear in the following 6 contexts: $LRRR$, $LRLR$, $RRLR$, $RRRLR$, $RRRR$, and RLR, each corresponding to a '1' in the Stefan matrix.

In the rightmost column of the Stefan matrix S there are always two 1's, one on top of the other. They come from the right and left neighbourhood of the critical point C. Drawing a horizontal line between them, we see that the segments above this line are located to the left of C, and those below the line to the right of C. Taking the letters a through d as denoting different states of an automaton, we can take the Stefan matrix as the definition of transfer function for the automaton, see Table 1.

Table 1. Transfer function for the NDFA associated with $(RLRRC)^\infty$.

	L	R
a	$c + d$	
b		d
c		$b + c$
d		a

The table reads, for example, state a on accepting a letter L goes into either state c or state d, etc. Taking the states as nodes, we draw the graph in Fig. 2 to visualize the table. This is an NDFA, because, first, there is no distinguished node to start with, one can start traveling along the graph from any node; second, at some nodes there is more than one choice of where to go next on seeing the same input letter.

In formal language theory, see, e.g., [17], there is a standard method to derive DFA from NDFA — the subset construction. It is simpler to continue with our example than formulating the rules. Instead of treating single states such as a or b, we take a certain combination (subset) of states as a new state and see what happens according to the transfer function. Let us start with $\{abcd\}$, i.e., the set of all single states. On seeing the letter L it goes into the set $\{cd\}$, while on seeing the letter R it remains unchanged. We put these observations into Table 2 for a new transfer function:

Table 2. Transfer function for the DFA associated with $(RLRRC)^\infty$.

	L	R
$\{abcd\}$	$\{cd\}$	$\{abcd\}$
$\{cd\}$		$\{abc\}$
$\{abc\}$	$\{cd\}$	$\{bcd\}$
$\{bcd\}$		$\{abcd\}$

The automaton, drawn in accordance with this table, is shown in Fig. 3. Now there is a starting node, representing the state $\{abcd\}$. Beginning with the starting node, encircled twice in the figure, there is a unique choice as to where to go next on seeing an R or an L in the input. Amongst all the DFA accepting the same language there is one with a minimal number of nodes. Wolfram [7] took the logarithm of the minimal number of nodes as a measure of the complexity.

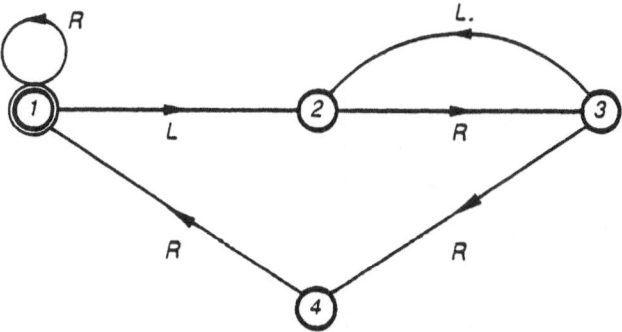

Fig. 3. Deterministic finite automaton corresponding to the period 5 kneading sequence $(RLRRC)^\infty$.

If we derive the Stefan matrix for a map with a $\rho\lambda^\infty$ type of kneading sequence, there will be a transient part in the final DFA. In order to deal with the stationary grammar, Grassberger [8] suggested to drop the transients. However, when we turn to infinite Stefan matrices and the infinite limits of the corresponding automata, sometimes the limit comes from the transient part of the finite automata. So some caution is required.

For any periodic kneading sequence $(\Sigma C)^\infty$ or eventually periodic kneading sequence $\rho\lambda^\infty$, at any admissible choice of the finite strings Σ, ρ, and λ, one always gets a finite automaton. Therefore, they belong to the lowest level of grammatical complexity — regular languages. In order to go beyond the regular level, one must turn to other types of kneading sequences, of which our knowledge is rather limited at present. A convenient way to look for a breakthrough is to construct a series of finite automata, then study the infinite limit.

Having mastered the construction of Stefan matrices and their relation to automata, both NDFA and DFA, we can work only at the transfer matrix level in our search for more complex limits. We continue with examples.

We first look at the $k \to \infty$ limit of $RL^k(RL^{k-1}R)^\infty$ type kneading sequences, which includes the period 3 band-merging point $RLL(RLR)^\infty$, studied in [10]. The limit is clear: RL^∞. The $(2k+2) \times (2k+2)$ Stefan matrix has a fixed structure with two simply growing parts. We show the matrix in Fig. 4. Consecutive '1's are drawn as a thick solid line and a single '1' is represented by a filled circle; all blanks are zero.

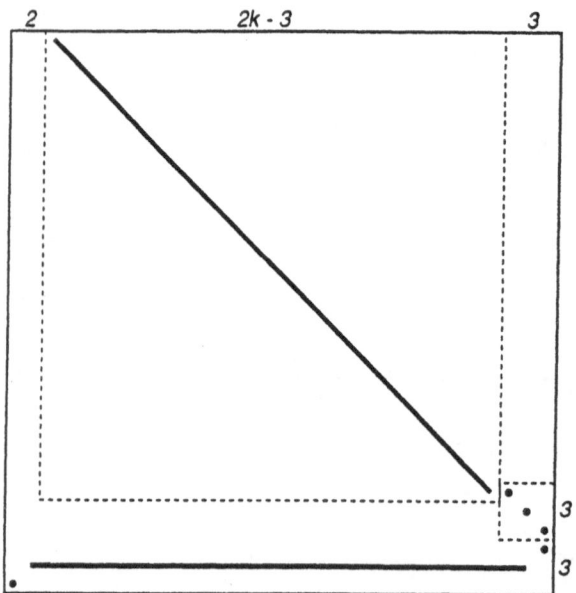

Fig. 4. The Stefan matrix for $RL^k(RL^{k-1}R)^\infty$.

We see that the structure of the matrix remains the same, only the size of one upper block and one horizontal line grows with k. This tells us that in the resulting automaton many similar nodes may be combined into one, and the grammar does not get more complex with growing k.

The next example is the limit of the Fibonacci sequence we mentioned before. Taking the letter R for Rabbit, and L for Little rabbit, and imposing a rule that after a time step a Rabbit gives birth to a Little rabbit while an existing Little rabbit grows up into a Rabbit, we get from a single R the following sequence:

$$R$$
$$RL$$
$$RLR$$
$$RLRRL \tag{11}$$
$$RLRRLRLR$$
$$RLRRLRLRRLRRL$$
$$\ldots$$

Note that the substitutions

$$R \to RL, \\ L \to R, \tag{12}$$

which generate the symbolic sequences in (11), do not satisfy the conditions of the generalized composition rule. Consequently, not all sequences in (11) are admissible. However, since we are interested in periodic orbits, it is always possible to make them shift-maximal by cyclic permutation. In this way we get the Fibonacci sequence. For a member of period F_n (the n-th Fibonacci number), the Stefan matrix is a $(F_n - 1) \times (F_n - 1)$ table. The case for $F_8 = 21$ is given in Fig. 5.

No matter how large the chosen Fibonacci number, this matrix always has the fixed structure shown above. Horizontally, from left to right, the block sizes are F_{n-2}, F_{n-3}, and $F_{n-2} - 1$; vertically, from top to bottom, the sizes are F_{n-3}, $F_{n-2} - 1$, and F_{n-2}. This implies that in the resulting automaton many nodes may be combined and the final effective automaton and its $F_n \to \infty$ limit cannot be more complex, a conclusion drawn in [10] with some effort by explicitly constructing the automata.

Our last example illustrates a case to which many authors assigned infinite complexity [8, 12], namely, the accumulation point of the period-doubling cascade. It is the result of repeatedly applying the substitutions (6) to (5). The symbolic dynamics notation of this limit, using the $*$-composition of [15], is simply $R^{*\infty}$. It can be reached either from the period 2^n orbits, or from the $2^n \to 2^{n-1}$ band-merging points. In the latter case all the symbolic sequences are of $\rho\lambda^\infty$ type and may be obtained by applying the substitutions (6) to the kneading sequence of the surjective map (8), i.e., to RL^∞. The construction of Stefan matrices is straightforward in both cases. We give the period 16 Stefan matrix in Fig. 6.

This matrix has a more complicated, yet quite regular, block structure. The number of blocks grows with n. For a period 2^n orbit, the block sizes are 2^{n-1},

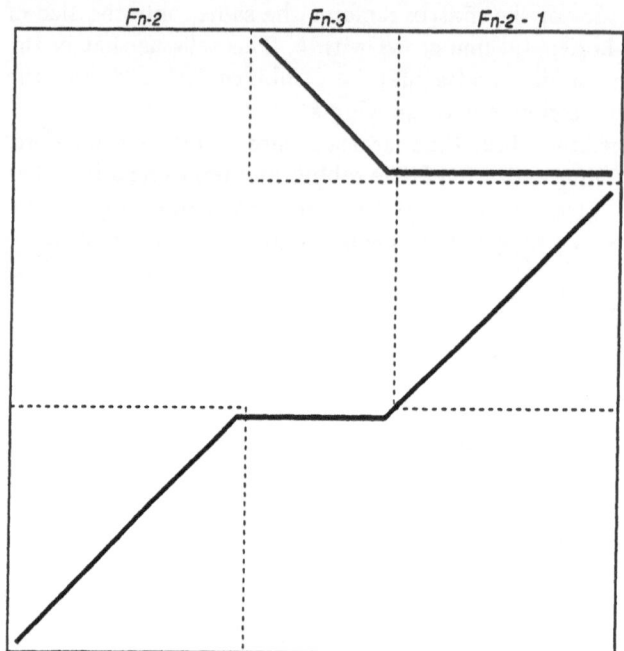

Fig. 5. The Stefan matrix for period F_n in the Fibonacci sequence.

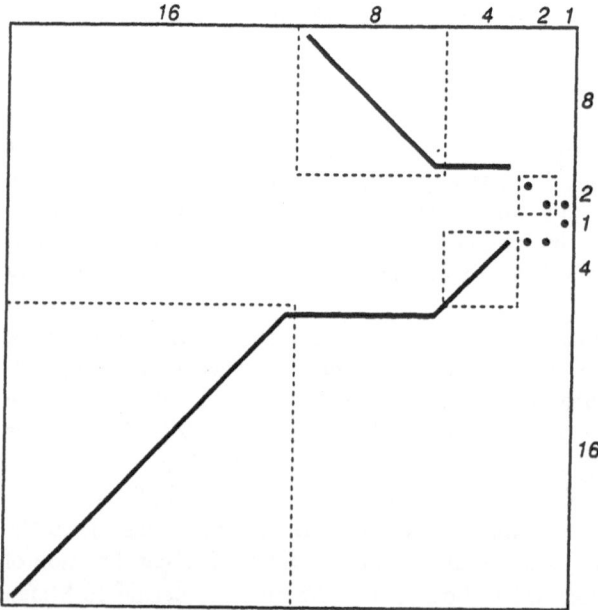

Fig. 6. The Stefan matrix for period 16 in the period-doubling cascade.

2^{n-2}, $\cdots$, 8, 4, 2, 1 along the horizontal, and the same numbers alternating along the vertical from top and bottom toward the two '1's in the rightmost column. Even when each block is represented by a single effective node in the resulting automaton, one still needs an infinite number of effective nodes to realize the automaton. No wonder some authors get infinite complexity for this case. It certainly goes beyond finite automaton, but the structure is too regular to be called "infinitely" complex. Is it reasonable that a quasiperiodic orbit at the accumulation of period-doublings is more complex than any chaotic orbit? In order to discover more complex orbits, one must look at other types of infinite Stefan matrices.

However, in doing so one should note that the overall structure of the Stefan matrices is subject to strong restrictions, originating from the underlying map. If one rotates the last three matrices anticlockwise by 90 degrees, there appears a kind of "monotonicity". To the left of the two central '1's, the location of '1's is "non-decreasing", while to the right it is "non-increasing". Of course, this observation holds only for unimodal maps.

Before concluding the paper, we would like to say a few words on the application of transfer matrices to experimental data. Based on the assumption that the system under investigation has been in a stationary state and the sampled data reflect a typical trajectory, one applies the standard phase space reconstruction technique and draws experimental Poincaré sections. The first return maps for one of the reconstructed coordinates may reveal the outline of an underlying map. If it is close to one-dimensional, as is usually the case with dissipative systems, then there is a good hope to introduce a partition, using a few letters. Then one can try to extract the grammatical rules from the symbolic sequence, obtained from the original time series. The simplest cases deal with two or three letters (in the presence of a discrete symmetry, as our experience with the Lorenz model shows [30]).

Suppose we are lucky enough to start with a symbolic dynamics of two letters, say, A and B, for Above and Below the average. The first thing to do is to count the occurrence of A and B, then the occurrence of the pairs AA, AB, BA, and BB, etc., and so forth. The absence of a pair, e.g., BB would imply a grammatical rule "two consecutive B's are not allowed in the language". This rule alone would cut the number of longer strings. The number of 3-letter strings would be reduced from 8 to 5; that of 4-letter strings from 16 to 8; that of n-letter strings from 2^n to F_{n+2} — the Fibonacci number again. This rule alone determines a smallest transfer matrix:

$$\begin{vmatrix} 1 & 1 \\ 1 & 0 \end{vmatrix}, \tag{13}$$

which, in turn, would give an upper bound for the topological entropy: logarithm of the golden mean $1.618\cdots$. If new rules were to be discovered during the counting of longer strings, they would change the number of allowed combinations of even longer strings, and yield larger approximants for the transfer matrices and better estimates for the entropy.

This way of thinking essentially follows the "pre-entropy" experiment of Kahlert and Rössler [31], and the best results so far, to our knowledge, have been obtained with the Belousov-Zhabotinskii reaction data by Lathrop and Kostelich [32], see also the recent work [33] along the line of Badii [9]. Nevertheless, the whole business of extracting grammatical rules and comparing the complexity of experimental chaotic attractors is still in its infancy.

As regards the application of symbolic dynamics to description of complexity in two and more dimensions, much more effort has to be made, since the symbolic dynamics itself has not been well-developed.

Acknowledgements

This work is a part of a project supported by the Chinese Natural Science Foundation and the Open Laboratories Project of Academia Sinica. The work has been continued and the paper written during a visit to the International Centre for Theoretical Physics, Trieste. A kind arrangement by the Kaiserlautern University encouraged me to summarize the work at its present stage. The author thanks all the above-mentioned institutions.

References

1. Livi R, Ruffo S, Ciliberto S, and Buiatti M (1988), eds., *Chaos and Complexity*, World Scientific.
2. Peliti L, and Vulpiani A (1988), eds., *Measures of Complexity*, Lecture Notes in Phys. **314**, Springer-Verlag.
3. J. Bernasconi, M. Droz, A. Malaspinas, and M. Ryter, eds., Proceedings of Conference on "Complex Systems", *Helv. Phys. Acta* **62** (1989), 461-630.
4. N. B. Abraham, A. M. Albano, A. Passamante, and P. E. Rapp, eds., *Measures of Complexity and Chaos*, Plenum, 1989.
5. Bohr H (1990), ed., *Characterizing Complex Systems — Interdisciplinary Workshop on Complexity and Chaos*, World Scientific.
6. Zurek W (1990), ed., *Complexity, Entropy, and Physics of Information*, Addison-Wesley.
7. S. Wolfram, *Commun. Math. Phys.* **96** (1984) 15.
8. P. Grassberger, *Int. J. Theor. Phys.* **25** (1986) 907; *Physica* **A140** (1986) 319; *Z. Naturforsch.* **43a** (1988) 671; *Helv. Acta Phys.* **62** (1989) 489; and in [2].
9. R. Badii, Europhys. Lett. **13** (1990) 599; and in [4], 313.
10. D. Auerbach, and I. Procaccia, *Phys. Rev.* **A41** (1990) 6602.
11. G. D'Alessandro, and A. Politi, *Phys. Rev. Lett.* **64** (1990) 1609.
12. J. P. Crutchfield, and K. Young, *Phys. Rev. Lett.* **63** (1989) 105.
13. J. Milnor, and W. Thurston, in *Dynamical Systems*, ed. by J. C. Alexander, *Lect. Notes in Math.*, No. 1342, Springer-Verlag(1988).
14. B.-L. Hao, *Physica* **D51** (1991) 161.
15. B. Derrida, A. Gervois, and Y. Pomeau, *Ann. Inst. H. Poincaré* **29A** (1978) 305.
16. J. P. Crutchfield, and N. H. Packard, *Physica* **D7**, (1983) 201.
17. J. E. Hopcroft, and J. D. Ullman, *Introduction to Automata Theory, Languages, and Computation*, Addison-Wesley, 1979.

18. M. Morse, and G. A. Hedlund, *Am. J. Math.* **60**, 815(1938); reprinted in *Collected Papers of M. Morse*, vol. 2, World Scientific, 1986.

19. A. M. Alekseev, and M. V. Yakobson, *Phys. Repts.* **75**, (1981) 287.

20. N. Metropolis, M. L. Stein, and P. R. Stein, *J. Combinat. Theory* **A15** (1973) 25.

21. Hao Bai-lin, *Elementary Symbolic Dynamics and Chaos in Dissipative Systems*, World Scientific, 1989.

22. W.-M. Zheng, and B.-L. Hao, "Applied Symbolic Dynamics", in *Experimental Study and Characterization of Chaos*, vol. 3 of *Directions in Chaos*, ed. by Hao Bai-lin, World Scientific, 1990, 363-459.

23. P. Collet, and J.-P. Eckmann, *Iterated Systems on the Interval as Dynamical Systems*, Birkhäuser, Boston, 1980.

24. B.-L. Hao, and W.-M. Zheng, *Int. J. Mod. Phys.* **B3**, (1989) 235.

25. Wei-mou Zheng, J. Phys. **A22** (1989), 3307.

26. Wei-mou Zheng, *Chaos, Solitons and Fractals*, **1** (1991) 243.

27. Wei-mou Zheng, "Symbolic dynamics for the Tél map", ASITP Preprint 92-001.

28. Hong Zhao, Wei-mou Zheng, and Yan Gu, "Determination of partition lines from dynamical foliations for the Hénon map", ASITP Preprint 92-002; Hong Zhao, and Wei-mou Zheng, "Symbolic analysis of the Hénon map at $a = 1.4$ and $b = 0.3$", ASITP Preprint 92-040.

29. Wei-mou Zheng, *Chaos, Solitions and Fractals*, **2**(1992)461

30. M.-Z. Ding, and B.-L. Hao, *Commun. Theor. Phys.*, **9** (1988) 375.

31. C. Kahlert, and O. E. Rössler, *Z. Naturforsch.* **39a** (1984) 1200.

32. D. P. Lathrop, and E. J. Kostelich, *Phys. Rev.* **A40** (1989) 4028; and in [4], 147.

33. M. Finardi, L. Flepp, J. Parisi, R. Holzner, R. Badii, and E. Brun, *Phys. Rev. Lett.* **68** (1992) 2898.

Instabilities in Nonlinear Dynamics: Paradigms for Self-Organization

R. K. Bullough

1 Introduction

This paper is concerned with aspects of nonlinear dynamics which may (but indeed may not) have bearing on our ultimate understanding of living matter. It is concerned with *dynamical self-organization* as well as with aspects of disorganization. A main theme is to show that rather simple causes can have complex effects. These complex consequences are all self-organized in so far as they arise as consequences of the equations governing a dynamics which is stimulated in very simple ways.

This paper looks first of all at ways in which instabilities in dynamical nonlinear wave systems lead to the spontaneous creation of coherent (or more-or-less coherent) pulses called generically 'solitons'. As used in this paper 'solitons' includes both true solitons, which are solutions of *integrable* dynamical systems [1], and those closely related 'solitary waves', solutions of non-integrable systems [1], which appear to display comparable properties. Solitons of this generic type form dynamically evolved patterns which *are* self-organized, show some considerable measure of coherence, and persist in time. Though that organization is relatively simple and primitive, the self-organized patterns can be both complicated and very stable against perturbations, including in these thermal perturbations which have a complex incoherent (chaotic) dynamics. It is hard to believe that the simple self-organizations described in this paper do not have *some* bearing on the complex self-organized dynamics apparently needed to describe living matter.

In their simplest description solitons and their related solitary waves move in one space dimension (x) and one time dimension (t). I call this $1+1$ dimensions. Integrability may be broken in $2+1$ dimensions or $3+1$ dimensions and the instabilities and resultant motions are more complicated. More-or-less coherent patterns can still emerge but these are also more complex and this is the study of the second part of the paper. In the final section, Sect.7, effects of Brownian motion (thermal perturbations) on the $1+1$ dimensional strictly integrable models are considered. These have significance to any role for the true solitons of integrable dynamical systems in $1+1$ in the dynamics of living matter.

Springer Series in Synergetics, Vol. 61
Eds.: R.K. Mishra, D. Maaß, E. Zwierlein **On Self-Organization**

In the spirit of some aspects of this meeting I make a few remarks on living matter and the roles of *complexity, understanding* and *consciousness* in living matter in the Sect. 2 next. But except for the return in the last section, Sect. 7, of the paper to some structural aspects of biological molecules, proteins, DNA which are essentially one dimensional, I do not try seriously to pursue this tantalising problem of living matter further in this paper.

2 Complexity and Living Systems: Self-Organization and Complexity

Certainly living systems would seem to be remarkable examples of *complex* self-organized systems. In this *intrinsically* complex context [2], and in some antithesis to this paper, self-organization is thought of as a process enabling creation, reproduction and perfection in its environment, of a complex dynamical system complex being of the essence and having its usual meaning, namely [3] 'complicated' or in an 'involved condition'. Complex is used qualitatively rather than quantitatively though it may be possible to attach a number for a threshold of this kind of complexity at which significant organization of this complex type can occur and below which it cannot. The dynamics of such systems may be governed by the currently accepted laws of physics, including quantum physics, or it may involve further aspects, or qualitatively new concepts, of which at present we have no understanding.

As examples of living systems human beings seem to have extraordinary capacities of intellect, of aesthetics, of belief, of creativity and imagination, which the known laws of physics apparently do not in themselves describe. What seems certain is, that over the last 10 000 years or so human beings have ascertained just these laws of physics; and, through their *understanding* of these, used them in the last 3 000 years (and more so during the last 300 years) to make the technological advances which have enabled them dramatically to change their relationship with the environment within which they have by self-organization organized themselves. This is both a local self-organization of the human organism *and* a global self-organization organized by cooperation between many organisms organized thus for good or ill. Such complex self-organizations may indeed involve thresholds of complexity for they involve reproduction at that complexity and the evolution of *that* may trace back some 7×10^9 years to the actual origin of life itself perhaps a much less complex event? Certainly life's reorganization about an oxygen cycle in the early Cambrian period some 5×10^8 years ago involved a global threshold for complexity.

'Understanding' likewise seems beyond the range of current physics: quantum theory, unlike classical dynamics, must take into account the role of the observer in relation to the observed (and it is the *human* observer since, apparently, no other observer on this planet conceives of quantum mechanics). Observation, in quantum mechanics, means measurement, and measurement means, in the simplest interpretation of quantum mechanics [4] a 'collapse of the wave packet' which breaks the causal evolution of the dynamics. Quantum mechanics is thus incomplete in this aspect and it can only *begin* to impinge on any idea we have of 'understanding'. Understanding apparently involves *consciousness* (a realization of its own existence?) and physics says nothing about this either. Considerable complexity is apparently involved in consciousness but can it be equated to it – however weakly? [2]. At most only above some critical threshold, perhaps?

3 Dynamics of the Integrable and Near Integrable Systems, Quantum and Classical

Against the background sketched in Sect. 2 this paper is concerned with self-organization at a naïve level. It is entirely physically based and is concerned with classical and quantum dynamics as these are currently understood. The dynamical systems are particularly simple for they are either integrable or close to being integrable and have the soliton solutions – though, as remarked, simple perturbations can induce complicated changes. I focus most attention first of all on Hamiltonian dynamical systems which are closed systems and within these the 'completely integrable' Hamiltonian systems which is what I mean by integrable. These are heavily constrained dynamical systems rare even amongst Hamiltonian systems yet they approximate to much of modern physics. As a definition [5] for $N < \infty$ degrees of freedom a classical completely integrable Hamiltonian system has N independent constants of the motion $I_k; k = 1, \ldots, N$; which are in involution, i.e., commute under the Poisson bracket $\{\ ,\ \}$: $\{I_k, I_\ell\} = 0$. Technically the system is equipped with a $2N$ dimensional phase space M to which the motion is confined: M is a $2N$ dimensional symplectic manifold [5] with a symplectic differential 2-form ϖ and the bracket $\{\ ,\ \}$ defined upon it: $d\varpi = 0$. If, [5], the manifold of level lines I_k = constant is also compact, motions become very simple and lie on N dimensional tori. The accessible phase space is then reduced to a $2N - N = N$ dimensional hyper-surface in M, the N dimensional tori. The motion on the $2N - 1$ dimensional energy surface $H = E$ is then not ergodic because of the $N - 1$ additional constraints I_k (H is itself a constant).

However, new features I consider are: (i) large systems, $N \to \infty$ (evidently $N < \infty$ for the physical realisation of living matter but 'consciousness' may be *unbounded*?); (ii) corresponding quantum systems; (iii) specifically *nonlinear* systems. I try to illustrate aspects of these features in this short paper.

It is (iii) (nonlinearity) which introduces apparent complexity into the observable motions viewed in the space-time framework. I also comment (iv) on large non-integrable nonlinear systems, but only in so far as these remain close to the integrable (completely integrable) Hamiltonian systems[4]. Antoniou [7] considers large ($N \to \infty$) non-integrable and non-Poincaré integrable Hamiltonian systems: they have no *analytic* invariants (constants of the motion) and no

---

[4] The non-integrable Ginsberg-Landau equation (36) in both 1+1 and 2+1 dimensions, and the equations in Sect. 6 after that, retain [6] structures which can be traced back to the constants of the motion of the integrable systems.

invariants. They thus embrace the chaotic dynamics of Brownian motion which I consider, necessarily briefly, in Sect.7 of this paper.

For $N < \infty$, generic classical Hamiltonian systems are neither integrable nor ergodic [8]. Crucial aspects of the dynamics are determined by 'fixed points': two such points are *stable* elliptic fixed points and *unstable* hyperbolic fixed points. Motions are on tori, called invariant tori, about elliptic points and stay near these points; motions move away from hyperbolic fixed points. Examples of both types are shown in Fig. 1.

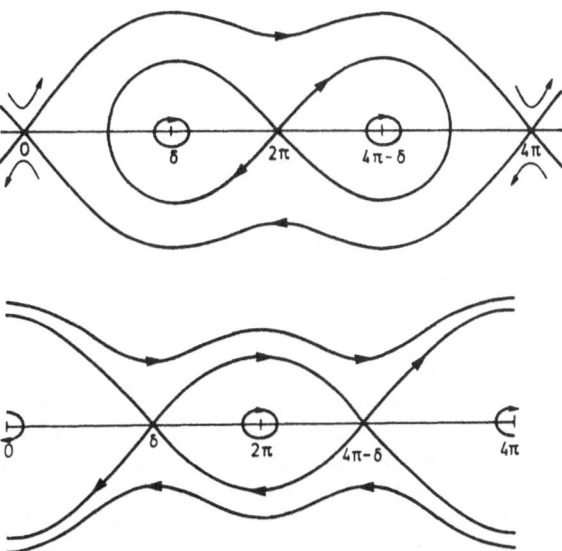

Fig. 1. Examples of hyperbolic and elliptic fixed points. The two figures actually show motions in the two-dimensional phase space for the 'double sine-Gordon equation', equation (31) below. In the upper figure there are hyperbolic points at $0, 2\pi, 4\pi$ and elliptic points marked at δ and $4\pi - \delta$ where $\delta = 2\cos^{-1}(-\frac{1}{4}\varepsilon)$, and $\varepsilon = 1$. In the lower figure the hyperbolic points are at $\delta, 4\pi - \delta$ and the elliptic at $0, 2\pi, 4\pi$.

Generic ergodic systems are Anosov systems with (for $N < \infty$) dense sets of hyperbolic fixed points [9]. But generic Hamiltonian systems have dense sets of hyperbolic *and* of elliptic fixed points. This is the KAM theorem (for $N < \infty$) [9]–[13], which shows that if the integrable Hamiltonian $H[I_1, \ldots, I_N]$ is perturbed by the non-integrable $\varepsilon H_1[I_1, \ldots, I_N; \phi_1, \ldots, \phi_N]$ (in which the ϕ_k are canonical to the I_k, as 'angle variables' [5]) then invariant tori, whose frequencies are sufficiently incommensurate persist densely about an elliptic point as ε increases from $\varepsilon = 0$. The motion is amazingly complex [9, 12]. An example essentially for $N = 2$ is shown in Fig.2.

The corresponding situation for large systems ($N \to \infty$) is open: the KAM theorem as it is described in [7] does not extend to $N \to \infty$ (apparently "resonances" appear "everywhere" and resonances destroy the KAM tori). Neverthe-

Fig. 2. Complexity of non-integrable Hamiltonian dynamics. Blown up to the bigger scale the square box would show features essentially identical to those in the larger figure. There is thus 'self-similar' structure on all possible scales and infinitely many tori. The plot is a plot of p against q found through the 'standard map' $p_{n+1} = p_n - \frac{K}{2\pi} \sin(2\pi q_n)$, $q_{n+1} = p_{n+1} + q_n$ (mod unity for p_n, q_n). This map is the discrete time form of the nonlinear pendulum equation $2\pi \ddot{q} = -\omega^2 \sin(2\pi q)$ reached for $K = O((\Delta t)^2)$ and $\Delta t \to 0$, and this pendulum equation is integrable. The invariant tori are therefore lines $p_n = \text{constant} = \alpha$ with $K = 0$; and with α irrational, some of these remain, distorted, as the invariant KAM tori for $K > 0$. The figure, taken from 'Nonlinear Science : from Paradigms to Practicalities' by David K. Campbell, Los Alamos Science, No.15 has $K = 1.1$ and is due to James Kadtke and David Umberger of the Los Alamos National Laboratory. This original figure is in colour and the complete trajectories in the chaotic regions can be identified by their colours.

less solitary waves can still be found. True solitons form certain generalisations (Sect.5) of elliptic fixed points for integrable systems, and solitary waves involve constants of the motion for large non-integrable systems[5].

Extensions to quantum completely integrable systems are [14] that for N quantum degrees of freedom there should be N independent operators $\hat{I}_k$, commuting under the commutator: $[\hat{I}_k, \hat{I}_\ell] = 0$. There is no Arnold theorem: motions do not lie on tori or generalised tori because the uncertainty principle means that precise trajectories in the phase space cannot be defined. However there *is* a mapping from the classical trajectories, which for integrable systems lie on tori, to

---

[5] Other results in infinite dimensional KAM theory are reported in a second note added in proof at the end of this paper.

their corresponding quantum paths. For a large infinity of degrees of freedom [15]

$$G(\phi(x,0), \phi(x,T); T) = \int \mathcal{D}\Pi\mathcal{D}\phi \exp \mathcal{S}[\Pi, \phi] \tag{1}$$

is a *quantum* 'propagator' (Green's function). The action S is the *classical* action on the phase space coordinatised by Π, ϕ:

$$S = \frac{i}{\hbar} \int_0^T d\tau \left[\int \Pi\phi_{,t}\, dx - H[\Pi, \phi] \right], \tag{2}$$

and H is the classical Hamiltonian; Π, ϕ are canonical, $\{\Pi, \phi\} = \delta(x - x')$. This propagator becomes the quantum partition function Z by formal Wick rotation in the time $t \to -i\tau : T = -i\beta$ (where $\beta^{-1} = k_B T$ with T now the *temperature* and k_B Boltzmann's constant) and

$$Z = Tr \int \mathcal{D}\Pi\mathcal{D}\phi \exp \mathcal{S}[\Pi, \phi] \tag{3}$$

in which Tr means 'trace'. The action S is now Wick rotated to

$$S = \frac{1}{\hbar} \int_0^{\beta\hbar} d\tau \left[i \int \Pi\phi_{,\tau}\, dx - H[\Pi, \phi] \right]. \tag{4}$$

Technically $\mathcal{D}\Pi\mathcal{D}\phi$ in (1) and (3) means functional (or path) integration [15]-[17] and is the measure for that integration, and I adopt the notation $\phi_{,t} = \partial\phi/\partial t$, a notation I use in the form ϕ_t systematically below. Explanation of this symbolism ((1) - (4)) for the non-expert is in e.g. [15]-[17]. For our purposes Z *is* the quantum partition function for a large Hamiltonian system ($N \to \infty$) integrable or non-integrable and the Helmholtz free energy is $F = -\beta^{-1}\ell n Z$.

There is the equivalent *Feynman* path integral [15]-[17] defined on a Hilbert space not a symplectic manifold. The integral in Π in (3) can be carried out for H quadratic in Π to leave

$$Z = Tr \int \mathcal{D}\phi \exp \mathcal{S}[\phi], \tag{5}$$

and

$$S[\phi] = \hbar^{-1} \int_0^{\beta\hbar} d\tau \int L[\phi, \phi_x, \phi_\tau]dx \tag{6}$$

is the *Euclidean* action (e.g. $\int L[\phi, \phi_x, \phi_\tau]dx = \int[\frac{1}{2}\phi_\tau^2 + \frac{1}{2}\phi_x^2 + V(\phi(x,\tau))]dx$ the "energy"). Then $Z = TrK \equiv \int \mathcal{D}\phi_0 K[\phi(x,0), \phi(x,0), \beta\hbar]$ where $\phi_o \equiv \phi(x,0)$ and $K[\phi(x,0), \phi(x,T), T]$ is the density matrix ρ [16] now for an infinite dimensional system. It is thus a Green's function [16] satisfying the Fokker-Planck-Kolmogorov equation

$$\frac{\partial K}{\partial T} - HK = \prod_x \delta(\phi(x,T) - \phi(x,0))\delta(T) \tag{7}$$

and

$$H \equiv \int dx \{ -\hbar \delta^2 / \delta \phi(x)^2 + \frac{1}{2} (\phi_\tau)^2 + V(\phi(x, \tau)) \} \tag{8}$$

in which the first term acts as an infinite dimensional Laplacian operator. This mathematical structure for infinite dimensional systems is developed in e.g. Ref [17] and its many references. Its mention here is relevant to the ideas of this paper since (7) is diffusive rather than wave-like. It shows that the measure $\mathcal{D}\phi$ in (5) is of Wiener type. It describes the (infinite dimensional) Gaussian chaos of Brownian motion.

For completely integrable Hamiltonian systems in one space and one time dimension $(1 + 1$ dimensions) we can calculate Z exactly [17] despite an evident complexity. Our methods exploit the classical integrability of these systems by making a canonical transformation to new simpler coordinates, the classical action-angle variables, and the 'simplicity' of these is the subject of Sect.5. The quantum integrability is nevertheless broken by a coupling to the infinite dimensional chaos provided (conceptually) by a heat bath at definite temperature β^{-1}. The dynamical system is open and in this one respect simulates the dynamics of living systems. It is the Wiener measure deriving from (5) which describes the chaos. There is a peculiar situation here since the chaos is ergodic and, depending on the status of the KAM theorem which we discuss again in Sect.5, generic large (infinite dimensional) Hamiltonian systems may not be ergodic and are certainly not ergodic in the integrable cases - except in fact through the breaking of the integrability induced by the action of the heat bath. We shall see in the Sect.7, in a brief sketch of the thermodynamics of integrable systems, that the classical tori still control this complicated thermal equilibrium (Z is of course evaluated at thermal equilibrium): a complex dynamical probabilistic self-organization nevertheless takes place about the steady classical trajectories in this equilibrium and that is one reason for exposing the reader to the equations (1) to (8) in this paper. Notice the notions of complexity and probabilistic character in this self-organization [2]. Let us now return next nevertheless to the simplicity of the wholly unperturbed motions of large integrable systems.

4 Some Integrable Models

The well-known classical integrable models in $1+1$ dimensions include [18] (recall the notation is $\phi_{xx} \equiv \partial^2 \phi / \partial x^2$ etc.)
1. The sine-Gordon (s-G) model

$$\phi_{xx} - \phi_{tt} = m^2 \sin \phi; \tag{9}$$

$\phi = \phi(x, t)$ is a real field (written $\phi \in \mathbb{R}$) and it has true soliton solutions (described at (27), (28) below); m is a real valued mass, or wave number.
2. The sinh-Gordon (sinh-G) model

$$\phi_{xx} - \phi_{tt} = m^2 \sinh \phi, \tag{10}$$

which is completely integrable but has no soliton solutions.

3. The nonlinear Schrödinger (NLS) models

$$- i\phi_t = \phi_{xx} - 2c\phi^*\phi^2 \tag{11}$$

in which ϕ is now complex ($\phi \in \mathbb{C}$) and $c \in \mathbb{R}$ is a coupling constant. For $c > 0$ the model is the *repulsive* NLS model with no soliton solutions; for $c < 0$ it is the attractive NLS model which has the soliton solutions

$$\phi = q \, sech[q(x - Vt)] \exp i[(q^2 - \frac{1}{4}V^2)t + \frac{1}{2}Vx] \tag{12}$$

with $q \in \mathbb{R}$.

4. The famous Korteweg de Vries (KdV) equation [1] always described in terms of the real field $u(x,t)$ (not $\phi(x,t)$!) and which in general form is $c^{-1}u_t + u_x + \alpha u_{xxx} + \delta u u_x = 0$ in which c is a speed (a sound speed). This can be scaled in the frame moving at c to

$$u_t + 6uu_x + u_{xxx} = 0. \tag{13}$$

Under vanishing *boundary conditions* at $x = \pm\infty$ this has the soliton solution

$$u(x,t) = 2\xi^2 sech^2\{\xi(x - 4\xi^2 t)\} \tag{14}$$

where ξ is a real parameter: (14) is both a solitary wave but also a true soliton, for, first of all, the amplitude $2\xi^2$ relates directly to the speed $4\xi^2$ at which this pulse travels up the x-axis - "bigger pulses travel faster" a property generic for other solitons (but not however of the soliton (12)). One can expect, and this is what happens, that arbitrary initial data $u(x,0)$ spontaneously *break up* into a succession of such solitons travelling at their different speeds. This is indeed the dynamical self-organization of solitons which is a substantial theme of this paper.

We list some other models and return to this dynamical self-organization below. The model (5.) is:-

5. The modified Korteweg de Vries (mKdV) equation

$$u_t + 6u^2u_x + u_{xxx} = 0. \tag{15}$$

This has soliton solutions which are *sech* functions not *sech²* functions as for KdV: they follow the NLS soliton (12) in this respect. These mKdV solitons retain the soliton property of spontaneous creation from arbitrary initial data $u(x,0)$.

Other classical integrable models with soliton solutions include [18]

6. The Landau-Lifshitz model in a vector field $\vec{S}$

$$\vec{S}_t = \vec{S} \times \vec{S}_{xx} + \vec{S} \times \Im . \vec{S} \tag{16}$$

where $\Im = diag\ (J_1, J_2, J_3)$, a diagonal second rank tensor, and $\times$ is vector product;

7. The Heisenberg ferromagnet

$$\vec{S}_t = \vec{S} \times \vec{S}_{xx} \tag{17}$$

which is (6.) with $\Im$ the unit tensor;

8. The classical massive Thirring model in the spinor field ψ

$$(i\partial_\mu \gamma^\mu + m)\psi + g^2\gamma^\mu \psi(\overline{\psi}\gamma_\mu \psi) = 0 \tag{18}$$

where $\overline{\psi} = \psi^*\gamma^0, \gamma^0 = \sigma_1, \gamma^1 = i\sigma_2$, with σ_1, σ_2 the two Pauli matrices, and g is a coupling constant;

9. The Toda *lattice*

$$q_{n,tt} = a(e^{-br_n} - e^{-br_{n-1}}) \tag{19}$$

where $a, b \in \mathbb{R}$ and $r_n \equiv q_n - q_{n-1}$. There are many other classical integrable *lattices* [17, 18] and see also [19, 20] in 1+1 dimensions. There are also classical integrable models in 2+1 dimensions of which an example is the Davey-Stewartson system in two fields ϕ, w

$$i\phi_t + \frac{1}{2}(\phi_{yy} + \alpha^2\phi_{xx}) = -\alpha^2\beta \mid \phi \mid^2 \phi + w\phi,$$
$$w_{xx} - \alpha^2 w_{yy} = 2\alpha^2\beta(\mid \phi^2 \mid_{xx}); \tag{20}$$

$\beta = \pm 1, \alpha = 1$ is DS-I; $\alpha = +i$ is DS-II [21]. There are no soliton solutions under vanishing boundary conditions at infinity imposed on both ϕ and w but so-called 'dromion' solutions exist for w, but not ϕ, non-vanishing at infinity [22]. For ϕ independent of $y, w = 2\alpha^2\beta \mid \phi \mid^2$ and the DS system in $2 + 1$ becomes the NLS system in $1+1$. It is relevant to the instability theories sketched in the two following Sects. 5,6, that

$$i\phi_t + \phi_{yy} + \phi_{xx} - 2c\phi^*\phi^2 = 0, \tag{21}$$

which is the two-dimensional NLS model, is *not* integrable and with $c < 0$ (attractive case) does not have stable soliton solutions: these blow up in a finite time but this leads to an interesting *sequence* of self-organizations (Sect. 6).

The classical integrable models in $1 + 1$ can be quantised to quantum integrable models: the NLS models, for example, have Hamiltonians

$$H = \int [\phi_x^*\phi_x + c\phi^{*2}\phi^2]dx \tag{22}$$

with Poisson bracket $\{\phi, \phi^*\} = i\delta(x - x')$. They are quantised by the Dirac canonical quantisation

$$i\hbar^{-1}[\phi, \phi^\dagger] = i\delta(x - x') \tag{23}$$

and are to be normally ordered for physical sense. Thus (11) normal orders to

$$-i\phi_t = \phi_{xx} - 2c\phi^\dagger\phi^2. \tag{24}$$

For $c > 0$ (no quantum solitons) this model is the quantum bose gas model of [23, 24]. It was solved by the method of Bethe ansatz in [23] and its quantum statistical mechanics given in [24]. In this analysis the partition function Z was not evaluated: an energy E and an entropy S were defined, and $F = E - \beta^{-1}S$ was obtained by a minimisation procedure. This calculation of F coincides with that found later [25] by evaluating the partition function Z. Notice that both calculations of the free energy F directly or indirectly associate an entropy S with this thermal equilibrium and thus with the equilibrium dynamics. This entropy derives from the Wiener measures introduced in Sect. 3, and is a quantitative measure of the chaos of Brownian motion (Sect.7).

5 The Simplicity of the Integrable Models

The dynamical self-organization exhibited by the solitons of KdV is generic for integrable systems with soliton solutions. This complex self-organization can be seen to be the result of a very simple underlying dynamics *expressed in the proper dynamical variables.*

This *simplicity* is well exemplified by the s-G model equation (9). The model is Hamiltonian and H can be taken in the form

$$H[\Pi, \phi] = \gamma_0^{-1} \int \left[\frac{1}{2}\gamma_0^2 \Pi^2 + \frac{1}{2}\phi_x^2 + m^2(1 - \cos\phi) \right] dx. \tag{25}$$

With the bracket $\{\Pi, \phi\} = \delta(x - x')$, Hamilton's classical equations of motion are

$$\phi_t = \{H, \phi\} = \gamma_0 \Pi$$
$$\Pi_t = \{H, \Pi\} = \gamma_0^{-1}(\phi_{xx} - m^2 \sin\phi), \tag{26}$$

so $\phi_{tt} = \phi_{xx} - m^2 \sin\phi$, which is the s-G equation (9).

This s-G model has three kinds of soliton solution, the kinks and antikinks

$$\phi(x, t) = 4\tan^{-1} exp\{\pm m(x - Vt)(1 - V^2)^{-\frac{1}{2}}\} \tag{27}$$

where V is a speed, and the breather solution

$$\phi(x, t) = 4\tan^{-1}(\tan\mu \sin\Theta_I sech\Theta_R)$$

with

$$\Theta_R \equiv m(\sin \mu)(x - Vt)(1 - V^2)^{-\frac{1}{2}}$$
$$\Theta_I \equiv m(\cos \mu)(t - Vx)(1 - V^2)^{-\frac{1}{2}}. \qquad (28)$$

Unless the initial data $\phi(x, 0), \phi_t(x, 0)$ correspond to *precisely* one of these solutions under the boundary conditions $\phi \to 0 \pmod{2\pi}$ as $x \to \pm\infty$(N.B. in (27) the kink has $\phi \to 0, x \to -\infty$ but $\phi \to 2\pi, x \to +\infty$ and the antikink, as written, has $\phi \to 2\pi, x \to -\infty$ and $\phi \to 0, x \to +\infty$, which is $\phi \to 0, x \to -\infty$ and $\phi \to -2\pi, x \to +\infty \pmod{2\pi}$) then that initial data is unstable to the emergence of a combination of such solitons characterized by the different speeds V. Thus in particular the kinks and antikinks, all self-organized, all travel at different speeds V, and the self-organized pattern at $x \to \infty$ as $t \to \infty$, for example, is a succession of these objects ordered according to decreasing V. Note that (27) means $\phi_x = \pm 2m(1 - V^2)^{-\frac{1}{2}} sech\{\pm m(x - Vt)(1 - V^2)^{-\frac{1}{2}}\}$ so that these *pulses* ϕ_x, with the boundary conditions $\phi_x \to 0$ as $x \to \pm\infty$, have larger amplitude for larger V and again "bigger pulses travel faster". Note that the speeds V are scaled so $0 < V < 1$: the s-G equation (9) is covariant with a 'velocity of light' $c = 1$. One can thus envisage two such kinks starting at $x = -\infty$ at $t = -\infty$ ordered by velocities $V_1 > V_2$ so that kink 1 overtakes kink 2 in a collision region near $x = 0$, subsequently reaching $x = +\infty$ at $t = +\infty$ in reverse order. All of this is actually contained in the exact analytical formula for the two-kink solution of s-G which is [1, 26]

$$\phi(x, t) = 4 \tan^{-1}\left[\frac{\sinh\{\frac{1}{2}(\Theta_1 + \Theta_2)\}}{(a_{12})^{\frac{1}{2}} \cosh\{\frac{1}{2}(\Theta_1 + \Theta_2)\}}\right] \qquad (29)$$

in which $\Theta_1 \equiv (x - V_1 t)(1 - V_1^2)^{-\frac{1}{2}}, \Theta_2 \equiv (x - V_2 t)(1 - V_2^2)^{-\frac{1}{2}}$. As $t \to \pm\infty, \phi(x, t) \sim 4 \tan^{-1} exp\{\Theta_1 + \delta_1^{\pm}\} + 4 \tan^{-1} exp\{\Theta_2 + \delta_2^{\pm}\}$ and $\delta_1^{\pm} = -\delta_2^{\pm} = \pm\frac{1}{2} \ell n \, a_{12}$. There are therefore 'phase shifts' $\delta_1^+ - \delta_1^- = -(\delta_2^+ - \delta_2^-) = \ell n \, a_{12}$. Otherwise the two kinks pass through each other without any change despite the strong nonlinearity of their interaction. In this picture the kinks, organized at $x \to -\infty$ as $t \to -\infty$, become apparently disorganized at $x = 0$ and reorganize themselves at $x = +\infty$ as $t \to +\infty$. However the analytical formula (29) shows there is a *very well organized* coherent behaviour throughout the whole motion.

One could also choose initial data to illustrate actual break-up for $x \to -\infty$. For any (x, t) initial data for s-G are $\phi(x, t), \phi_t(x, t)$ and generic initial data break up for times greater than t into some set of kinks, antikinks, breathers and yet something else – namely weak propagating oscillatory motion [1]; we shall call this oscillatory motion 'phonons' in this paper. All of the solitons, kinks, antikinks and breathers, move through the phonons and each other without loss of speed, shape and coherence: but each of these 'phase shift' each other.

The condition on the phase shifts $\delta_1^+ - \delta_1^- = -(\delta_2^+ - \delta_2^-)$ for the two kinks alone conserves total phase shift, and this proves to be true of all of the phase shifts in the more general case: conservation of phase shifts is thus a 'conservation law', a constant of the dynamical motion. This, and the whole of the collision properties, rest on the existence of infinitely many conserved quantities. These are the constants of the motion for a large completely integrable Hamiltonian system. Thus for the s-G model there is a canonical transformation of the Hamiltonian (25) which writes this in terms of action-angle type variables [5] in the form [1, 18, 27]

$$H[p] = \sum_{i=1}^{N_k}(M^2 + p_i^2)^{\frac{1}{2}} + \sum_{j=1}^{N_{\bar{k}}}(M^2 + \bar{p}_j^2)^{\frac{1}{2}} + \sum_{\ell=1}^{N_b}(4M^2\sin^2\Theta_\ell + \hat{p}_\ell^2)^{\frac{1}{2}}$$

$$+ \int_{-\infty}^{\infty} \omega(k)P(k)dk. \quad (30)$$

All of the N_k momenta p_i, the $N_{\bar{k}}$ momenta $\bar{p}_j$, the N_b momenta $\hat{p}_\ell$, the N_b internal momenta Θ_ℓ and all of the $P(k)$ are constants of the motion: they are action variables $(P(k))$ for the 'phonons' and generalised action variables $(p_i, \bar{p}_j, \hat{p}_\ell, \Theta_\ell)$ for the kinks, antikinks, and breathers respectively. The function $\omega(k) \equiv (m^2 + k^2)^{\frac{1}{2}}$, the dispersion relation of the *linearised* s-G which is the Klein-Gordon (K-G) equation $\phi_{xx} - \phi_{tt} = m^2\phi$, and $M \equiv 8m\gamma_0^{-1}$. The number $\gamma_0 > 0$ plays the role of coupling constant (see (25)). Under Hamiltonian (30) rather than (25) all of the motion can be seen to be very simple: $\int_{-\infty}^{\infty}\omega(k)P(k)dk$ is the Hamiltonian of a large (infinite) dimensional torus: $H(k) \equiv \omega(k)P(k)$ means for canonical $Q(k)$, i.e. $\{P(k), Q(k')\} = \delta(k - k'), Q(k)_{,t} = \partial H/\partial P = \omega(k), P(k)_{,t} = -\partial H/\partial Q = O$; and $Q(k)$ is an angle variable $Q(k) = \omega(k)t + \delta(mod\,2\pi)$. Then $Q(k)$ describes motion on a circle (a 1-torus) at fixed radius $\propto (P(k))^{\frac{1}{2}}$. Thus $\int_{-\infty}^{\infty}\omega(k)P(k)dk$ describes the *large dimensional* torus. Note that these very simple motions are coherent motions describing the motions of the phonons.

For the rest, the solitons, these traverse generalised tori opened into sheets coordinatised by $p_i, q_i; \bar{p}_j, \bar{q}_j; \hat{p}_\ell, \hat{q}_\ell$ with $-\infty < p_i, q_i < \infty, \{p_i, q_j\} = \delta_{ij}$, etc. These are therefore tori which it takes an infinite time to 'go round'. The motion of the phonons is evidently stable motion about an elliptic fixed point and lies on invariant tori. The motion of the kinks and antikinks generalizes this to stable motions which can be seen (cf.[28] and below) to join one unstable hyperbolic fixed point to another by a motion about an elliptic fixed point: this is the reason for the infinite time going round the torus (the elliptic fixed point).

As for the Θ_ℓ, $4\gamma_0^{-1}\Theta_\ell(0 < \Theta_\ell < \frac{1}{2}\bar{a})$ is canonical to an angle variable $\Phi_\ell(0 \leq \Phi_\ell < 8\pi) : \{4\gamma_0^{-1}\Theta_\ell, \Phi_m\} = \delta_{lm}$. Thus the l_{th} breather, an each of the other $N_b - 1$ breathers, carries with its motion along $\hat{q}_l$ a further $1 - torus$ (of periode $8\,\bar{a}$) describing its internal motion.

This then is the 'self-organization' of these integrable models: they are simply strapped by the large infinity of constants of the motion which make large completely integrable Hamiltonian systems completely integrable.

I am not able to discuss here the breakdown of that self-organization via a KAM theorem since (Sect. 3) there is no KAM theorem for infinite dimensional systems available[6]. Poincaré proved [7] that for non-integrable systems

---

[6] Some recent results on infinite dimensional KAM theory are given in a second note

with $N < \infty$ under periodic boundary conditions all analytical invariants are destroyed by resonances if these occur. The situation becomes different in large systems ($N \to \infty$) because although resonances can still destroy the tori which are infinite dimensional [7] the solitons can survive as solitary waves.

An example is the 'double sine-Gordon' equation [28]-[30]

$$\phi_{xx} - \phi_{tt} = m^2(\sin\phi + \varepsilon\frac{1}{2}\sin\frac{1}{2}\phi), \quad \varepsilon > 0. \tag{31}$$

The double s-G is not integrable [29, 30], but it becomes the s-G for $\varepsilon = 0$. For $\varepsilon = 1$ it has the solitary wave solutions made up of two bound s-G kinks [29]

$$\phi = 4\tan^{-1}(e^{\Theta-\Delta}) + 4\tan^{-1}(e^{\Theta+\Delta}); \tag{32}$$

$\Delta = \ell n(\sqrt{5} + 2)$, and $\Theta = m(x - Vt)(1 - V^2)^{-\frac{1}{2}}$ and the kink is evidently a 4π−kink. The upper figure in Fig.1 shows how a trajectory associated with this solution connects two hyperbolic singular points much as does the soliton of the s-G. Created alone this solution is a constant of that motion showing no change at any t. Moreover at the level of computer accuracy achieved [29, 30] it survives in collision with other similar 4π−kinks: [29] shows results of computer calculations for two colliding 4π−kinks, and [30] shows this for three. However, these 4π−kinks can be made to oscillate in the parameter Δ [29], an internal degree of freedom induced by the *self-organized* binding of the two s-G kinks shown in (32): this binding is induced by the perturbation $\frac{1}{2}\varepsilon m^2 \sin\frac{1}{2}\phi$ and nothing else. An extreme case of this self-organized oscillation can be interpreted [28, 31] as an actual "leap-frogging" of two 2π-pulses - a remarkable example of self-organized dynamics. Collisions can also excite or change the magnitude of this oscillation. The ultimate stability of this self-organized oscillatory state is not decided since [29] phonons radiate continuously during the short times these systems have been followed on the computer.

A perturbation theory analytical in the perturbation parameter ε has been given precisely when radiation of phonons is ignored [29]. We can try to follow the analytical behaviour of the 4π−kink solution (32) as $\varepsilon \to 1$ from $\varepsilon = 0$ by putting it into the alternative form for $V = 0$ which is [29]

$$\phi = 4\tan^{-1}\left[(1 + \frac{4}{\varepsilon})^{\frac{1}{2}}\sinh\{(1 + \frac{\varepsilon}{4})^{\frac{1}{2}}x\}\right] \tag{33}$$

with $-2\pi < \phi < +2\pi$. The behaviour is singular (i.e. non-analytic) at $\varepsilon = 0$, the s-G limit, and the reason is that for $V = 0$, the ordinary differential equation $d^2\phi/dx^2 = m^2(\sin\phi + \varepsilon\frac{1}{2}\sin\frac{1}{2}\phi), \varepsilon > 0$, which is 4π-periodic in ϕ, becomes 2π-periodic for $\varepsilon = 0$. The singular points and trajectories for this ordinary differential equation are shown in the upper figure in Fig. 1 and the singular points change smoothly from $\varepsilon = 0$. For $\varepsilon > 0$ these points are hyperbolic at $-2\pi, 0, +2\pi$ and elliptic at $\delta - 2\pi$ and $2\pi - \delta$ where $\delta = 2\cos^{-1}(-\frac{1}{4}\varepsilon)$. For $\varepsilon = 0$ the hyperbolics at $-2\pi, 0, 2\pi$ are unchanged and $\delta = \pi$ means elliptics at $\pm\pi$.

---

added in proof at the end of this paper.

However the trajectories cannot change smoothly: for $\varepsilon > 0$ the 4π-kink (33) corresponds to a trajectory between hyperbolics at -2π and $+2\pi$ (4π-kink); and for $\varepsilon = 0$ this becomes a double representation[7] of the 2π-kink of s-G described by trajectories between hyperbolics at $-\pi$ and $+\pi$ and $+\pi$ and $-\pi + 4\pi = 3\pi$. For $\varepsilon > 0$ this double representation is described by *three* trajectories, that between -2π and $+2\pi$ (4π-kink) and the two different trajectories between the hyperbolic at 0 and itself (0π pulses [29]). Transition from $\varepsilon = 0$ to $\varepsilon > 0$ appears smoother in the time frame (evolution in t at fixed x, the lower figure in Fig. 1): the hyperbolics at $-2\pi, 0, 2\pi$ are now elliptics and the elliptics at $\delta - 2\pi, 2\pi - \delta$, hyperbolics. Trajectories connecting hyperbolics are then $\delta - 2\pi$ to $2\pi - \delta$ and $2\pi - \delta$ to $\delta - 2\pi + 4\pi$, which are distinct trajectories, and these become a double representation $-\pi$ to $+\pi$ and $+\pi$ to $+3\pi$ for $\varepsilon = 0(\delta = \pi)$. The essential point of this analysis is that the 2π-kink of the integrable s-G persists as the 4π-kink of the double s-G by a smooth change of its singular point structure as the perturbation $\varepsilon = 0$ changes to $\varepsilon > 0$. If this behaviour has large enough measure it points to a rather complicated form for the KAM theorem for large, infinite dimensional, systems still to be formulated.

The stability of the solutions (33) or (32) for arbitrary t is important. It is shown in [29], via the perturbation theory developed about the two-kink solution of s-G there, that close-to the stationary solution (33) and for small enough ε the frequency of the internal oscillation has period $(\frac{1}{2}\varepsilon)^{\frac{1}{2}}$ and the 'wobbling' [29] 4π-pulse with this internal oscillation is stable and long-lived providing, as in the calculations, all effects of radiated phonons are explicitly dropped from the analysis. The evidently all time existence of solutions (32) for the large perturbation with $\varepsilon = 1$ and its apparent stability under collisions albeit for short times as calculated numerically suggests that dense sets of stable solutions (32) labelled by the parameter V can persist as $\varepsilon = 0$ increases to $\varepsilon > 0$. The perturbation theory of this remains equivocal because radiation has so far been ignored. 'Radiation' represents the large dimensional tori for $\varepsilon = 0$, and perturbation affects all of these tori for $\varepsilon > 0$ [29]. But how far any of these tori could remain for $\varepsilon > 0$ is not yet known[8]. Still the existence of the coherent solitary waves (32) together with the possible long time existence of the coherently wobbling 4π-pulses as established for short times as well as the topological condition (Sect.7) that both (32) and the coherent 'wobbler' satisfy boundary conditions $\phi \to 0, x \to -\infty, \phi \to 4\pi, x \to +\infty$ indicate that the double s-G is not ergodic and so is neither integrable nor ergodic. This may be true in a sufficient measure for generic large, non-integrable, systems. And this is one of the reasons why I consider that, despite the Poincaré theorem as described in [7], KAM theory has not yet been formulated for large non-integrable systems. Certainly from the simpler viewpoint of this paper coherent solitary waves exist amongst the

---

[7] The *stationary* 4π-kink cannot become two 2π-kinks of s-G for $\varepsilon = 0$ since these 2π-kinks cannot both be brought to rest simultaneously. The *oscillating* 4π-pulse appears to have the required property.

[8] The results in the second note added in proof at the end of this paper may suggest that a large majority of these could remain for small $\varepsilon > 0$.

Hamiltonian chaos of large non-integrable Hamiltonian systems.

Note that KAM theorems concern *self-disorganization* - of coherent pulses into chaos. The sequence is from simple motions on tori (generalised tori) about stable elliptic fixed points to complicated chaotic motions involving dense sets of unstable hyperbolic points. This *is* Hamiltonian chaos and is ergodic. Open systems with damping can display other chaotic motions. Some of these are sketched in the Sects. 6, 7 following.

6 Instabilities and Self-Organization

The NLS equations (11) arise in many physical contexts. An early application was made in [32] where an intense laser field was shown to break up inside an irradiated medium into intense *filaments* of light. The action is as sketched in the Fig.3.

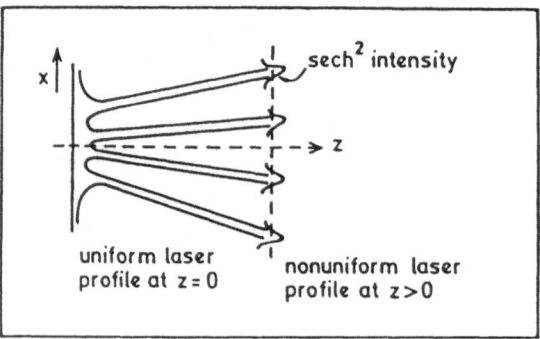

Fig. 3. Filamentation of intense laser light in a medium. In two dimensions x, z the equation is the NLS equation (11) with $c < 0$ and z replacing time t.

In the steady state motion along z becomes the effective time variable t and the input laser light has the profile in x as sketched in Fig. 3 at $z = 0$. But for $c < 0$ as can be arranged (see below) the soliton solutions (12) of the NLS equation (11) mean that with increasing z the light profile breaks up into soliton forming channels with light intensity $\propto | \phi |^2$ given by $q^2 sech^2[q(x - Vz)]$: these are the filaments and such filaments have been seen in many different physical circumstances.

This so-called modulational instability along x is a spontaneous self-organization into these filaments. In most situations two transverse directions x and y are involved and the two dimensional NLS equations, equation (21), with z replacing t, is no longer integrable. The consequent instability is however still more interesting. The Fig.4 shows consecutive behaviours, as found by numerical integration, in the 2-dimensional transverse profile for a harmonic wave of frequency ω entering and traversing an optical ring cavity: in the steady state distance z

along the path in the ring cavity again acts as time and the nonlinearity is a
saturable nonlinearity of the form

$$\alpha(\Delta\omega - i\gamma_0)\Phi/[(\Delta\omega)^2 + \gamma_0^2 + |\Phi|^2 \gamma_0\gamma_1^{-1}], \tag{34}$$

'saturable' because it tends to zero as $|\Phi| \to \infty$. In (34) $\Delta\omega = \omega - \omega_0$ is the 'de-
tuning', the difference between ω and the natural resonance ω_0 of the nonlinear
medium inside the ring cavity. The expression (34), developed in powers of the
dependent variable Φ yields a nonlinearity of the form $A\Phi - B\Phi|\Phi|^2$ and the
equivalent NLS model to this order is

$$-i\Phi_z = \Phi_{xx} + \Phi_{yy} + A\Phi - B\Phi|\Phi|^2. \tag{35}$$

If $\phi = \Phi e^{-iAz}$, (35) becomes the 2D NLS model in ϕ and if ϕ is independent of
$y, -i\phi_z = \phi_{xx} - B\phi|\phi|^2$ so that, if B is restricted to its real part, $B = 2c$ for
the integrable 1D NLS models. For solitons ($c < 0$) $\Delta\omega$ must thus be negative.
The saturable form of the NLS model in 2D (x and y) and one time (z) is non-
integrable because of the dependence on x *and* y, because of the saturable form
(34) of the non-linearity, *and* because (34) has the complex number $\Delta\omega - i\gamma_0$ (not
a real number) as coefficient. For $\Delta\omega < 0$ it still has remarkable modulational
instabilities leading to 2D solitary-wave type solutions which however evolve
with increasing z.

The Fig.4 shows detail of the actual modulational instability in the transverse
profile in x and y for different z. For small enough z, because $Re(B) < 0$, the
system 'self-focusses' light of intensity $\propto|\Phi|^2$ to a bright central spot: this is the
transverse profile of a harmonic wave of frequency ω traversing the cavity and,
with changing input intensity, this wave displays optical bistability associated
with the z-direction. The Fig. 5 sketches typical optical bistability.

As indicated in the caption to Fig.5 the system is bistable (has two possible
output states for given input) between switch-up (on the right) and switch-down
(on the left). The S-shaped curve is evidently a form of first order phase transition, with hysterisis, induced by the changes in input intensity. If the input
intensity is fixed, then for small enough z there is self-focussing in the x, y plane.
But, for increasing z, as shown in Fig.4 field gradients in this plane at the edge
of the focussed beam now induce surrounding circular filaments first of all. Subsequently actual filaments (spots) of NLS type push symmetrically into the ring
through a further instability. The vertical strip at the right of the first (top
left) figure in Fig. 4 is a central slice across the whole transverse profile. Subsequently (bottom left) these filaments (bright spots) interact and a further stage
in the process is shown top right. At the bottom right with increased z there is
a still further modulational instability. Random patterns due to computer noise,
which breaks the symmetry, subsequently emerge, together with recurrent modulational instabilities. This complicated sequence of self-organizations is indeed
non-integrable: it was solved numerically and the computer time on a Cray II
for this sequence was six hours! [33].

It may be noticed that, because A and B in (34) are complex rather than
real numbers, equation (34) is actually a form of the complex Ginsberg-Landau

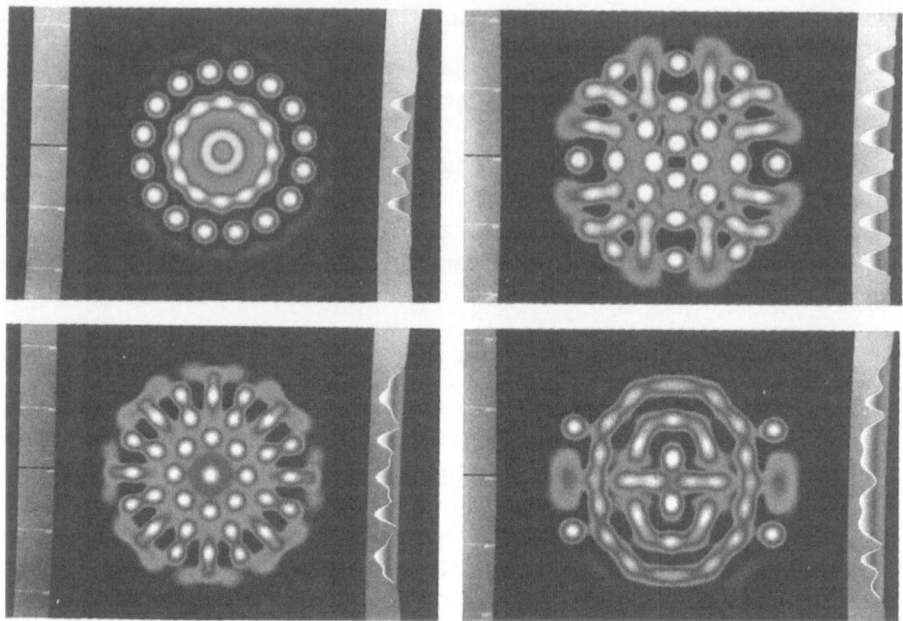

Fig. 4. Modulational instabilities in x and y induced by the saturable nonlinearity equation (34). The evolution of these transverse instabilities is described in the text. The figure is from [33].

equation (G-L equation) in 2D [6]. In 1D this is

$$\Phi_t = R\Phi + (1 + i\nu)\Phi_{xx} - (1 + i\mu)\Phi \mid \Phi \mid^2 \tag{36}$$

in which we now assume R is a *real* parameter. It is a "control" parameter or a "pumping" parameter as occurs in the laser equations described below. The complex numbers $(1 + i\nu)$ and $(1 + i\mu)$ in (36) arise out of the physics and are simply scaled to these forms without changing that physics. If $R = 0$ and $1 + i\nu \rightarrow i, 1 + i\nu \rightarrow 2ic$ (set $\mu\nu^{-1} = 2c$ and let $\nu, \mu \rightarrow \pm\infty$ also rescaling the time t) we regain the NLS equations (11). This is a wave equation, Hamiltonian and integrable. But the complex numbers $1 + i\nu, 1 + i\mu$, whatever their scaling if they remain complex, change the behaviour of the equation very considerably. It becomes dissipative, with the possibility of energy loss, and a number of new nonlinear features emerge in the dynamics. In particular it now has the feature that all motions ultimately move into a small region in the configurational space whatever the initial motions. That space is not a phase space but a Hilbert space simply. The region in the configurational space is called an "attractor" since motions are attracted into it, and it is a global attractor since all motions are ultimately attracted into it. Below we become concerned with so-called "strange" attractors in which the motion is chaotic but not Hamiltonian chaotic. Motions on global attractors may also include such chaotic motions depending on the initial conditions.

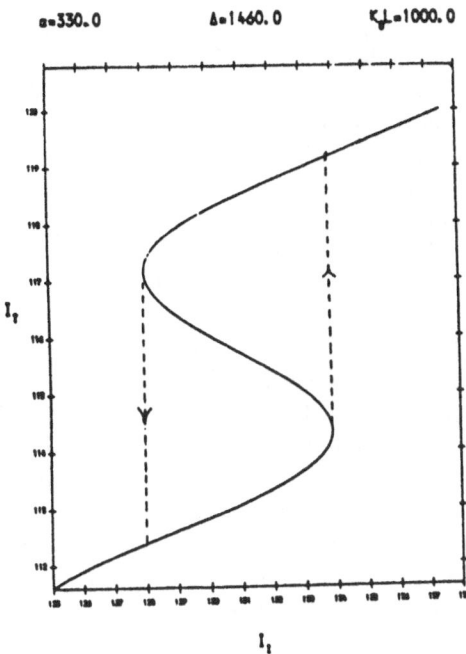

$\alpha = 330.0$ $\Delta = 1460.0$ $k_0 L = 1000.0$

Fig. 5. Input-Output plot of an optical bistable system: the downward sloping part of the S-shaped curve is unstable and the curve followed is the hysterisis loop marked by the arrows. Between switch-up and switch-down the system *is* bistable. The output intensity, the transmitted intensity I_T, is the ordinate, the input intensity I_I the abscissa. Parameters α, Δ are proportional to the number density of atoms and the detuning $\Delta\omega$ respectively: L is a path length and $k_0 = \omega c^{-1}, c$ the velocity of light *in vacuo*.

If we return from (36) to the 2D G-L equation in x *and* y, the blow-up of the 2D solitary wave of the NLS equation (21) means that the corresponding G-L equation has a global attractor when, but only when, it is "attractive" that is μ and ν must have opposite signs. These equations arise as a reduction of the Navier-Stokes equations of hydrodynamics applied to the problem of heat convection. Recent work on the 2D complex G-L equation shows that it can display turbulence analogous to fluid turbulence and characterized as both "soft" and "hard" turbulence [6].

I make these points to show how less than dramatic changes in the equations, induced by the particular physics in question, can lead to dramatic changes in the dynamics. Similarly dramatic changes arose from small perturbations of large integrable Hamiltonian systems (Sect.5). Modulational instabilities in transverse profiles of the types described arise by physical demands which break transverse translational symmetry and so accompany most intense field optics. The s-G system (9) is an extreme form of the dynamical description of a physical phenomenon called self-induced transparency (SIT) [1, 28], [34]-[36]. In SIT optical pulses typically in the 10^{-9} sec regime [35] traverse very dilute but resonant media (dilute is about 10^{12} particles cc^{-1} [35]). Transverse translational symmetry in x and y means that *plane-wave* pulses travel along z governed in a rough sense [28, 34] by the s-G $\Phi_{zz} - \Phi_{tt} = m^2 sin \, \Phi$. The electric field is then essentially Φ_z, not Φ itself, and spontaneously self-organized break-up of the field into sech pulses (derivatives of the kinks and antikinks, Sect. 5) with different speeds $V_n > V_{n-1} > \ldots > V_1$ occurs. These sech pulses integrate to the values $\pm 2\pi$ in 'area', characteristic of 2π-kinks or 2π-antikinks. These $\pm 2\pi$ optical pulses [28, 34, 35] rotate the 'spin' [28, 34] of a resonant atom by $\pm 2\pi$ and thus take such atoms from their initial ground state to an excited state and back to their ground state. This spin (spin-$\frac{1}{2}$) description for a single resonant 2-level atom has been very fruitful in laser physics [28, 34]. Evidently in SIT no energy is extracted from the optical pulse to remain lost in the atoms. Hence the 'self-induced', i.e. self-organized, 'transparency' of SIT.

In practice, even in SIT, transverse translational symmetry must be broken because the input light must have a finite transverse profile. Typically this is Gaussian in x, y and then the sech pulses along z lose amplitude, and then speed, across their transverse profile and a complicated 3-dimensional pulse has to evolve. The relevant equations are actually the self-induced transparency equations [37] in three space dimensions and these can be put [38] in the form

$$\lambda_z + [i(\Delta\omega)\tau_p + \tau_p T_2^{-1}]\lambda = \mathcal{E}\mathcal{N}$$

$$N_z + (N - N_0)\tau_p T_1^{-1} = -\frac{1}{2}(\mathcal{E}^*\lambda + \mathcal{E}\lambda^*)$$

$$-iF\nabla_T^2\mathcal{E} + \mathcal{E}_t = \int_{-\infty}^{\infty} g(\Delta\omega)\lambda(x, y, z, t; \Delta\omega)d(\Delta\omega) \qquad (37)$$

in which $\mathcal{E}$ is the electric field and λ is another complex field, both fields in x, y, z, t. The field λ also depends on the detuning $\Delta\omega$ and so-called 'inhomogeneous broadening' is included through the $g(\Delta\omega)$. This could arise through the different speeds at which atoms travel inside a gas for example. The number $F^{-1} = 2\omega c^{-1}d_{beam}^2(\alpha/2\pi)\tau_p$ is the 'Fresnel number' $d_{beam}^2/\lambda L_0$ for propagation distance $L_0 = (2\alpha\tau_p)^{-1} = \frac{1}{2}c\tau_p^{-1}\tau_c^2$ in which τ_c is a *coherence* time [38]: $\nabla_T^2 \equiv \partial^2/\partial x^2 + \partial^2/\partial y^2$. The time τ_p is a pulse characteristic time $\sim nano-secs$ $(\tau_p = \frac{1}{2}\hbar\mu^{-1}E_0^{-1}$ where E_0 is the field amplitude and μ is a quantum mechanical matrix element for the atomic transition between ground state and excited state at the frequency $\omega_0)$: N_0 is the steady state inversion of the atom - if $| c_e |^2$ is the probability for the atom in its excited state and $| c_g |^2$ that for the ground state the inversion is $N = | c_e |^2 - | c_g |^2$: T_2 is a time damping the coherent propagation of the pulse, and T_1 is the associated damping of the inversion. The scaling of the propagation (along z) is $z \Leftrightarrow (t - zc^{-1})\tau_p^{-1}$ (c = velocity of light *in vacuo*), $t \Leftrightarrow \alpha\tau_p z$. The equations (37) are integrable if and only if F, T_1^{-1} and T_2^{-1} are all zero [36, 37].

None of these details really matter for present purposes. Transverse effects are included in (37) through the $-F\nabla_T^2\mathcal{E}$. If $\mathcal{E}$ in (37) is written $\mathcal{E}e^{i\phi}$ and $\lambda = \frac{1}{2}i(Q - iP)e^{i\phi}$ where $\mathcal{E}, Q, P, \phi$ are all real quantities which vary slowly on the slow time-scale of the $10^{-9}sec$ pulses (light oscillates at $10^{15}Hz$, 10^6 times faster than the $10^{-9}sec$ scale) then

$$- F\nabla_T^2 \mathcal{E} + \phi_t \mathcal{E} \propto \int_{-\infty}^{\infty} Q(\Delta\omega)g(\Delta\omega)d(\Delta\omega) \tag{38}$$

[38]. On resonance $Q = 0$ and $\lambda = \frac{1}{2}Pe^{i\phi}$. If $\phi = 0$ but $\phi_t \neq 0$ initially, then *any* variation across the transverse profile means $\nabla_T^2 \mathcal{E} \neq 0$, so $\phi_t \neq 0$ and the 'chirp', the slowly varying phase ϕ in the evolution along the z-direction, spontaneously grows. Then, from $Q_z = -P(\Delta\omega\tau_p + \phi_z) - QT_2^{-1}\tau_p$ from (37), Q grows and generates ϕ_t until ϕ_t is distributed across the profile. One can show [38] that there is transverse energy flow, either inwards (self-focussing) or outwards (self-defocussing) though self-focussing was observed in the actual experiments in SIT where a pulse with trailing sides developed (N.B. "bigger pulses travel faster" even in this transverse profile). All of these features are 'self-organized' by the nonlinear equations (37). The evolution of the transverse modulational instability of a one dimensional close to integrable system in z and t which can be followed through in this example is not so much different from the examples built around the NLS model as given above. Experts may be able to pick their way through the different equations to see why they must behave similarly, but it is not easy to give one generic example.

In the same vein is the behaviour inside lasers of wide aperture, both 2-level atom and Raman lasers in particular [39]. Assuming there is a single longitudinal mode along z in the laser cavity oscillating at frequency ω_c one can take equations for the 2-level atom laser in the form [39] (compare (37) for SIT)

$$\frac{\partial \mathcal{E}}{\partial t'} + \kappa\mathcal{E} - iF\nabla_T^2 \mathcal{E} = \frac{i\omega_c}{2\varepsilon_0}P$$

$$\frac{\partial P}{\partial t'} + (\gamma_\perp + i(\omega_0 - \omega_c))P = \frac{i\mu^2}{\hbar}\mathcal{E}N$$

$$\frac{\partial N}{\partial t'} + \gamma_\parallel(N - N_0) = -\frac{2i}{\hbar}(\mathcal{E}^*P - \mathcal{E}P^*) \tag{39}$$

where κ is the cavity damping constant, $\gamma_\perp = T_2^{-1}$ is the dephasing rate (damping of the coherent evolution) and $\gamma_\parallel = T_1^{-1}$ is the energy loss decay (corresponding damping of the inversion). The number N_0 is the steady population inversion in the absence of any field $\mathcal{E}$ and is now supposed *positive*. With $P \propto i\lambda$ one can see that equations (39) for the laser are very similar to the SIT equations (37). The 'coherent' dephasing from inhomogeneous broadening has been omitted from (39) while cavity damping through κ is included. Otherwise the two 'time' variables z, t in (37) become the single time variable t' in (39) a form which follows from the assumption of a single *longitudinal* mode (along z) in the laser cavity. We shall see that the important difference between the SIT equations (37) and the laser equations (39) is however that $N_0 < 0$ in the SIT equations and $N_0 > 0$ in the laser equations. This fact leads to critically different behaviours.

The system (39) is conveniently *scaled* (by $t' = t\gamma_\perp^{-1}, \varepsilon = (i\hbar\gamma_\perp/2\mu)e, P = (\hbar\gamma_\perp\varepsilon_0\kappa/\omega_c\mu)p, N - N_0 = -(2\varepsilon_0\kappa\hbar\gamma_\perp/\omega_c\mu^2)n$ to

$$e_t - if\nabla_T^2 e = -\sigma e + \sigma p$$
$$p_t + (1 + i\Omega)p = (r - n)e$$
$$n_t + bn = \frac{1}{2}(ep^* + e^*p) \tag{40}$$

where $\sigma = \kappa\gamma_\perp^{-1}, \Omega = (\omega_0 - \omega_c)\gamma_\perp^{-1}, b = \gamma_\parallel\gamma_\perp^{-1}, f = F\gamma_\perp^{-1}$, and $r = \frac{\omega_c\mu^2 N_0}{2\varepsilon_0\hbar\kappa\gamma_\perp}$ with $N_0 > 0$ is interpreted as a 'control parameter'. This scaling shows that equations (39) are a complex valued form of the equations of the Lorenz attractor [34] derived from the Bénard instability occurring in the thermal convection between two plates heated from below [40]. This equivalence was first recognised by Haken [41]. Apart from their complex valued character, the new feature in (39) and (40) is the transverse term in ∇_T^2. In [40] I extended the Lorenz equations to the case of a laser with saturable absorber: *these* (real valued) equations also arise in the description of thermosolutal convection [40]. With transverse effects included the laser-with-absorber equations become in the thermosolutal form [40]

$$a_t - if\nabla_T^2 a = \sigma(-a + r_T b - r_S d)$$
$$b_t + b = (1 - c)a; d_t + \tau d = (1 - e)a$$
$$c_t = (-c + \frac{1}{2}(ab^* + a^*b)); e_t = (-\tau e + \frac{1}{2}(ad^* + a^*d)). \tag{41}$$

If the second control parameter $r_S = 0$, these complex valued equations decouple to an alternative form of the real valued Lorenz equations [40] when the specific transverse effects are excluded. Here we can only show what transverse effects do in the case of the simple, laser, system (40) which has the single control parameter r.

It is now well known that, as the control parameter r of the Lorenz equations increases, the system undergoes a sequence of period doubling coherent oscillations leading to a limit point beyond which a chaotic 'strange attractor' develops (cf. the comparable analysis for the laser with saturable absorber and thermosolutal convection in [40]). The simple coherent behaviour breaks down beyond the limit point, to a very complicated motion characterised by everywhere exponentially diverging trajectories: this motion, which is confined to within some finite fraction of the whole configuration space leads to the fundamental impossibility of defining actual trajectories. These trajectories are both classical and causal, but any infinitesimal shift in the data at $t = 0$ leads to widely departing trajectories for all $t > 0$. This is a further example of the *breakdown* of coherence (self-disorganization driven here by the value of a single number, the control parameter r). Such dynamical behaviours may be relevant to e.g. human cardiac arhythmias and the behaviour of animal populations, as well as to the physical problem of fluid turbulence.

However we are more concerned here with transverse patterns formed at smaller r [9]. The system (40) has the trivial solution $(e, p, n) = (0, 0, 0)$ and the

---

[9] This analysis closely follows Moloney, Ref. [39].

lasing solution $(e, p, n) = (\bar{e}e^{i\theta}, \bar{p}e^{i\theta}, \bar{n})$ with θ a phase, $\theta(x, t) \equiv \mathbf{k.x} - \omega t$ (and $\mathbf{x} = (x, y), \mathbf{k} \equiv (k_x, k_y)$). The nontrivial lasing solution is

$$\bar{e} = \sqrt{b(r - r_c)}, \bar{p} = (1 - \frac{i\Omega}{\sigma + 1})\sqrt{b(r - r_c)}, \bar{n} = r - r_c \qquad (42)$$

with the dispersion relation $\omega = (\sigma\Omega + fk^2)/(\sigma + 1)$: r_c is thus the critical value of r beyond which lasing occurs. If $r < r_c$ the *equilibrium* solution is the trivial solution: if $r > r_c$ the system moves to the lasing state.

Linear stability analysis about the trivial solution $e = 0 + \delta e, p = 0 + \delta n, n = 0 + \delta n$ leads to

$$(\delta e, \delta p, \delta n) = (V_1, V_2, V_3)e^{\lambda t + i\mathbf{k.x}} \qquad (43)$$

and (V_1, V_2, V_3) is a column vector V_0 satisfying $AV_0 = 0$ with the matrix A given by

$$A = \begin{bmatrix} \lambda + \sigma + ifk^2 & -\sigma & 0 \\ -r & \lambda + 1 + i\Omega & 0 \\ 0 & 0 & \lambda + b \end{bmatrix}. \qquad (44)$$

Solutions for λ are then $\lambda = -b$ with $V_1 = V_2 = 0, V_3 = 1$, the usual isolated atom decay, and the two roots of

$$\lambda^2 + (\delta + 1 + ifk^2 + i\Omega)\lambda + (\sigma + ifk^2)(1 + i\Omega) - \sigma r = 0. \qquad (45)$$

If $\lambda = \mu - i\nu(\mu, \nu real), \mu > 0$ determines the growth rate and ν is a frequency of oscillation about the cavity frequency ω_c. The trivial equilibrium state $(0, 0, 0)$ is stable for smaller r. But as r increases $(0, 0, 0)$ becomes unstable and there is a critical curve $r = 1 + (\Omega - fk^2)^2/(\sigma + 1)^2$, the neutral stability curve, beyond which lasing occurs. There is now a critical difference in behaviour for detunings $\Omega > 0$ and $\Omega < 0$ When $\Omega < 0$ the lowest lasing threshold is at $k = 0$, a steady oscillation in the transverse profile without spatial modulation and the critical r for this is $r_c = 1 + \Omega^2(\sigma + 1)^{-2}$ while the frequency ν is $\nu_c = \sigma\Omega(\sigma + 1)^{-1}$: this $\nu_c = \sigma\omega_0 + \omega_c(\sigma + 1)^{-1} - \omega_c$ so the laser oscillates at $\sigma\omega_0 + \omega_c(\sigma + 1)^{-1}$.

However, when detuning $\Omega > 0$, the lowest lasing threshold occurs for $k = \pm\sqrt{\Omega/f}$, meaning travelling plane wave solutions in the x, y plane, and $r_c = 1$. The frequency is $\nu_c = \Omega$, and this means the laser oscillates at ω_0, the atomic frequency. Thus the laser *chooses* transverse spatial structure to compensate initial detuning and maximises emission by lasing *at the gain centre* ω_0. The neutral stability curve has *bifurcated* with the two equilibrium k values $k_c = \pm\sqrt{\Omega/f}$, a form of phase transition. If, for $\Omega > 0$, the transverse travelling wave is in the x-direction there is a growth *band* in the (k_x, k_y) plane forming an annulus in this plane. The bifurcation is 'super-critical' namely a simple pitchfork bifurcation [40] and in 1D the travelling wave at $k_c = \sqrt{\Omega/f}$ appears to be globally attracting (i.e. *all* motions move towards it much as in the complex Ginsberg-Landau case).

Close to these thresholds one can derive 'order parameter' equations patterned on those for phase transition theory in the Ginsberg-Landau context.

Thus for $\Omega < 0$ one finds a two-dimensional form of the G-L system (39) (integrable and with soliton solutions in $1+1$ in the limit described here). For $\Omega > 0$ one has to take into account waves travelling in both directions in the laser cavity and various people have derived such equations. In their simplest form and without any pumping these are [42, 43]

$$F_z + V_g^{-1} F_t - (i/2k)\nabla_T^2 F = i\eta(|F|^2 + G|B|^2)F$$
$$-F_z + V_g^{-1} F_t - (i/2k)\nabla_T^2 F = i\eta(G|F|^2 + |B|^2)B. \qquad (46)$$

The number $G = 2$ takes into account a standing wave pattern set up in the cavity: $G = 1$ smooths this out as can occur in practice. In one space dimension (x) these equations are integrable in the steady state with soliton type solutions for $G = 0$ and $G = 1$ [44]. Moloney [39] gives the coupled Newell-Whitham-Segur system more complicated than (46) and involving the pumping parameter r.

Examples of optical patterns created from systems like (46) are described in [45]. Firth in [42] explains how hexagonal patterns tend to develop in x and y when G in (46) is zero and in the F case there is both quadratic nonlinearity F^2 and the cubic nonlinearity $|F|^2 F$. The patterns and their stability beyond critical onset for the system (39) are likewise of considerable interest. When $\Omega > 0$ (travelling wave solution) a dynamical stability analysis about criticality can be described in terms of slowly evolving quantities $\bar{e}, \bar{p}, \bar{n}$ and phase θ. The phase 'slaves' the amplitudes, that is [46] derivatives of the amplitudes are so small that they can be eliminated from the coupled phase and amplitude equations. Ref. [39] derives $\theta_t = \omega + \alpha(k)\theta_{xx} + \beta(k)\theta_{yy}$ for the evolution of the phase θ this way. This allows a classification of possible instabilities - Benjamin-Feir ($\alpha(k) < 0$ for all k); Eckhaus ($\alpha(k) < 0$ for $k > \hat{k}$); zig-zag ($\beta(k) < 0$, absent of course in the 1D-transverse profile case). Together these form a very rich 'phenomenology' of pattern forming instabilities and remarkable patterns can emerge. Some of these patterns are computed and shown in [39]. The Fig. 6 shows patterns derived from the Raman laser equations studied in Ref. [39], much as the 2-level atom laser equations (39) have been studied here.

Evidently it is possible to trace essential similarities underlying the sequence of equations from the strictly integrable relativistic (9) and non- relativistic (11) in $1+1$ dimensions, through (34), (35), (36) to (37) with (38), (39) with (40), (41) and (46) and its generalisations in [39].

The nonlinearities are much the same in each of these equations [10]: however, key changes reflecting changes in the physics occur at (34), (35) (the saturable nonlinearity, the increase in dimension $x \to (x, y)$, the advent of complex coefficients, and at (36) and (39) - (41) where pumping described by simple pumping parameters r is included. Pumping, in particular drives a dynamical system far from simple equilibrium and can create remarkably complex looking patterns and some of these are those shown in the Fig. 6.

A conclusion from these still relatively simple examples is that complex patterns of a certain 'coherence' can be induced in these simple ways. The pumping

---

[10] Sine-Gordon (9) has a development $m^2(\phi - \frac{1}{6}\phi^3)$ in the nonlinear sine function which can be related to the complex valued nonlinearity $\phi|\phi|^2$ of the NLS model.

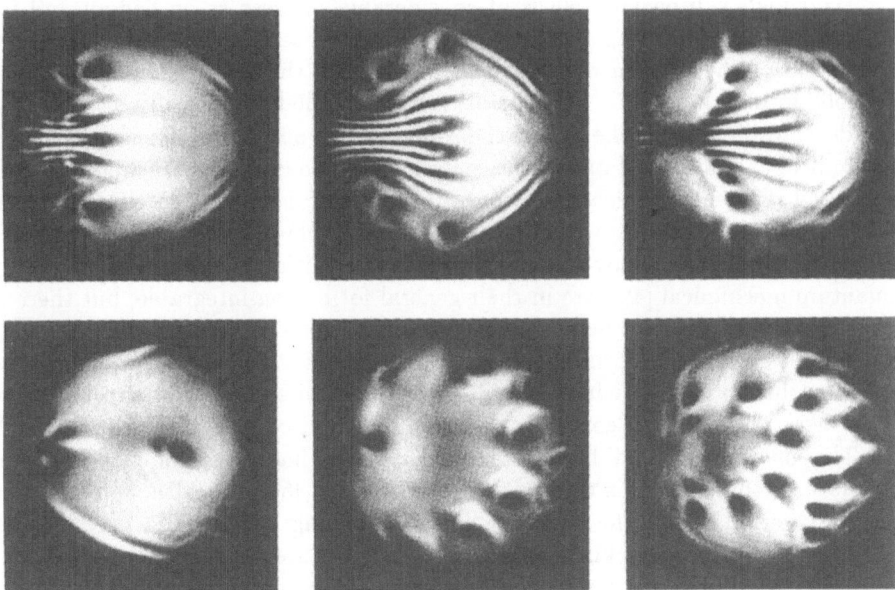

Fig. 6. Transverse patterns computed for the wide angle Raman laser in [39].

in the far from equilibrium systems constituting living matter is substantially more complicated; and this may be one further aspect of the complex character of living matter only briefly considered as such in Sect.2.

7 Effects of Brownian Motion

Effects of Brownian motion, the creation of disorder and entropy and of order and 'negative entropy', some correlation between body temperatures and 'intensity of life', were all considered by Schrödinger in his essay 'What is Life' [43]. In the Sects. 3-6 of this paper we have been concerned mostly with the self-organization of ordered (coherent) or partially ordered (partially coherent) structures. Mostly in terms of KAM theorems (Sects. 3, 5) we have also been concerned with *dis*organization and dynamical chaos which is Hamiltonian chaos; in Sect. 6 we have also been concerned with strange attractor chaos. With mammalian body temperatures of about $310^o K$ the disordering effects of Brownian motion must be important. I describe in this last section some effects of Brownian motion, namely infinite dimensional Gaussian Hamiltonian chaos, on the coherent, ordered solitons in $1 + 1$ dimensions. [11]

---

[11] Correlation functions can be used to give quantitative measures of 'coherence' 'partial coherence', 'order' etc. as used in this paper. So can appropriate definitions of entropy (Sect.4): for example equations (49, 50, 51) below will give explicit expressions for the *thermodynamic* entropy of s-G classical solitons at any temperature,

Davydov's soliton [48]-[50] is of considerable interest as an energy and information carrier forming a natural and apparently realisable $1 + 1$ dimensional excitation of the protein α-helix. The α-helix has three independent channels [49, 50], formed by $\cdots H - N - C = 0 \cdots$ units bonded sequentially by hydrogen bonds, spiralling about the axis of the α-helix: Davydov's excitations are natural dynamical excitations of one of these essentially one dimensional channels. It is supposed that energy released by hydrolysis of ATP is transfered to the protein as an excitation of Davydov type: this may be self-trapped [49] or propagate as a Davydov soliton. Davydov's equations of motion, either semiclassical or wholly quantum mechanical [49], are in their general forms non-integrable; but there is a particular limit where they become the attractive NLS equation, equation (11) with $c < 0$ in classical form or equation (24) with $c < 0$ in quantum form. The soliton solution of (11) with $c = -1$ is the classical soliton (12) and this is of 'breather' type in that it contains the internal oscillation of frequency $(q^2 - \frac{1}{4}V^2)$.

The 'one dimensional' biological molecule DNA has been modelled in a similar way by the classical s-G equation, equation (9) [51, 52]. Although the classical s-G has the breather solutions (28) satisfying the boundary conditions $\phi \rightarrow 0, x \rightarrow \pm\infty$, the kinks and antikinks (27) are *topological* solitons stabilised by the boundary conditions $\phi \rightarrow 2\pi, x \rightarrow +\infty$ and to $-\infty$. Consequently breathers, and kinks and antikinks, have very different properties in the face of the onslaught by Brownian motion. In particular, classical breathers break up into classical phonons under the action of classical Brownian motion. On the other hand quantum mechanics stabilises the s-G breathers at the lower temperatures at which quantum Brownian motion, (i.e. quantum statistical mechanics, Sect.3) applies. The classical soliton (12) of the NLS is of breather type and this too breaks up into phonons under classical Brownian motion. But the corresponding quantum soliton can likewise be stabilised by quantum mechanics.

These results have all been found by application of the methods briefly sketched in Sect.3 [17, 53]. For example, for the classical attractive NLS model coupled to a heat bath we find [25, 53] that the classical Helmholtz free energy $F = -\beta^{-1}\ell n \ Z$, calculated in a thermodynamic limit achieved under periodic boundary conditions of period L with $L \rightarrow \infty$ at finite density, has a free energy density in that limit

$$\lim_{L \to \infty} F L^{-1} = (2\pi\beta)^{-1} \int_{-\infty}^{\infty} dk \quad \ln(\beta\epsilon(k)). \tag{47}$$

The excitation energies $\epsilon(k)$ in this expression are given by

$$\epsilon(k) = \omega(k) + (2\pi\beta)^{-1} \int_{-\infty}^{\infty} dk'[\partial\Delta(k - k')/\partial k] \ln(\beta\epsilon(k')) \tag{48}$$

with $\omega(k) = k^2$ and $\Delta(k - k') = -2c(k - k')^{-1}$ with $c < 0$. The important point is that there is only *one* contribution to F, and experts will see that the

results which extend to the quantum theories. 'Quantum coherence' in biological systems is discussed by Ke-Hsueh Lu, at this meeting - especially concerning biophoton emission from living matter.

$\beta^{-1}ln(\beta\epsilon(k))$, which is the classical free energy of a single classical oscillator of frequency $\epsilon(k)$ (energy $\hbar\epsilon(k)$ with $\hbar$ set to unity) is of phonon type associated with the weak oscillatory solutions of the NLS model. There is no contribution from the breather-like solitons (12) and these have broken up into the phonons.

In some contrast the similar calculation for classical s-G yields two contributions to F and [53, 54]

$$\lim_{L\to\infty} FL^{-1} = (2\pi\beta)^{-1} \int_{-\infty}^{\infty} \omega(x)\ln(\beta\epsilon(x))dx - 2(2\pi\beta)^{-1}$$

$$\int_{-\infty}^{\infty} E(x)e^{-\beta\tilde{E}(x)}dx \tag{49}$$

which is conveniently expressed in terms of 'rapidity' x such that $\omega(x) = m\cosh x$, $E(x) = M\cosh x$ (so that if momenta $p(x) = m\sinh x$, $P(x) = M\sinh x$, $\omega(x) = (m^2 + p^2)^{\frac{1}{2}}$, $E(x) = (M^2 + P^2)^{\frac{1}{2}}$, the proper relativistic expressions with velocity c set equal to unity). The excitation energies, and there are *two* namely $\epsilon(x)$ and $\tilde{E}(x)$, are given by the coupled system

$$\epsilon(x) = \omega(x) + (2\pi\beta)^{-1} \int_{-\infty}^{\infty} (\partial\Delta/\partial x)\ln(\beta\epsilon(x'))dx'$$

$$+ 2(2\pi\beta)^{-1} \int_{-\infty}^{\infty} (\partial\Delta_k/\partial x)e^{-\beta\tilde{E}(x)}dx', \tag{50}$$

$$\tilde{E}(x) = E(x) - 2(2\pi\beta)^{-1} \int_{-\infty}^{\infty} (\partial\Delta_{kk}/\partial x)e^{-\beta\tilde{E}(x')}dx'$$

$$- 2(2\pi\beta)^{-1} \int_{-\infty}^{\infty} (\partial\Delta_k/\partial x)\ln(\beta\epsilon(x'))dx'. \tag{51}$$

The classical phase shifts Δ, Δ_k and Δ_{kk} are all given in Refs. [53, 54] and are evidently phonon-phonon, kink-phonon and kink-kink phase shifts. One can see that the breather-type solitons (28) of s-G have broken up into phonons, but the topological solitons, the kinks and antikinks, survive the Brownian motion: they remain as stable generalised elliptic fixed points now 'dressed' however by the phonons and the other solitons (this is the interpretation of the coupled equation (51) for $\tilde{E}(x)$). Likewise the phonons are dressed by the other phonons and the solitons (eqn. (50) for $\epsilon(x)$). Both kinks and antikinks behave in the same way under Brownian motion (the factors 2 in eqns. (49), (50), (51)).

For the purposes of Davydov's soliton these results would seem to mean that at body temperatures ($310°K$), where *classical* statistical mechanics would seem to apply, his coherent breather-like soliton (strictly a breather-like solitary wave in general) will break up into phonons: these phonons are Gaussian chaotic through the action of the heat bath providing the Brownian motion as explained in Sect.3. Of course if a pure, undressed, Davydov soliton can be excited (by ATP hydrolysis?) on the α-helix at some initial time $t = 0$ it will take some time $t > \tau > 0$ to approach thermal equilibrium and break up. Estimates of the time constant τ are discouraging ($\tau \sim 10^{-14}sec$ has been suggested [55] by other methods, but I am unable yet to confirm these estimates by extending our statistical mechanical methods to the approach to thermal equilibrium [56, 57]).

It may have some relevance to these problems that certain trans-cell-membrane proteins act quite differently transferring excitations essentially structurally rather than dynamically, perhaps as a hopping process. In this context the quantum integrable Hubbard model of hopping fermions, or the boson hopping models solved in [58, 59] and their free energies calculated, may be more relevant.

What is now known in some detail [60]-[62] is that of three trans-membrane proteins from Halobacteria, called the bacteriorhodopsin family [60], two act as ion pumps: bacteriorhodopsin acts as a light driven proton pump and halorhodopsin acts as a light driven chloride ion pump, while sensory rhodopsin acts as a phototactic receptor. Each of these three consist of one continuous protein folded into seven columnar trans-membrane helical pieces grouped to embrace a roughly cylindrical trans-membrane channel, and this structure is a feature also of the larger family of rhodopsins [60]. Rhodopsin occurs in the visual receptor cells of the retina. In the case of bacteriorhodopsin from Halobacteria, studied by high resolution electron cryo-microscopy methods [61, 62], there is a light sensitive element, retinal, transverse to and some half-way up the trans-membrane channel, and photons, absorbed by the retinal, pump protons through this channel via successive attachments, i.e. hoppings, between aspartic acid sites. The proton *gradient* created this way seems to be essential for the synthesis of ATP from ADP [63]. This synthesis appears to take place in a second complex of proteins, ATP synthetase (F_0F_1 synthetase [63] – mitochondrial ATP synthetase is a complex of 14 different proteins embebbed in the inner membrane of the cell). Protons return to the interior of the cell through this complex and the free energy accumulated in the proton gradient is apparently used to synthesise the ATP. There seems to be no oppotunity for a Davydov soliton mechanism to act anywhere in this sequence of steps as described, but the energy stored in the ATP can again be released through its hydrolysis. However, the seven helix structure of the bacteriorhodopsins and rhodopsins seems to be typical for information transfer: the much larger family of eukaryotic G-protein coupled receptors, with widely different functions [63], seems to exhibit the same structural features.

In so far as the classical Davydov soliton may have only the very short lifetime $\tau \sim 10^{-14}$ sec in approach to thermal *equilibrium* it is significant that living systems are plainly not in thermal equilibrium. Evidently they are at least being pumped to some 'far from equilibrium' state in the sense of Sect.6. In the case of the 2-level atom laser (Sect.6) pumping creates coherent excitations and the Benjamin-Feir instability in particular leads to NLS type solitons. In living systems adjacent material must introduce very substantial damping so the natural excitations of the protein α-helix as computed by Davydov must be significantly damped and Brownian motion introduces substantial further damping. The excitations and dynamics of the damped systems described in Sect.6 show that once pumping and damping are introduced together the patterns formed (solitary waves created) become substantially different from the solitons and solitary waves of integrable and non-integrable, but strictly Hamiltonian, systems. Thus for any dynamics of living systems all of the remarks made in this section can only point at widely open questions.

The role of quantum theory rather than classical theory may also be important. Although the classical breathers of s-G break up into phonons in thermal equilibrium at body temperatures, the quantum breathers act quite differently [53]. If at such temperatures *quantum* statistical mechanics is still relevant, we find [53] that at thermal equilibrium quantum s-G has quantum breathers stabilised by quantum mechanics and there are no quantum phonons. The quantum stabilisation of the breathers arises from quantum constraints [15] which impose the Bohr quantisation conditions on the breather mass spectrum. From eqn.(30) the classical breathers have mass $2M \sin \Theta_\ell, 0 < \Theta_\ell < \frac{1}{2}\pi$. This becomes quantised in the quantum s-G through the Bohr conditions [15, 53, 64]

$$4\gamma_o''^{-1} \oint \Theta_\ell d\Phi_\ell = 2\pi n_\ell \hbar \qquad (52)$$

where classical $4\gamma_o''^{-1}\Theta_\ell, \Phi_\ell$ are canonical and $0 < \Phi_\ell < 8\pi$. The quantity $\gamma_o'' \equiv \gamma_o[1 - \gamma_o/8\pi]^{-1}$ is the renormalised coupling constant γ_o, and n_ℓ in (52) is a positive integer. Since Θ_ℓ is a constant of the motion (Sect. 5 and eqn.(30)),

$$\Theta_\ell = n_\ell \hbar \gamma_o''/16 \qquad (53)$$

and $\Theta_\ell < \frac{1}{2}\pi$ means $n_\ell < N_\ell \equiv [8\pi/\gamma_o'']_P$, with $[--]_P$ integral part. There are thus $N_\ell - 1$ distinct quantum breathers with masses $2M \sin(n_\ell \hbar \gamma_o''/16), n_\ell = 1, 2, \cdots, N_\ell - 1$ for each of the labels ℓ appearing in eqn. (30), and it is the dynamics of these *distinct* quantum breathers which undergoes quantum Brownian motion. To these quantum breathers must be added quantum kinks and quantum antikinks each with mass M: these come in pairs in thermal equilibrium and form breather-like kink-antikink pairs of total mass $2M$. There are no quantum phonons in this quantum statistical mechanics their role being taken over by the breathers: this is of course opposite to the *classical* statistical mechanics of s-G just described. The NLS model is a non-relativistic form of the s-G: there are no topological kinks and quantised breather-like solitons and nothing else contribute to the quantum free energy: (47) shows only phonons contribute to the classical free energy. The situation in the quantum NLS is complicated by the fact that there is no stable quantum mechanical ground state unless the total number of particles is restricted.

For living matter it now becomes an important dynamical question to ascertain whether, in effect, pumping can lower the *effective* temperatures so that these quantum solitons can survive at body temperatures. In the laser pumping in effect produces *negative* temperatures (inverted populations): there is then the thresholds to coherent laser oscillation in the presence of both pumping and damping described in Sect. 6. Analysis of the quantum s-G or quantum NLS models at negative temperatures is not yet carried out but it can be done. The stabilising character of quantum mechanics was recognized by Schrödinger [47] where the quantum mechanical binding of atoms, capable of creating a genetic code, was seen as strongly resistant to Brownian motion on energetics grounds.

The molecule of DNA has also been modelled by the classical Toda lattice, eqn.(19) [65]. The lattice is driven by a Gaussian random force inducing Brownian motion at temperatures β^{-1}, and because of the 'fluctuation -dissipation theorem' [66] a corresponding damping term enters the Toda equation of motion. It is possible to count the equivalent bare Toda solitons in the system in thermal equilibrium [65] and this shows an interesting $\beta^{-\frac{1}{3}}$ (a one third power $T^{\frac{1}{3}}$ in the temperature T) dependence. This analysis necessarily assumes that Toda lattice solitons survive the action of the Brownian motion in thermal equilibrium: Toda solitons are neither breather-like nor topological. But we have shown that this view is wholly correct in so far as we have derived [67] the same $\beta^{-\frac{1}{3}}$ dependence by the methods sketched in Sect. 3.

The problem is of some considerable technical interest since the result depends on the fact [67] that, whilst the free energy density $lim\ FL^{-1}$ is made up of phonon and soliton contributions, the solitons couple only to *themselves* in the expressions (like (50) and (51)), for their excitation energies. The solitons thus "dress" each other in thermal equilibrium in this example but the phonons play no role in this dressing. As noted these solitons are neither topological nor breather-like. They are essentially the KdV solitons eqn.(14) and evidently robustly survive as such in thermal equilibrium. Such classical solitons, dynamically self-organized from arbitrary initial data as they are, may thus yet be seen to play a significant role in living systems beset by Brownian motion – whether they are in actual thermal equilibrium or are in some far from equilibrium pumped state. The quantum statistical mechanics of the Toda lattice has still to be worked out.

References

1. 'Solitons', eds. R.K. Bullough and P.J.. Caudrey, Springer Topics in Current Physics 17 (Springer-Verlag: Heidelberg, 1980).
2. V.V. Nalimov 'Self-organization as a creative process: philosophical aspects'. This volume.
3. 'The Concise Oxford Dictionary' (Oxford University Press : Oxford, 1951) Fourth Edn. p.243.
4. For example B. de Witt 'Quantum mechanics and reality', Physics Today, September 1970 (American Institute of Physics) and references, where the 'Copenhagen collapse' - the 'conventional interpretation of quantum mechanics'- is described as the second of three still current, alternative, and debated interpretations of the measurement process in quantum mechanics. Measurement leads to the "von Neumann infinite regression catastrophe". E.P.Wigner for example then argued that it is the entry of the measurement signal to the *human consciousness* which triggers decision on the actual outcome of an observation. De Witt argues that the 'Copenhagen collapse' is equally anthropocentric. The strictly causal interpretation of measurement which does not change the formalism of quantum mechanics due to Everett, Wheeler and Graham (referenced) raises bizarre problems concerning reality; and in so far as it apparently involves an infinite splitting into imperfect copies of each observer, each however undetectable within the formalism by the other, the problems of human 'consciousness' and its 'complexity' become accentuated rather than in any way resolved.
5. V.I. Arnold 'Mathematical Methods of Classical Mechanics' (Springer-Verlag: Berlin, 1978).

6. C.R. Doering, J.D. Gibbon, D.D. Holm and B. Nicolaenko, Nonlinearity 41, 279-309 (1988); and more particularly J.D. Gibbon, M.V. Bartuccelli C.R. Doering 'Weak and Strong Turbulence in the GGL Equation' in "Nonlinear Processes in Physics", A.S. Fokas, D.J. Kaup, A.C. Newell and V.E. Zakharov eds. (Springer-Verlag, Berlin 1993) pp. 275-278 and references.

7. I.E. Antoniou and I. Prigogine, 'Intrinsic irreversibility and integrability of dynamics', Preprint, and paper by I.E. Antoniou, this volume.

8. L. Markus and K.R. Meyer 'Generic Hamiltonian Dynamical Systems are Neither Integrable nor Ergodic', Mem. Am. Math. Soc. 144, (1974).

9. A.S. Wightman, 'The Mechanism for Stochasticity in Classical Dynamical Systems'in:"Perspectives in Statistical Physics" (North Holland: Amsterdam, 1981) pp. 343-363.

10. A.N. Kolmogorov, 'Preservation of Conditionally Periodic Movement with Small Change in the Hamiltonian Function', Dokl. Acad. Nauk, SSSR 98, 527-531 (1954).

11. V.I.Arnold, 'Proof of Theorem of A.N. Kolmogorov on the Invariance of Quasi-periodic Motions Under Small Perturbations of the Hamiltonian' Russian Math. Surveys 18: 5, 9-36 (1963)

12. V.I. Arnold, 'Small Denominators and Problems of Stability of Motion' in:" Classical and Celestial Mechanics", Russian Math. Surveys 18: 6, 85-191, (1963).

13. J.K. Moser, Nachr. Akad. Wiss. Göttingen 1 (1962).

14. R.K. Bullough and J. Timonen,'Quantum groups and quantum complete integrability' in :"Differential Geometric Methods in Theoretical Physics", Springer. Lecture Notes in Physics 375, C. Bartocci, U. Bruzzo and R. Cianci eds. (Springer-Verlag : Heidelberg, 1991) pp. 71-90.

15. R.K. Bullough and J. Timonen "Quantum solitons on quantum chaos : coherent structures, anyons, and statistical mechanics" in :"Microscopic Aspects of Nonlinearity in Condensed Matter" A. R. Bishop et al, eds (Plenum Press : New York, 1991) pp. 263-280.

16. R.P. Feyman, and A.R. Hibbs, 'Quantum Mechanics and Path Integrals', (McGraw-Hill Book Co: New York, 1965).

17. R.K. Bullough and J. Timonen, 'Soliton statistical mechanics', Physics Reports (Elsevier : Amsterdam). To appear 1993.

18. L.D. Faddeev and L.A. Takhtadjan, 'Hamiltonian Methods in the Theory of Solitons' (Springer-Verlag : Berlin, 1987).

19. A.G. Izergin and V.E. Korepin, Nucl. Phys. B 205, [F55] 401 (1982).

20. N.M. Bogoliubov and R.K. Bullough, J. Phys. A: Math. Gen. 25, 4057-4071 (1992).

21. A. Davey and K. Stewartson, Proc. Roy. Soc. London, A 338, 101 (1974).

22. M. Boiti, J. Léon, L. Martina and F. Pempinelli, Phys. Letts. 132A 1, 432 (1988).

23. E.H. Lieb and W. Liniger, Phys. Rev. 130, (1963).

24. C.N. Yang and C.P. Yang, J. Math. Phys. 10, 1115 (1969).

25. R.K.Bullough, D.J. Pilling and J. Timonen, J. Phys. A: Math. Gen 19, L955-960 (1986).

26. P.J. Caudrey, J.C. Eilbeck, and J.D. Gibbon, Nuovo Cimento 25, 497 (1975).

27. R.K. Dodd and R.K. Bullough, Physica Scripta 20, 512-530 (1979).

28. R.K. Bullough, 'Solitons' in: "Interaction of Radiation with Condensed Matter Vol I" IAEA-SMR 20/51 International Atomic Energy Agency, Vienna 1977, pp. 381-469.

29. R.K. Bullough, P. J. Caudrey and H.M. Gibbs, Chapter 3 of Ref [1], and references.

30. R.K. Dodd and R.K. Bullough, Proc. Roy. Soc. London A, 351 , 499-523, (1976).

31. R.K. Bullough, P.J. Caudrey, J.D. Gibbon, S. Duckworth, H.M. Gibbs, B. Bölger and L. Baede, Optics Comm, 18, 200 (1976).

32. R.Y. Chiao, E. Garmire and C.H. Townes, Phys. Rev. Lett. 13, 479 (1969).

33. From the 'Remarkable World of Nonlinear Systems' - a publication by the UK's SERC (1989). The calculations are by J.V. Moloney, Heriot-Watt University (now at Arizona) with A.C. Newell and D.W. McLaughlin, Tucson, Arizona (and DWM now at Princeton Univ.).

34. R.K. Bullough, 'Optical solitons, chaos and all that: quantum optics and nonlinear phenomena' AAPPS Bulletin, Vol. 2, No.2, 1992, pp. 23-37.

35. R.E. Slusher and H.M. Gibbs, Phys. Rev. A 5, 1634 (1972) and references.

36. G.L. Lamb Jr. and D.W. McLaughlin, Chap.2 in Ref. [1].

37. P.J. Caudrey, J.D. Gibbon, J.C. Eilbeck and R.K. Bullough, J. Phys. A 6, 1337-1347 (1973).

38. R.K. Bullough, P.W. Kitchenside, P.M. Jack and R. Saunders, Physica Scripta 20, 364-381 (1979).

39. J.V. Moloney 'Patterns and weak turbulence in wide aperture lasers' in: "Future Directions of Nonlinear Dynamics in Physical and Biological Systems", P.L. Christiansen, J.C. Eilbeck, R.D. Parmentier and A.C. Scott eds. (Plenum Press : New York, 1993. To appear.

40. R.K. Bullough, 'Conceptual models of co-operative processes in nonlinear dynamics' in :"Nonlinear Electrodynamics in Biological Systems", W. Ross Adey and Albert F. Lawrence eds (Plenum Press : New York, 1984) pp. 347-392.

41. H. Haken, Phys. Letts. A53, 77 (1975).

42. W. Firth, 'Transverse Nonlinear Optics' to appear in : Proc Second Intl. Summer School on Nonlinear Optics, O. Keller ed. (Nova Science Publ. Inc : New York). To appear in 1993.

43. R.K. Bullough, S.S. Hassan, M.N.R. Ibrahim, N. Nayak and G.P. Hildred 'Analytical and numerical results in a nonlinear refractive index theory of optical bistability'. To be published.

44. It seems integrability for G=1 is due to S. Manakov, c. 1974, see [42].

45. A. Petrossian, M. Pinard, J.Y. Courtois and G. Grynberg, Europhys. Letts. 18, 689-695 (1992).

46. H. Haken, 'Some Aspects of Synergetics' in : "Synergetics A Workshop" H. Haken ed. (Springer-Verlag : Berlin, 1977) pp. 10-11.

47. E. Schrödinger, 'What is Life?' (Cambridge U P : Cambridge, 1967) Combined reprint (1967) with 'Mind and Matter'.

48. A.S. Davydov and N.I. Kislukha, Physica Status Solidi B 59, 465-470 (1973), refs. in Refs.[49, 50], and A.S. Davydov, 'Problem of the transport of biological energy in living systems'. This meeting.

49. Alwyn C. Scott, 'Davydov's Soliton' in : "Molecular Theories of Cell Life and Death" Sungchal Ji ed. (Rutgers University Press : New Brunswick, N.J. 1991) pp. 264-281.

50. A.C. Scott, Phil. Trans. Roy. Soc. London, A, 313, 423-436 (1985).

51. S.W. Englander, N.R. Kallenbach, A.J. Heeger, J.A. Krumhansl and S. Litwi, Proc. Nat. Acad. Sci., USA 77, 7222 (1980).

52. S. Yomosa, Phys. Rev. A, 27, 2120 (1983). See also Asok Banerjee and Henry M. Sobell in:"Nonlinear Dynamics", Ref [40].

53. R.K. Bullough, Y.Z. Chen and J. Timonen 'Soliton Statistical Mechanics - Thermodynamic Limits for Quantum and Classical Integrable Models' in : "Nonlinear

World Vol. 2", V.G. Bar'yakhtar et al. eds. (World Scientific: Singapore, 1990) pp. 1377-1422.

54. J. Timonen, M. Stirland, D.J. Pilling, Yi. Cheng and R.K. Bullough, Phys. Rev. Lett. 56, 2233-6 (1986).

55. J.P.Gettingham and J.W. Schweitzer, Phys. Rev. Lett. 62, 1752 (1985).

56. R.K. Bullough 'Remarks on the Thermalisation of Solitons on Biological Molecules' in : "The Living State II", R.K. Mishra ed. (World Scientific : Singapore, 1985) pp. 458-466.

57. R.K. Bullough, J Timonen and D.J.Pilling in: "Coherence, Cooperation and Fluctuations", Proc. of the Symposium on the Occasion of the Sixtieth Birthday of Professor Roy J. Glauber, Harvard University, October 19 , 1985. F. Haake, L.M. Narducci and D.F. Walls eds. (CUP, Cambridge, 1986).

58. N.M. Bogoliubov, R.K. Bullough and G.D. Pang, 'Exact solution of a q-boson hopping model' Phys. Rev. B. 47, 17, 11495-8 (1993).

59. R.K. Bullough, N.M. Bogoliubov and G.D.Pang 'The quantum Ablowitz- Ladik equation as a q-boson system' in:"Future Directions", Ref [39].

60. R. Henderson and G. Schertler, Phil. Trans. Roy. Soc. London B 326, 379-389 (1990).

61. R. Henderson, J.M. Baldwin, T.A. Ceska, F. Zemlin, E. Beckmann and K.H. Downing, J. Mol. Biol. 213, 899-929 (1990).

62. Sriram Subramaniam, Mark Gerstein, Dieter Oesterhelt and Richard Henderson, 'Electron diffraction analysis of structural changes in the photocycle of bacteriorhodopsin' EMBO J. In Press, Oct. 1992.

63. For example C.K. Mathews and K.E. van Holde, 'Biochemistry' (The Benjamin/Cummings Publ. Co. Inc: Redwood City, California, 1990) pp.528-529, pp. 525 - 527, pp. 797 - 801.

64. R.H. Dashen, B. Hasslacher and A. Neveu, Phys. Rev. A11, 3424 (1975).

65. V. Muto, A.C. Scott and P.L.Christiansen, Physica, D44, 75 (1990).

66. R. Kubo, Rep. Progr. Phys. 29, 255 (1966).

67. R.K. Bullough, Yu-zhong Chen and J.T. Timonen, 'Thermodynamics of Toda lattice models: application to DNA' Physica D 000,000 (1993).

Note 1:

A structural mechanism, rather than an energy transport mechanism involving Davydov's soliton, is apperently preferred in the process of infection of a cell by an invading virus. In two reports on the influenza virus ('Breakthrough on Flu Viruses ' reported by Warren E. Leary, New York Times, Friday 21st May, 1993, and 'A Spring-Loaded Mechanism for the Conformational Change of Influenza Hemaglutinin' by Chavela M. Carr and Peter S. Kim in Cell, Friday 21st May 1993) it is postulated that a conformational change in Hemaglutinin induced by a change in pH from 7 to 5 allows the transfer of a segment of protein (the hydrophilic "fusion peptide") over a distance of some 100 Å allowing its fusion to the endosome membrane and the subsequent rupture of that membrane. The mechanism is apparently driven by triplets of "super-coiled" (i.e. coiled-coiled) -helices. A 'hair-pin' structure of the coiled-coiled -helices at pH7 opens up from 80 Å to 135 Å at pH5; and the fusion peptide springs up with this extended system of -helices to reach the endosome membrane. The postulated mechanism is based on coiled-coil propensity and other physico-chemical data including the temperature dependence of circular dichroism data which support the idea of a "kinetic" control from a metastable state at pH7 to a stable one at pH5 for the conformational change. This is rather than a thermodynamic control through a sequence of stable equilibrium states. The complete structure of Hemaglutinin at pH5 has been solved by X-ray methods at Harvard University's Departement of Biochemistry and Molecular Biology, and its forthcoming report by D.C. Wiley and colleagues should serve to endorse or otherwise this postulated mechanism – and I believe this endorsement can only be partial. Nevertheless, a dynamical theory of the whole mechanism could be very rewarding for future studies of the dynamics of biological molecules.

Note 2:

Reference the infinite dimensional KAM theorem, in a current preprint 'The nonlinear Klein – Gordon equation on an interval as a perturbed Sine – Gordon equation' A.I. Bobenko and S.B. Kuksin apparently show that for the nonlinear Klein - Gordon equation $u_{tt} - u_{xx} = -mu + f(u)$, $m > 0$, with $u = u(t, x)$, $0 < c < \pi$ and Neumann boundary conditions $u_x(t, 0) = u_x(t, \pi) \equiv 0$, and with $f(u) = \chi u^3 + O(|u|^4)$ and $\chi \neq 0$ viewed as a perturbation of the Sine – Gordon equation $u_{tt} - u_{xx} = -m \sin u$, where $\sin u = u - \frac{1}{6}u^3 + O(|u|^5)$, most of the small amplitude finite gap solutions of Sine – Gordon persist. The argument appeals to infinte dimensional KAM theory as described in S.B. Kuksin 'Nearly integrable infinite dimensional Hamiltonian systems' in Lecture Notes in Mathematics, Springer-Verlag (to appear) and 'KAM theory for partial differential equations' in Proc. First European Congress of Mathematics (Paris 6 - 10 July, 1992) (to appear).

Uncertainty Principle, Coherence and Structures

Ke-Hsueh Li

1 Introduction

Recently, problems concerning order and disorder, coherence and incoherence, structure and chaos etc. have been lively discussed not only among physicists. As physicists, we should frequently be asked: what is the coherence which is used today in almost all branches of the sciences? how is this coherence defined exactly? what is the connection between structure and coherence etc. In this paper we try to show that 1) the concept of coherence is really rooted in the uncertainty principle of Heisenberg, the coherence space-time is identical to the uncertainty space-time for states of minimum fluctuations; 2) within the uncertainty (coherence) space-time we find structures of matter waves by using the general time-dependent Schrödinger equation, where the boundary condition plays an important role; 3) from this point of view we have a solid physical basis for understanding the principle of the "order-from-order" mechanism in the sense of Schrödinger [15].

2 Uncertainty Principle and Coherence

The uncertainty principle for a system of material particles follows from the wave–corpuscle duality of matter. The dynamic state of a classical particle is defined at every instant by specifying precisely its position $r(x, y, z)$ and its momentum $p(p_x, p_y, p_z)$. In contrast, since the wave function $\Phi(r)$ has a certain spatial extension, one cannot attribute a precise position to a quantum particle; one can only define the probability of finding the particle in a given region of space when one carries out a measurement of its potition.The same consideration is also valid for the observation of momentum. The wave function actually represents the probability amplitude including all information about the system under consideration. The probability amplitude is the basic element or alternative (in the sense of Feynman) for analysing the interference and interaction between different particles and fields in quantum mechanics. Thus we should not forget Feynman's formulation of the uncertainty principle [1]: "Any determination of the alternative taken by a process capable of following more than one alternative destroys the interference between alternatives." Heisenberg's original statement of the uncertainty principle was not given in the form we have cited above. For a one-dimensional case, from Heisenberg's original paper we have [2]

$$\Delta x \Delta p \geq \hbar \tag{1}$$

with the de Broglie relation

$$\Delta p = \hbar \Delta k = \hbar \Delta \frac{2\pi}{\lambda} = \hbar \frac{\Delta \lambda}{\lambda_0^2} \tag{2}$$

where λ_0 denotes the mean value of the wavelength under consideration.

From (2) and (1) we have

$$\Delta x \geq \frac{1}{2\pi} \frac{\lambda_0^2}{\Delta \lambda} = L \tag{3}$$

where L is the coherence length of the matter wave [3, 4]. With the exception of the equality sign on the right hand side, the relation (3) had already been deduced by Heisenberg himself. Just as momentum is a wave number and cannot be localized in space, so energy is a frequency and cannot be localized in time. Thus there exists, as suggested by the principle of relativity, a time-energy uncertaity relation

$$\Delta t \Delta E \geq \hbar \tag{4}$$

where $\Delta E = \hbar \Delta \omega$ according to Planck's relation.
Because of

$$\hbar \Delta \omega = \hbar \left(2\pi \Delta \frac{v}{\lambda} \right) = \hbar v \left(2\pi \frac{\Delta \lambda}{\lambda_0^2} \right) = \hbar \frac{v}{L} \tag{5}$$

we have from (4)

$$\Delta t = \frac{L}{v} = \tau \tag{6}$$

where $v = c$ for electromagnetic waves, and τ represents the coherence time.

One should be aware of the important fact that the Planck constant $\hbar$ has been eliminated from the equations (3) and (6). Therefore the concept of coherence length and time (coherence space-time) can be used certainly for macroscopic scales, as we see in classical optics for her analysing interference phenomena [3, 4].

Here we will not get involved in the difficulty of interpretation of relation (4). Readers who are interested in that interpretation are refered to P.Busch [5]. In the well known Feynman diagrams the electromagnetic interaction(or e.m. force) can be described by virtual photons exchanged between particles. The existence of virtual photons would be limited by the time- and space-interval of the uncertainty relation. The temporal and spatial range of interaction is limited by

$$\Delta t = \frac{\hbar}{\Delta E} \tag{7}$$

where E is the energy of a virtual photon and the range of interaction forces is

$$R = \Delta x = c\Delta t = \frac{c\hbar}{\Delta E} = \frac{c\hbar}{h\Delta v} = \frac{1}{2\pi}\frac{\lambda_0^2}{\Delta\lambda} = L \qquad (8)$$

Here we obtain the coherence length of photons [6] L. The uncertainty relation of Heisenberg is the only fundamental condition to determine the space-time region of interactions between particles and fields. The important fact which may be traced back to the original papers of Heisenberg is that the uncertainty relation is only an alternative approach to coherence properties of fields and particles.

The interference between different probabilitiy amplitudes, hence the coherence property of probability distributions (wave packets) constitutes the essential point of quantum theory [2].

3 Structures in Coherence Volume

In classical optics, the coherence length L is equivalent to the breadth of the wave packet [7]. Similarly, in quantum mechanics the uncertainty region Δx in the position of a particle corresponds to the breadth of the wave function within which the matter is statistically distributed in the sense of Born [8]. The description of a quantum particle by the wave packet can be established by using the superposition principle. According to this principle Schrödinger first introduced the coherence state [9] which represents a bridge between the microscopic quantum description and macroscopic intuition. The same consequence of coherence has been derived from the uncertainty principle of Heisenberg as we discussed above.

If the Hamiltonian of Schrödinger's wave equation is subjected to some boundary condition, e.g. potential well, then we have essentially a standing wave as the solution of the Schrödinger equation due to the interference structure which forms within the coherence volume. This fact can be generally shown as follows [16]:

We rewite the general time-dependent Schrödinger equation

$$i\hbar\frac{\partial}{\partial t}\Psi = \left(-\frac{\hbar^2}{2m}\nabla^2 + V\right)\Psi \qquad (9)$$

with the Hamiltonian:

$$H = -\frac{\hbar^2}{2m}\nabla^2 + V \qquad (10)$$

in the form

$$\hbar^2\nabla^2\Psi = 2m\left[V - i\hbar\frac{\partial}{\partial t}(\ln\Psi)\right]\Psi \qquad (11)$$

with the momentum operator

$$\mathbf{p} = \frac{\hbar}{i}\nabla \quad \text{and} \quad p^2 = -\hbar^2\nabla^2 . \tag{12}$$

If we consider the equation (11) as an eigenvalue problem and Ψ as an eigenfunction of (9) i.e.

$$i\hbar\frac{\partial}{\partial t}\Psi = H\Psi = E\Psi \tag{13}$$

we have then the appropriate eigenvalue of the momentum operator

$$|p| = \sqrt{2m(E - V)} \tag{14}$$

where E is the well-defined energy of the whole system in a certain state. We see here that the very important information carried by the wave function actually arises from the boundary condition represented by potential V and the total energy E of the system.

On account of the relation

$$\Delta|p| = \sqrt{\langle|p|^2\rangle - \langle|p|\rangle^2} \tag{15}$$

and the constraint $\frac{\langle V\rangle}{E} < 1$, we obtain from equations (1) and (3)

$$\Delta x \geq \hbar\sqrt{\frac{2E}{m}}\frac{1}{\Delta V} = L \tag{16}$$

3.1 Structure of the Hydrogen Atom as an Example

Taking for V a simple Coulomb potential of hydrogen atom, we get [8]

$$\langle V\rangle = \left\langle\frac{-ke^2}{r}\right\rangle = -\frac{ke^2}{n^2 a_0}$$

$$\langle V^2\rangle = \frac{2k^2 e^4}{(2l + 1)n^3 a_0^2} \tag{17}$$

where n is the principal quantum number and l is the angular momentum quantum number, a_0 the radius of the first Bohr orbit.

If

$$|E| = \frac{mk^2 e^4}{2\hbar^2}$$

is constant, then the stationary problem of this subsection brings equation (16) into the form:

$$\Delta x \geq \frac{n^2 a_0}{\sqrt{\frac{2n}{2l+1} - 1}} = L_{n,l} \tag{18}$$

where $L_{n,l}$ denotes the coherence length of an electron wave with different quantum numbers n and l. For minimum uncertainty states we have [16]

$$\Delta x = L_{n,l} \tag{19}$$

Herein we see that the uncertainty (coherence) length generally may not be smaller than the Bohr radius. Taking $n = 1$, $l = 0$ for the first Bohr orbit, we have

$$\Delta x = a_0 = L_{1,0} \tag{20}$$

the coherence length of the electron in the state $n = 1$, $l = 0$. The general coherence length of different orbitals of the hydrogen atom can be described by the equation (18) or (19).

A similar problem of electron structures in the hydrogen atom has been discussed in all elementary text books of quantum mechanics, but it is not evidently shown that the structures can be formed only within the uncertainty (or coherence) volume. Interference is a general phenomenon of matter waves and coherence is an inherent feature of them.

3.2 Building of More Complex Structures

The covalent chemical bond building between two atoms is usually said to be formed by the "sharing of electrons". That is essentially a consequence of the uncertaity principle. As two atoms approach each other, the outher electrons, which are not at localized points in space but instead must be thought as a "smear" of charge around the atom, cease to be assignable to a particular atom. This means that the uncertainty(coherence) volume of the electron under consideration is now enlarged; therefore, the uncertainty in the momentum, and thus also the minimum possible momentum, decreases. Since momentum and energy are related by

$$\frac{1}{2}mv^2 = E, \qquad mv = p \qquad \text{and} \qquad E = \frac{p^2}{2m} \tag{21}$$

the energy of the system also decreases. But a decrease in the energy of the system is an increase in the binding force, and therefore the atoms are held together.

The understanding of this force depends on a quantum mechanical explanation that was first provited by Heitler and London in 1927. Because of the interesting display of coherence phenomena in this case we will briefly illustrate its important concerns: Let us consider simple molecular orbitals of the hydrogen molecular ion. Its Hamiltonian is usually given by

$$H = -\frac{\hbar}{2m}\nabla^2 + \frac{e^2}{4\pi}\left(-\frac{1}{r_a} - \frac{1}{r_b} + \frac{1}{R}\right) \tag{22}$$

where r_a and r_b are the distances of the single electron from the nuclei a and b, separated by R. The lowest energy solution, a binding orbit, can be modelled as

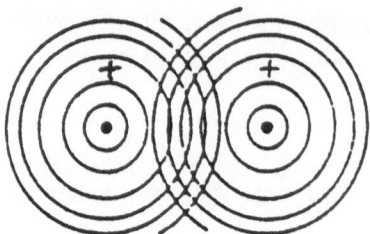

Fig. 1. Constructive overlap between 1s-orbitals

the linear combination $\Psi_a + \Psi_b$ of hydrogen 1s-orbitals centred on each nucleus, see Fig. 1

The antibonding orbital can also be modelled as $\Psi_a - \Psi_b$ by using the same two atomic orbitals, but this time superimposed with opposing amplitudes, as shown in Fig. 2

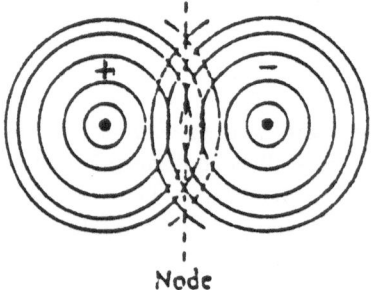

Fig. 2. Destructive overlap between 1s-orbitals

In the first case there is constructive interference where the two wave functions symmetrically overlap and their amplitudes augment each other. This models the accumulation of amplitude in the internuclear region which we know is an important feature of the bonding process. In the antibonding combination the amplitudes tend to cancel each other and there is exact cancelation to form a nodal plane half way between the two nuclei, just as in the exact solution. The destructive interference where the two wave functions antisymmetrically overlap with opposite phase models the annihilation of the amplitude in the internuclear region [10].

Now let us replace the electron by a photon and assume that the two atoms are sufficiently separated so that the overlap of electron wave functions is impossible. In this case the symmetric state describes the fact that if a photon is emitted by one atom, one must be emitted by the other. That is just the most superradiant state, a cooperative coherent spontaneous radiation process. On the other hand, the antisymmetric state describes the fact that a photon released by one atom must be absorbed by the other and vice versa. That is only possible if one atom is initially excited and the other not and both the atoms must be within the uncertainty length (coherence length) of the photons. This latter case corresponds, therefore, to the most subradiact state of the two atom-system in which the photon can be trapped for a long time. All these cases have been discussed in detail by different authors since the fundamental pioneering work of Dicke [11].

In the discussion in this subsection we have aimed to show that there really are coherent structures within the uncertainty volume. These structures are formed by the interference of matter waves constrained by boundary conditions. The existence of structures does not contradict the state indeterminacy of the uncertainty principle.The indeterminacy actually means the delocalization of the particle in position and momentum compared with the localization in the classical sense. Due to delocalization within the uncertainty volume the particle loses its classical picture.

3.3 Pauli Principle and Coherent Structures

For two-, and more than two-, electron problems we must also take the Pauli principle into account. Then the problem is of course complicated: For example, the problem of the helium atom or the hydrogen molecule. But, the general principle: the structure builds only within the coherence volume, remains intact, because the Pauli principle itself represents destructive interference within the uncertainty volume; outside that the electrons are independent. In order to illustrate the main feature of this case let us consider two free electrons with parallel spins. To the first electron we assign the velocity v_1 and position x_1 and to the second the velocity v_2 and position x_2. Hence we have the wave functions [12]:

$$\Psi_1(x_1) = A_1(x_1) \exp(2\pi i \frac{m}{h} v_1 x_1)$$
$$\Psi_2(x_2) = A_2(x_2) \exp(2\pi i \frac{m}{h} v_2 x_2) \tag{23}$$

for the two electrons respectively. The probability to find an individual electron at the position x_i within a unit volume is

$$\langle \Psi_i^* \Psi_i \rangle_{V_u} = \int_{V_u} |A_i(x_i)|^2 dx_i \tag{24}$$

where V_u denotes the unit volume of the configuration space. Due to the validity of the Pauli principle the total wave function of the whole system must be antisymmetrized by reason of the degeneracy of states described by equation (23).

$$\Psi(x) = \Psi_1(x)\Psi_2(x+R) - \Psi_1(x+R)\Psi_2(x) \tag{25}$$

where we have redefined variables as follows:

$$x = x_1; \qquad x_2 = x_1 + R = x + R$$

Then we have the probability distribution to find both electons in a unit configuration volume determined by the coordinates x_1 and x_2:

$$\int_{V_u} \Psi^*(x)\,\Psi(x)dx =$$

$$\int_{V_u} |A_1(x)|^2 |A_2(x+R)|^2 dx + \int_{V_u} |A_1(x+R)|^2 |A_2(x)|^2 dx$$

$$- \int_{V_u} A_1^*(x+R)A_2^*(x)A_1(x)A_2(x+R)\exp(-2\pi i\frac{m}{h}(v_1-v_2)R)dx$$

$$- \int_{V_u} A_1^*(x)A_2^*(x+R)A_1(x+R)A_2(x)\exp(2\pi i\frac{m}{h}(v_1-v_2)R)dx \quad (26)$$

The last two terms represent the correlation function of electron wave packets just as that of light wave trains. Because of the continuity of the spatial coordinate x, the integral over unit volume is actually an ensemble average in it. In addition we must be aware that the centre of the first electron is at the position x and that of the second one is at $x+R$, where they have the maximum value of the wave packet.

On account of the calculation method for correlation functions [13, 14] we have

$$\int_{V_u} \Psi^*(x)\,\Psi(x)dx =$$

$$2\int_{V_u} |A_1(x)|^2 |A_2(x+R)|^2 dx \left\{1 - e^{-\frac{R}{L}}\cos 2\pi\frac{m}{h}(v_1-v_2)R\right\} \quad (27)$$

where we have inserted into eq. (27) the identity

$$\int_{V_u} |A_1(x)|^2 |A_2(x+R)|^2 dx \equiv \int_{V_u} |A_1(x+R)|^2 |A_2(x)|^2 dx \quad (28)$$

for two identical particles. The mathematical structure of eq. (27) represents really a destructive interference of two electron wave functions.

In the equation (27) we introduced L as the coherence length of the electron wave packet. For $R > L$, the distance between two electrons is greater than the coherence length of wave packet, the exponential term $\exp(-R/L) \to 0$, then the interference term in the bracket disappears. In such a case eq. (27) represents the probability distribution of two independent electrons. :

$$\Psi^*(x)\Psi(x) = 2|A_1(x)|^2 |A_2(x+R)|^2 = 2|A_1(x_1)|^2 |A_2(x_2)|^2 \quad (29)$$

If $R \leq L$, we have a situation where the velocity v_1 cannot be permitted to take the same value of the velocity v_2 if the conservation of particles is not to be violated. For different v_1 and v_2 the destructive interference pattern can in principle be recognized.

But, we must be aware that the case $R \leq L$ means that the distance between two electrons falls into the uncertainty region, so that the difference of the rigorous positions of two electrons cannot be recognized due to the delocalization of the particle within the uncertainty region. Therefore, it is reasonable that the two electron positions x_1 and x_2 may be considered as equal in such a case. At the same time the exponential factor $\exp(-R/L)$ can be approximated to unity, especially for $R \ll L$,

$$e^{\frac{-R}{L}} \rightarrow 1, \qquad \text{for} \qquad R \ll L$$

So thus we fall again into a situation violating the conservation law of particles. On account of this fact we can conclude that the Pauli principle, by antisymmetrization of the wave function of the whole system, forbits two electrons with parallel spins from staying within a region smaller than the uncertainty (coherence) volume. Outside the coherence volume the two electrons are independent. By deriving this statement we have not considered the interaction between the two electrons. The repulsion of one another of electrons in this case comes only from the antisymmetrization of the electron wave function. The physical background would be represented by the destructive interference of electron wave packets.

4 Some Conclusions

In the present paper the identity of coherence space-time with the uncertainty space-time for states of minimum fluctuations in the sense of Heisenberg has been demonstrated where the coherence space-time can be used both for microscopic quantum systems and also for macroscopic scales. In particular, the building of structures, interference patterns of matter waves within the coherence volume has been illustrated using Schrödinger equation with boundary constraints. Thus we get the picture that a quantum particle within its uncertainty space-time is uncertain only in the sense of delocalization, but not indeterminate in its structure of space-time. The Schrödinger wave function has a certain delicate spatial-temporal structure determined by constraints such as space-time boundary conditions appearing in the Hamiltonian of the problem. The information of the wave function can be, at least partly, represented by the interference pattern in the coherence (uncertainty) volume. The question is: what happens outside the coherence volume? If two particles are separated by a distance greater than the coherence length, the two wave packets cannot overlap each other, so that they are independent. This fact has been shown by a special example of two electrons with parallel spins which must be subjected to the Pauli principle. The Pauli principle itself presents an argument via the requirement of antisymmetrization of the wave function that the structure builds only within the coherence volume (here we have destructive interference); outside that the two electrons are independent of each other and no structure can be identified. We have also shown that the complex structure can be built from many particles or particle systems by the expansion of the coherence volume of the electron wave at the cost of lowering the kinetic energy, for example of the covalent bond of molecules. In reality, the overlap of electron wave packets indicates the expansion of the coherence volume due to the delocalization of electrons within it. Of course we must not forget the positive role of photons in the formation of complex systems. Here we have to consider that different matter waves in different coherence space-times display cooperative effects for building a harmonic and dynamic complex of systems under different constraints and interactions. Now, we come with the above point of view to the hard problem, namely [15], "It appears that there are two different 'mechanisms' by which orderly events can be produced: the 'statistical mechanism' which produces 'order from disorder' and the new one, producing 'order from order'. To the unprejudiced mind the second principle appears to be much simpler, much more plausible. No doubt it is. That is why physicists were so proud to have fallen in with the other one, the 'order-from-disorder' principle,which is actually followed in Nature and which alone conveys an understanding of the great line of natural events, in the first place of their irreversibility. But we cannot expect that the 'laws of physics' derived from it suffice straightaway to explain the behavior of living matter, whose most striking features are visibly based to a large extent on the 'order-from-order' principle. You would not expect two entirely different mechanisms to bring about the same type of law – you would not expect your latch-key to open your neighbour's door as well."

We think that we have presented in this part a solid physical basis for the principle of the 'order-from-order' mechanism.

References

1. R.P. Feynman and A. Hibbs, Quantum Mechanics and Path Integrals (McGraw-Hill Book Company, 1965).
2. W. Heisenberg, Physikalische Prinzipien der Quantentheorie, (Bibliograph. Institut, Mannheim,1958).
3. M. Born and E. Wolf, Principles of Optics 5th edition (Pergamon Press, Oxford, 1975) p. 511.
4. M. Garbuny, Optical Physics, (Academic Press, New York and London 1965), p.303.
5. P. Busch, Foundations of Physics, 20, (1990) p.1 and p.33.
6. R.T. Weidner and R.L. Sells, Elementare moderne Physik, (Friedr. Vieweg+Sohn, Braunschweig, Wiesbaden 1982) p.419-436.
7. E. Goldin, Waves and Photons, An Introduction to Quantum Optics, (John Wiley + Sons, Inc.,1982).
8. A. Messiah, Quantum Mechanics vol.1, (North-Holland Pub. Comp. Amsterdam, 1965).
9. E. Schrödinger, Der stetige Übergang von der Mikro- zur Makromechanik. Die Naturwissenschaften, Heft 28, 9.7.1926, p.665.

10. P.W. Atkins, Molecular Quantum Mechanics 2nd ed. (Oxford University Press, Oxford, New York, 1983) p. 250-283.
11. R.H. Dicke, Coherence in Spontaneous Radiation Process, Phys. Rev. 93, (1954), p. 99-110.
12. W. Finkelnburg, Einführung in die Atomphysik (Springer-Verlag, Berlin, Göttingen, Heidelberg, 1954)
13. H. Haken, Synergetics, An Introduction (Springer-Verlag, Berlin, Heidelberg, New York, 1977)
14. H. Risken, Fokker-Planck Equation (Springer Berlin Heidelberg New York Tokyo, 1984).
15. E. Schrödinger, What is Life? (Cambridge University Press, 1945), p.80.
16. F.A. Popp, K.H.Li and Q.Gu, Recent Advances in Biophoton Research and its Applications (World Scientific, Singapore New Jersey London Honkong, 1992) p. 113 and p. 431.

Intrinsic Irreversibility of Unstable Dynamical System

Ioannis E. Antoniou and Shuichi Tasaki

1 Introduction

Self-organization is characterized by the irreversible emergence of structure and functions in open systems far from equilibrium [1] - [4]. The basis for self-organization is the entropy change of the system. The change in entropy ΔS during a time interval Δt can be decomposed into two parts [1, 4]:

$$\Delta S = \Delta_e S + \Delta_i S \qquad (1)$$

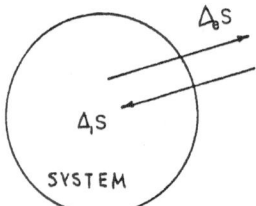

Fig. 1. Entropy change

$\Delta_e S$ is the entropy exchange with the outside world or the environment and $\Delta_i S$ is the internal entropy production due to irreversible processes, like diffusion, heat conduction or chemical reactions, in the interior of the system.

According to the second law of thermodynamics the internal entropy increases monotonically in time.

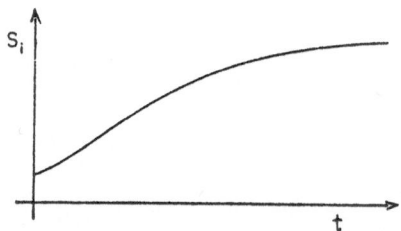

Fig. 2. The Second Law of Thermodynamics

The second law is just an expression of the unique privileged direction of time for all actual processes in the Universe we observe. Irreversibility as "the fundamental difference between the two directions of time, the future and the past, is one of the most important and striking properties of the phenomenal world" [5]. However, irreversibility is absent at the fundamental level because dynamics in the conventional formulations in terms of Hamilton's or Schrödinger's equation is a temporally reversible theory. Historically the bridge between dynamics and thermodynamics was constructed by appealing to plausible probabilistic assumptions like Boltzmann's Stosszahlansatz or coarse-graining projections or approximations [4, 6] or to the irreversible act of measurement [5]. All these approaches are based however on assumptions extraneous to dynamics and they cannot be considered as the final answer to the problem. According to Ilya Prigogine we should look for intrinsically irreversible representations of dynamics where the dynamical group splits into two distinct semigroups corresponding to the two directions of time. We have recently shown that such representations indeed exist and can be constructed for highly unstable dynamical systems [7] - [10]. The construction provides a generalized spectral decomposition of the evolution operator. The eigenvalues correspond to the decay rates, the Lyapounov times or the diffusion coefficients, while the eigenvectors acquire meaning in suitable functional spaces beyond the conventional Hilbert space formulations.

Our systematic method for the construction of the spectral decomposition of evolution operators applies to arbitrary linear operators with continuous spectrum and is a generalization of perturbation theory, of the partitioning technique for matrices as well as of the intertwining wave operator method of scattering theory. The method [10] is a continuation of the Brussels–Austin work [11] - [16] whence many ideas originate. The algorithm is briefly presented in section 3 after a discussion on the spectral theory of dynamical systems. The method is applied to two representative examples of unstable systems, namely the baker transformation and the Friedrichs model. The intrinsic irreversibility of these models is discussed in section 6.

2 Dynamical instability and spectrum

Liouville established in 1855 [17] that a Hamiltonian system with N degrees of freedom can be integrated through a canonical transformation to a solvable system, if N constants of motion in involution exist. Despite the fact that all one-dimensional systems can be integrated, the integration program faces difficulties for systems with $N \geq 2$ degrees of freedom. Attempts to construct canonical transformations using pertubation theory formulated as in the one-dimensional case lead to divergences, due to the famous problem of small denominators. Most interesting problems of mechanics starting from the three-body problem (Sun, Earth and Moon or Jupiter) do not possess analytic constants of motion. Poincaré's theorem [18, 19] made clear that the reason for the non-existence of analytic invariants, or equivalently of analytic canonical transformations to a solvable system, is the presence of resonances.

Poincaré considered Hamiltonians of the form :

$$H = H_0(J_1...J_N) + \lambda V(J_1...J_N, \alpha_1...\alpha_N). \qquad (2)$$

H_0 is the Hamiltonian of a solvable system and V is the potential energy of interaction depending on both the actions J_k and the angles α_k, $k = 1, ..., N$. Perturbation terms contain frequency denominators

$$\frac{1}{\sum_{k=1}^n \omega_k n_k} ,$$

where $\omega_k = \frac{\partial H_0}{\partial J_k}$ are the frequencies of the system and n_k are integers. The appearance of resonant denominators where $\sum_{k=1}^n \omega_k n_k = 0$, prevents the analytic integration of the system.

Poincaré's theorem was a turning point in the development of mechanics [20]. The role of resonance became clear, however, only in the late 1950's from the Kolmogorov, Arnold and Moser theorems [17]. The main result is that, if the frequencies $\omega_1 \cdots \omega_N$ are sufficiently far from resonance, then such divergences can be avoided and the motion is in principle controllable in the long term. It should be emphasized however, that the KAM theorem did not solve the problem of integration of Poincaré's non-integrable systems, but only established conditions which guarantee that sufficiently far from resonances, reasonable answers may be found through suitable perturbation schemes. Let us also remark here that the resonance divergence is a manifestation of the general fact [21] that not only the integrability question is undecidable but also the actual integration of integrable systems is a non-computable problem.

Furthermore, KAM's result cannot be generalized to large systems ($N \to \infty$), because the small denominators $n_1\omega_1 + \cdots + n_N\omega_N$ cannot be bounded away from zero as N tends to infinity [22], in contrast to the case of finite degrees of freedom N. In large systems, resonances are present everywhere, as the frequencies become arbitrarily close to resonance and Fourier sums are replaced by Fourier integrals. The limiting case $N \to \infty$ corresponds to large Poincaré non-integrable systems. Large Poincaré systems are a special class of dynamical systems with continuous spectrum for which the resonant denominators in the secular perturbation terms are arbitrarily close to zero for a continuous set of spectral values. Typical cases of such denominators are

$$\frac{1}{\sum_{k=1}^\infty n_k \omega_k} , \quad \text{or} \quad \frac{1}{\omega - \omega_\alpha} . \qquad (3)$$

The first type appears in Hamiltonian systems as $N \to \infty$, while the second type appears when certain discrete or continuous modes ω_α are coupled to a continuum mode ω. We emphasize that the term "large" refers to the continuous spectrum while the term "Poincaré" refers to the difficulties in any perturbational treatment of the problem expressed by Poincaré in his famous theorem [18, 19].

A dynamical system $S_t, t \in (-\infty, +\infty)$ on the phase space Γ has a continuous spectrum if the Frobenius-Perron operator U_t

$$U_t \rho(y) = \rho\left(S_t^{-1} y\right) , \tag{4}$$

has a continuous spectrum on the Hilbert space of square integrable phase functions.

For Hamiltonian systems, the generator of the unitary group U_t is the Liouville operator L :

$$U_t = e^{-iLt} ,$$

$$L = i\{H, \}_{\text{P.B.}} .$$

The idea of using operator theory for the study of dynamical systems is due to Koopman [23] and was extensively used thereafter in statistical mechanics [11] and ergodic theory [24] - [26] because the dynamical properties are reflected in the spectrum of the density evolution operators.

Dynamical systems with continuous spectrum evolve in a mixing manner, as first pointed out by Koopman and von Neumann [24]. They observed that, for dynamical systems with continuous spectra, "the states of motion corresponding to any set become more and more spread out into an amorphous everywhere dense *chaos*. Periodic orbits, and such like, appear only as very special possibilities of negligible probability". This result was refined later into the ergodic hierarchy of mixing, Lebesgue, Kolmogorov, exact and Bernoulli systems [25, 26] which is the basis of modern theory of "chaos". To our knowledge, the paper by Koopman and von Neumann contains the first use of the term "chaos" in the context of dynamical systems. The baker transformation is a simple representative example of Kolmogorov systems.

It is important to notice that non-integrability due to resonances appears also in large quantum systems with a continuous spectrum. Here, divergences due to resonances appear in the constructive perturbative solutions of the eigenvalue problem [15]. These divergences are manifestations of the general non-computability of the eigenvalue problem for self-adjoint operators in Hilbert space [27]. Large Poincaré systems, classical or quantum, appear quite commonly in our dynamical models of natural phenomena, as they include the models of the kinetic theory and Brownian motion, the interaction between matter and light as well as interacting fields and quantum many-body systems involving collisions. The Friedrichs model is a simple representative example.

3 The spectral decomposition algorithm

For unstable systems with continuous spectrum, the conventional spectral theory in Hilbert space not only suffers from divergences [15], but also gives spectral resolutions which do not reflect the dynamical processes involved. One would like to have the Lyapounov times or the lifetimes of the unstable states in the spectrum. Such spectral decompositions can be constructed for unstable systems with a systematic method [10]. The key idea is the construction of an intermediate operator Θ which is intertwined with the system operator U:

$$U\Omega = \Omega\Theta \ , \quad \text{or} \quad U = \Omega\Theta\Omega^{-1} \ . \tag{5}$$

The intertwining relation (5) was obtained by Prigogine, Henin, George and Rosenfeld [12]. Recently, Petrosky and Prigogine [16] pointed out that the intertwining relation can be used for the construction of the spectral decomposition of the Liouville operator. The method as reformulated by us [10] may also be considered as a generalization of the partitioning technique of matrices and of the intertwining wave operator method of scattering theory.

The construction begins with the choice of a suitable biorthonormal system $|\varphi_\sigma\rangle$, $\langle\tilde{\varphi}_\sigma|$. The system operator U is then decomposed into the diagonal U_0 and the nondiagonal part U_1:

$$U_0 = \sum_\sigma \langle\tilde{\varphi}_\sigma| \, U \, |\varphi_\sigma\rangle \, |\varphi_\sigma\rangle\langle\tilde{\varphi}_\sigma| \ = \sum_\nu \omega_\nu P_\nu, \tag{6}$$

$$U_1 = \sum_{\tau\neq\sigma} \langle\tilde{\varphi}_\tau| \, U \, |\varphi_\sigma\rangle \, |\varphi_\tau\rangle\langle\tilde{\varphi}_\sigma| \ , \tag{7}$$

ω_ν labels the identical diagonal elements $\langle\tilde{\varphi}_\sigma| \, U \, |\varphi_\sigma\rangle$ (degenerate eigenvalues) of U_0 and $P_\nu \equiv \sum_{\sigma(\omega_\sigma=\omega_\nu)} |\varphi_\sigma\rangle\langle\tilde{\varphi}_\sigma|$ is the ω_ν-eigenprojector ($\nu = \sigma$ if there is no degeneracy). The intermediate operator Θ is decomposable with respect to a complete family of projectors P_ν. After constructing the spectral decomposition of Θ, the intertwining operator Ω provides the spectral decomposition of the operator U.

The intermediate operator Θ as well as the similarity operator Ω are obtained [10] from the auxiliary "creation" and "destruction" operators C_ν and D_ν:

$$\Theta \equiv \sum_\nu \left(P_\nu U P_\nu + P_\nu U C_\nu P_\nu\right) = \sum_\nu \Theta_\nu \ , \tag{8}$$

$$\Omega \equiv \sum_\nu \left(P_\nu + C_\nu\right) \ , \tag{9}$$

$$\Omega^{-1} \equiv \sum_\nu \left(P_\nu + D_\nu C_\nu\right)^{-1}\left(P_\nu + D_\nu\right) \ . \tag{10}$$

The creation and destruction operators C_ν and D_ν are constructed iteratively as solutions of the following nonlinear equations for the components $P_\mu C_\nu$ and $D_\nu P_\mu$, ($\nu \neq \mu$):

$$[U_0, P_\mu C_\nu]_- = (P_\mu C_\nu - P_\mu)U_1(P_\nu + C_\nu) \ , \tag{11}$$

$$[U_0, D_\nu P_\mu]_- = (P_\nu + D_\nu)U_1(P_\mu - D_\nu P_\mu) \ , \tag{12}$$

or

$$(\omega_\mu - \omega_\nu)P_\mu C_\nu = (P_\mu C_\nu - P_\mu)U_1(P_\nu + C_\nu) \ , \tag{13}$$

$$(\omega_\nu - \omega_\mu)D_\nu P_\mu = (P_\nu + D_\nu)U_1(P_\mu - D_\nu P_\mu) \ . \tag{14}$$

If there is resonance we use the time-ordering boundary condition or regulariza-
tion rule, which is a natural generalization of scattering or radiation processes.
The matrix elements of $P_\mu C_\nu$ and $D_\nu P_\mu$ are regularized according to the process
they represent. Increase of correlations or emission takes place in the future and
we use the forward propagators. Decrease of correlations or absorption takes
place in the past and we use the backward propagators.

$$\langle \alpha | \, C_\nu \, | \beta \rangle = \frac{\lambda}{\omega_\alpha - \omega_\beta + i\epsilon_{\alpha\beta}} \langle \alpha | \, (C_\nu - Q_\nu) L_1 (P_\nu + C_\nu) \, | \beta \rangle, \qquad (15)$$

$$\langle \beta | \, D_\nu \, | \alpha \rangle = \frac{\lambda}{\omega_\beta - \omega_\alpha + i\epsilon_{\beta\alpha}} \langle \beta | \, (P_\nu + D_\nu) L_1 (Q_\nu - D_\nu) \, | \alpha \rangle, \qquad (16)$$

where the sign of the infinitesimal $\epsilon_{\alpha\beta}$ for the transition $\beta \to \alpha$ is given by

$$\epsilon_{\alpha\beta} = \begin{cases} -\epsilon \, , & \text{Forward propagation,} \\ +\epsilon \, , & \text{Backward propagation.} \end{cases} \qquad (17)$$

Let us remark in passing that the equations (11–16) are a nonlinear generalization of the Lippmann–Schwinger equations for the Möller wave operators
of scattering and may provide non-unitary intertwining operators even if the
scattering asymptotic condition [28] between U_0 and U fails.

The algorithm for the construction of spectral decomposition of the operator
U is the following:

1) Choose a convenient initial biorthonormal system $|\varphi_\nu\rangle$, $\langle \widetilde{\varphi}_\nu |$ and decom-
   pose the operator U into the diagonal part U_0 and the nondiagonal part or
   perturbation U_1 (6,7).
2) Construct the creation and destruction operators C_ν and D_ν iteratively as
   solutions of the equations (13,14), or (15,16) if resonances exist, starting
   with $C_\nu^{[0]} = D_\nu^{[0]} = 0$.
3) Construct the intermediate operator Θ from formula (8) and find the spectral
   decomposition of Θ by solving the eigenvalue problem in each P_ν subspace.
   As Θ is not Hermitian, we expect that Θ may have a generalized Jordan
   decomposition.
4) Obtain the spectral decomposition of U from the spectral decomposition of
   Θ using the similarity Ω (5,9,10).

Note that the intertwining relation (5) follows if C_ν and D_ν satisfy the nonlinear equation (11,12) and that the equality $\Omega\Omega^{-1} = \Omega^{-1}\Omega = I$ can be proved
from the following relations [10]:

$$(P_\nu + D_\nu)(P_\mu + C_\mu) = \delta_{\nu\mu}(P_\nu + D_\nu C_\nu), \qquad (18)$$

$$\sum_\nu (P_\nu + C_\nu)(P_\nu + D_\nu C_\nu)^{-1}(P_\nu + D_\nu) = I. \qquad (19)$$

The resulting spectral decompositions are meaningless in Hilbert space, but
they acquire meaning in suitable linear functional spaces [8] - [10] as will be

illustrated for two representative examples of systems with continuous spectra, namely, the baker transformation [9] and the Friedrichs model [8].

4 The baker transformation

The β-adic, $\beta = 2, 3, \cdots$, baker transformation B on the unit square $\Gamma = [0, 1) \times [0, 1)$ is a two-step operation: 1) squeeze the 1×1 square to a $\beta \times 1/\beta$ rectangle and 2) cut the rectangle into β $(1 \times 1/\beta)$-rectangles and pile them up to form another 1×1 square:

$$(x, y) \longmapsto B(x, y) = (\beta x - r, \frac{y + r}{\beta}) \quad (\text{for } \frac{r}{\beta} \le x < \frac{r + 1}{\beta}, \, r = 0, \cdots \beta - 1) . \quad (20)$$

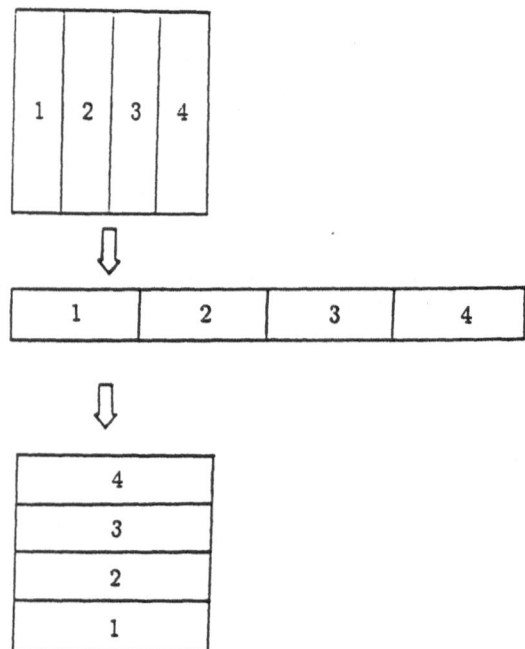

Fig. 3. The baker transformation ($\beta = 4$)

The dyadic baker transformation is the well-known model of Hopf introduced in 1930's as one of the first examples of mixing transformations. The name baker reflects the close resemblance to the action of a baker rolling dough.

The β-adic baker transformation is isomorphic to the β-adic equal weight Bernoulli shift and has positive Kolmogorov–Sinai entropy $\log_2 \beta$. The model is a prototype of chaotic behavior. The probability densities $\rho(x, y)$ evolve according to the Frobenius–Perron operator U [26]:

$$U\rho(x,y) \equiv \rho(B^{-1}(x,y)) \tag{21}$$
$$= \rho(\frac{x+r}{\beta}, \beta y - r), \quad (\text{for } \frac{r}{\beta} \leq y < \frac{r+1}{\beta}, \ r = 0, \cdots \beta - 1).$$

The Frobenius–Perron operator U is unitary on the Hilbert space L^2 of square integrable densities over the unit square and has countably degenerate Lebesgue spectrum on the unit circle plus the simple eigenvalue 1 associated with the equilibrium (as is the case for all Kolmogorov automorphisms).

By suitably choosing the initial biorthonormal system in terms of the Bernoulli polynomials we obtain [9] a Jordan spectral decomposition of the Frobenius–Perron operator U.

$$U = |f_{00}\rangle\langle\tilde{f}_{00}| + \sum_{\nu=1}^{\infty}\left\{\sum_{r=0}^{\nu}\frac{1}{\beta^{\nu}} |f_{\nu,r}\rangle\langle\tilde{f}_{\nu,r}| + \sum_{r=0}^{\nu-1} |f_{\nu,r+1}\rangle\langle\tilde{f}_{\nu,r}|\right\}. \tag{22}$$

The vectors $|f_{\nu,r}\rangle$ and $\langle\tilde{f}_{\nu,r}|$ form a Jordan basis for the Frobenius–Perron operator U

$$U |f_{\nu,r}\rangle = \begin{cases} \frac{1}{\beta^{\nu}} |f_{\nu,r}\rangle + |f_{\nu,r+1}\rangle, & (r = 0 \cdots \nu - 1) \\ \frac{1}{\beta^{\nu}} |f_{\nu,r}\rangle, & (r = \nu) \end{cases} \tag{23}$$

$$\langle\tilde{f}_{\nu,r}| U = \begin{cases} \frac{1}{\beta^{\nu}}\langle\tilde{f}_{\nu,r}| + \langle\tilde{f}_{\nu,r-1}|, & (r = 1 \cdots \nu) \\ \frac{1}{\beta^{\nu}} \langle\tilde{f}_{\nu,r}|, & (r = 0) \end{cases} \tag{24}$$

$$\langle\tilde{f}_{\nu,r}| f_{\nu',r'}\rangle = \delta_{\nu\nu'}\delta_{rr'}, \tag{25}$$

$$\sum_{\nu=0}^{\infty}\sum_{r=0}^{\nu} |f_{\nu,r}\rangle\langle\tilde{f}_{\nu,r}| = I. \tag{26}$$

While the Frobenius–Perron operator U is unitary in the Hilbert space L^2 and thus has a spectrum on the unit circle $|z| = 1$ in the complex plane, the spectral decomposition (22) of U includes the numbers $1/\beta^{\nu} < 1$ which are not in the Hilbert space spectrum. The spectral decomposition (22) also shows that the Frobenius–Perron operator has Jordan-block parts despite the fact that it is diagonalizable in the Hilbert space.

The principal vectors $f_{\nu,r}$ and $\tilde{f}_{\nu,r}$ are linear functionals over the tensor product spaces $L^2_x \otimes \mathcal{P}_y$ and $\mathcal{P}_x \otimes L^2_y$ correspondingly. $\mathcal{P}_x$, $\mathcal{P}_y$ are the spaces of polynomials in the first and second variables. The spaces $L^2_x \otimes \mathcal{P}_y$ and $\mathcal{P}_x \otimes L^2_y$ inherit the strict inductive limit topology from the space of polynomials and they are complete, barrelled locally convex topological spaces dense in L^2 [9, 29].

5 The Friedrichs model for resonance scattering

The model of Friedrichs [30] implicitly discussed also by Dirac [31] is the simplest model involving quantum transitions due to resonance scattering and it is solvable in a non-analytic way. The model served as a prototype test model for several proposals on the decay problem in quantum mechanics [32] - [35]. On the basis of the time ordering rule, Petrosky, Prigogine and Tasaki [15] obtained the explicit formulas for the eigenvectors with complex eigenvalues, as well as a complex spectral decomposition for the Friedrichs Hamiltonian. These eigenvectors were previously constructed by Sudarshan, Chiu and Gorini [32] using analytic continuation techniques.

In the Friedrichs model a discrete state $|1\rangle$ is coupled with a continuum of states $|\omega\rangle$, $0 \leq \omega < \infty$, corresponding to field modes. The Hamiltonian operator is:

$$H = H_0 + \lambda V, \tag{27}$$

$$H = \omega_1 |1\rangle\langle 1| + \int_0^\infty d\omega\, \omega|\omega\rangle\langle\omega| + \lambda \int_0^\infty d\omega V_\omega \left(|\omega\rangle\langle 1| + |1\rangle\langle\omega|\right). \tag{28}$$

The spectrum of the solvable Hamiltonian H_0 consists of a purely absolutely continuous part $[0, \infty)$ and a simple point of eigenvalue ω_1 embedded in the continuum.

The perturbation λV causes transitions between the eigenvalue ω_1 and the continuum only. The embedded eigenvalue ω_1 becomes unstable under the influence of the perturbation, due to resonance with the continuum and dissolves in the continuum. The spectrum of the Friedrichs Hamiltonian H consists of the absolutely continuous spectrum of the solvable Hamiltonian H_0 only.

Friedrichs model satisfies the asymptotic condition of scattering theory [28]. The unitary wave operators $\Omega_{\text{OUT} \atop \text{IN}}$ provide the Friedrichs [30] outgoing and incoming solutions to the eigenvalue problem [34]

$$H\left|f_\omega^{\text{OUT} \atop \text{IN}}\right\rangle = \omega\left|f_\omega^{\text{OUT} \atop \text{IN}}\right\rangle, \tag{29}$$

$$\left|f_\omega^{\text{OUT} \atop \text{IN}}\right\rangle \equiv \Omega_{\text{OUT} \atop \text{IN}}\left|\omega\right\rangle. \tag{30}$$

The Friedrichs solutions expressed in terms of the basis $|1\rangle$, $|\omega\rangle$, have the form:

$$\left|f_\omega^{\text{OUT} \atop \text{IN}}\right\rangle = |\omega\rangle + \frac{\lambda V_\omega}{\eta_\pm(\omega)}\left(|1\rangle + \int_0^\infty d\omega' \frac{\lambda V_{\omega'}}{\omega - \omega' \pm i0}|\omega'\rangle\right) \tag{31}$$

$1/\eta_\pm(\omega)$, $0 \leq \omega < \infty$ are the boundary functions of the partial resolvent $1/\eta(z)$ with respect to the state $|1\rangle$, from above and below the real axis correspondingly:

$$\left\langle 1\left|\frac{1}{z - H}\right|1\right\rangle \equiv \frac{1}{\eta(z)} \tag{32}$$

$$\eta(z) = z - \omega_1 + \int_0^\infty d\omega' \frac{\lambda^2 |V_{\omega'}|^2}{\omega' - z} \tag{33}$$

$$\eta_\pm(\omega) \equiv \eta(\omega \pm i0) = \omega - \omega_1 + \int_0^\infty d\omega' \frac{\lambda^2 |V_{\omega'}|^2}{\omega' - \omega \mp i0}. \tag{34}$$

Due to the cut along the semiaxis $[0, \infty)$, the resolvent formula (33) defines two analytic functions $1/\eta_\pm(z)$ on the upper and lower half-plane correspondingly. It is assumed that the interaction is such that the function $1/\eta_+(z)$ has a meromorphic extension to the lower half-plane with a simple pole at z_1 and that similarly $1/\eta_-(z)$ has a meromorphic extension to the upper half-plane with a simple pole at the complex conjugate z_1^* of z_1. Conditions which establish the meromorphic structure of the partial resolvent are discussed for example by Exner [35].

There are two difficulties with the Friedrichs eigenvectors:

a) They are not analytic in the coupling parameter as expected from the resonance [15].
b) The unstable state $|1\rangle$ is lost in the continuum. Evolution is simply a shift. The quantum transitions and decay events are not explicitly manifested in this representation.

The general consensus is that the dissolved eigenvalue appears either in the spectral concentration of H or as a pole of the resolvent operator or of the scattering operator analytically continued to the complex plane. Either way the eigenvalue "influences" the evolution and produces decay. A complete set of right and left eigenvectors were constructed by Sudarshan, Chiu and Gorini [32]. These solutions were also obtained by Petrosky, Prigogine and Tasaki [15] using the time ordering rule and perturbation theory.

Our method together with the time ordering rule gives also the same generalized eigenvectors. The right and left eigenvectors associated with the complex eigenvalue z_1 are:

$$|f_1\rangle = |1\rangle + \int_0^\infty d\omega \frac{\lambda V_\omega}{[z_1 - \omega]_+} |\omega\rangle, \tag{35}$$

$$\langle \tilde{f}_1| = \frac{1}{\eta_+'(z_1)} \left(\langle 1| + \int_0^\infty d\omega \frac{\lambda V_\omega}{[z_1 - \omega]_+} \langle \omega| \right). \tag{36}$$

The eigenvalue z_1 satisfies the equation:

$$z_1 = \omega_1 - \int_0^\infty d\omega \frac{\lambda^2 |V_\omega|^2}{[\omega - z_1]_+}. \tag{37}$$

Eq.(37) means that z_1 is the simple zero of the meromorphic extension $\eta_+(z)$ to the lower half-plane of the analytic function $\eta(z)$ (33). The real part $\tilde{\omega}_1$ of the complex eigenvalue

$$z_1 = \tilde{\omega}_1 - i\frac{\gamma}{2}, \qquad \gamma > 0 \tag{38}$$

gives the level shift $\tilde{\omega}_1 - \omega_1$, while the imaginary part gives the decay rate γ.

The eigenvectors f_1 and $\tilde{f}_1$ cannot exist in Hilbert space because they correspond to complex eigenvalues of the self-adjoint operator H. However, they acquire meaning as linear functionals over the test function spaces Φ_+ and Φ_- correspondingly defined as [8]:

$$\Phi_{\pm} \equiv \left\{ \phi \mid \langle \omega \mid \phi \rangle \quad \text{are in} \quad \theta[\mathcal{S} \cap \mathcal{H}_{\pm}^2] \right\}, \tag{39}$$

where $\langle \omega \mid \phi \rangle$ is the spectral representation of ϕ in the continuous eigenvectors $|\omega\rangle$ of H_0, $\mathcal{H}_{\pm}^2$ are the Hardy functions from above and below correspondingly, $\mathcal{S}$ is the Schwartz class of infinitely differentiable, rapidly decreasing functions on the real line and θ is the Heaviside unit step function. The function spaces $\theta[\mathcal{S} \cap \mathcal{H}_{\pm}^2]$ were introduced by Bohm and Gadella [36], who also studied their properties.

6 Intrinsic irreversibility

The spectral decomposition (22) of the Frobenius–Perron operator of the baker transformation and the generalized eigenvectors (35,36) of the Friedrichs model give rise to time asymmetric evolutions in the extended functional spaces.

For the baker transformation, the unitary evolution U^n, $n = 0, \pm1, \pm2 \cdots$ of densities associated with the Frobenius–Perron operator (21) can be extended to the dual of $L_x^2 \otimes \mathcal{P}_y$ for positive times $n > 0$ only and to the dual of $\mathcal{P}_x \otimes L_y^2$ for negative times $n < 0$ only [9]. The time symmetric unitary evolution splits therefore into two semigroups, the forward semigroup on the dual of $L_x^2 \otimes \mathcal{P}_y$ and the backward semigroup on the dual of $\mathcal{P}_x \otimes L_y^2$. This is the essence of intrinsically irreversible representations of dynamics provided by the principal vectors $f_{\nu,r}$ and $\tilde{f}_{\nu,r}$ (23,24). Irreversibility emerges naturally as the selection [4, 14] of the semigroup corresponding to our observations.

The split of the unitary Frobenius–Perron group into two semigroups is a manifestation of the evolution of the forward and backward K-partitions correspondingly. The forward K-partition of the baker transformation consists of the vertical lines and becomes progressively refined in the future, while the backward K-partition consists of the horizontal lines and becomes progressively refined in the past [4]. The intrinsic irreversibility of the baker transformation and of Kolmogorov systems in general was formulated by Misra, Prigogine and Courbage [4, 37] in terms of the forward and backward Markov semigroups associated with averaging over the cells of the forward and backward K-partitions correspondingly.

For the Friedrichs model we obtain similar results [8]. The extension of the unitary group $U_t = \exp(-iHt)$ generated by operator H to the duals of Φ_+ and Φ_- (39), splits also into two semigroups, the forward semigroup, $t > 0$, on the dual of Φ_+ and the backward semigroup, $t < 0$, on the dual of Φ_-. The asymmetry in the time evolution associated with f_1 and $\tilde{f}_1$ was also noticed [15] as divergence in the formal expressions for the time evolution of f_1 in the past and $\tilde{f}_1$ in the future. This divergence can now be understood from the fact that f_1 can evolve only towards the future and $\tilde{f}_1$ only towards the past. Because of the complex eigenvalues, the evolution of f_1 and $\tilde{f}_1$ is:

$$e^{-iHt}|f_1\rangle = e^{-i\widetilde{\omega}_1 t}e^{-\frac{\gamma}{2}t}|f_1\rangle , \qquad t > 0, \qquad (40)$$

$$e^{-iHt}|\tilde{f}_1\rangle = e^{-i\widetilde{\omega}_1 t}e^{+\frac{\gamma}{2}t}|\tilde{f}_1\rangle , \qquad t < 0. \qquad (41)$$

The state f_1 decays in the future and the state f_1 decays in the past.

Irreversibility emerges naturally as the selection [4, 14] of the semigroup corresponding to our observations. In this case decay is observed in our future, therefore the extension to the dual of Φ_+ is the relevant one.

In both systems we can define an entropy functional through the generalized eigenvectors (23,24) and (35,36). For the baker transformation:

$$S^B_{[\rho]} = -\sum_{\nu=0}^{\infty}\sum_{r=0}^{\nu}\langle\rho|\ \tilde{f}_{\nu,r}\rangle\langle\tilde{f}_{\nu,r}|\ \rho\rangle, \qquad (42)$$

for every admissible density ρ.

For the Friedrichs model:

$$S^F_{[\rho]} = -\mathrm{Tr}\Big(\rho\ |\tilde{f}_1\rangle\langle\tilde{f}_1|\ \Big) - \mathrm{Tr}\rho, \qquad (43)$$

for every admissible density operator ρ.

7 Concluding remarks

1. The application of our method to the characteristic prototypes of mixing systems, namely the Kolmogorov systems and large Poincaré systems is only the beginning. The application to non-invertible chaotic systems like the expanding part of the baker transformation [38] is also possible. More realistic models will be considered in future publications.

2. In both baker transformation and Friedrichs model, the dynamical properties are reflected in the spectrum because the eigenvalues include the powers of the Lyapounov time and the lifetime of the excited state. In the case of the baker transformation the eigenvalues are just the Pollicott–Ruelle resonances [39, 40], which describe the approach to equilibrium of the autocorrelation functions. In the case of the Friedrichs model, the imaginary of the complex eigenvalue z (38) gives the decay rate of the survival amplitude of the unstable state. In this sense, the spectral decompositions obtained are the natural ones for discussing the approach to equilibrium.

3. The trajectories are excluded from the domain of the spectral decomposition (22) of the baker transformation. This remark shows the intrinsically probabilistic character of unstable dynamical systems. The construction for the Friedrichs

model shows the intrinsic irreversibility at the dynamical level, but does not deal with the probabilistic character of the macroscopic evolutions [4]. The only possibility for a coherent world view is to have both irreversibility and probability at the fundamental level. This can be achieved for large Poincaré systems at the level of the Liouville space of density functions or density operators. The time ordering on the basis of dynamics of correlations [11, 16] leads to generalized eigendensities of the Liouville operator associated with complex eigenvalues. These eigendensities are not reducible to trajectories or wave functions, they live in appropriate extended Liouville spaces and they provide an intrinsically irreversible and probabilistic representation of dynamics [41].

Acknowledgements

Several fruitful discussions with Prof. I. Prigogine motivated to a large extent this work. We are grateful to Profs. C. George, K. Gustafson, L.P. Horwitz, T. Petrosky, J. Reignier, Z. Suchanecki and E.C.G. Sudarshan for discussions and useful comments. We also thank Profs. D. Maass, R. Mishra and E. Zwierlein who organized the Poincaré centenary symposium on "Self-organization as a paradigm in science" for their invitation, hospitality and several fruitful discussions during the conference at Kaiserslautern. We acknowledge the financial support of the Belgian Government (under the contract "Pole d'attraction interuniversitaire"), the European Communities Commission (contract n° 27155.1/BAS.), the U.S. Department of Energy, Grant N° FG05-88ER13897, and the Robert A. Welch Foundation.

References

1. G. Nicolis and I. Prigogine: "Self-organization in Non-equilibrium Systems", Wiley, New York, 1977.
2. H. Haken: "Synergetics, Non-equilibrium Phase Transitions and Self-organization in Physics, Chemistry and Biology", Springer-Verlag, Berlin, 1977.
3. G. Nicolis and I. Prigogine: "Exploring Complexity", Freeman, New York, 1989.
4. I. Prigogine: "From Being to Becoming", Freeman, New York, 1980.
5. J. von Neumann: "Mathematical Foundations of Quantum Mechanics", Princeton Univ. Press, Princeton, 1955.
6. P. Davies:"The Physics of Time Asymmetry", Surrey Univ. Press, U.K., 1974.
7. I. Prigogine: Phys. Reports **219** 93, 1992.
8. I. Antoniou and I. Prigogine: Physica **A 192** 443, 1993.
9. I. Antoniou and S. Tasaki: Physica **A 190** 303, 1992.
10. I. Antoniou and S. Tasaki: Int. J. Quantum Chemistry **45**, 1993.
11. I. Prigogine: " Non-Equilibrium Statistical Mechanics", Wiley, New York, 1962.
12. I. Prigogine, C George, F. Henin and L. Rosenfeld: Chemica Scripta 4 5, 1973.
13. A. Grecos, T. Guo and W. Guo: Physica **80 A** 421, 1975.
14. I. Prigogine and C. George: Proc. Natl. Acad. Sci. USA **80** 4590, 1983.
15. T. Petrosky, I. Prigogine and S. Tasaki: Physica **A 173** 175, 1991.
16. T. Petrosky and I. Prigogine: Physica **A 175** 146, 1991.

17. V. Arnold: "Mathematical Methods of Classical Mechanics", Springer, Berlin, 1978.
18. H. Poincaré: "Les Methodes Nouvelles de la Mécanique Céleste", Vol. I (1892), Dover, New York, 1957.
19. E. Whittaker: "A Treatise on the Analytical Dynamics of Particles and Rigid Bodies", 4th ed. Cambridge Univ. Press, London, 1937.
20. L. Brillouin: "Scientific Uncertainty and Information", Academic Press, New York, 1964.
21. N. da Costa and F. Doria: Int. J. Theor. Phys. 30 1041, 1991.
22. J. Pöschell: "On small divisors with spatial structure", Habilitationsschrift, Universität Bonn, 1989.
23. B. Koopman: Proc. Nat. Acad. Sci. USA 17 315, 1931.
24. B. Koopman and J. von Neumann: Proc. Nat. Acad. Sci. USA 18 255, 1932.
25. I. Cornfeld, S. Fomin and Ya. Sinai: "Ergodic Theory", Springer-Verlag, Berlin, 1982.
26. A. Lasota and M. Mackey: "Probabilistic Properties of Deterministic Systems", Cambridge Univ. Press, U.K., 1985.
27. M. Pour-el and J. Richards: "Computability in Analysis and Physics", Springer, Berlin, 1989.
28. See any book on scattering. For example, R. Newton:"Scattering Theory of Waves and Particles", 2nd ed., Springer, Berlin, 1982; W. Amrein, J. Jauch and K. Sinha:"Scattering Theory in Quantum Mechanics", Benjamin, Massachusetts, 1977.
29. F. Treves:"Topological Vector Spaces, Distributions and Kernels", Academic Press, New York, 1967.
30. K. Friedrichs: Comm. Pure Appl. Math. 1 361, 1948.
31. P.A.M. Dirac:"The Principles of Quantum Mechanics", Oxford University Press, London, 1958.
32. E.C.G. Sudarshan, C. Chiu and V. Gorini: Phys. Rev. D18 2914, 1978.
33. G. Parravicini, V. Gorini and E.C.G. Sudarshan: J. Math. Physics 21 2208, 1980.
34. L. Horwitz and J.P. Marchand: Rocky Mnt. J. Math. 1, 225, 1971.
35. P. Exner:"Open quantum systems and Feynman integrals", Reidel, The Netherlands, 1985.
36. A. Böhm and M. Gadella: "Dirac kets, Gamow vectors and Gelfand triplets", Springer Lect. Notes on Physics 348, Berlin, 1989.
37. B. Misra, I. Prigogine and M. Courbage: Physica 98 A 1, 1979.
38. I. Antoniou and S. Tasaki: J. Phys. A: Math. Gen. 26 73, 1993.
39. M. Pollicott, Invent. Math. 81 413, 1985.
40. D. Ruelle, Phys. Rev. Lett. 56 405, 1986.
41. I. Antoniou, S. Tasaki and Z. Suchanecki, in preparation.

Self-Organization as a Creative Process Philosophical Aspects

V. V. Nalimov

"Are life and mind indeed important in the working of existence?"

(Wheeler, 1988, p. 125)

1 Introduction

In the Russian Philosophical Encyclopaedia (1989) we read:

Self-organization is a process which enables creation, reproduction and perfection of a complex dynamic system organization. Self-organization processes can be realized only in systems of high complexity with a great many of elements interconnected by probabilistic character and not by a hard one ... Their distinctive property is described by a purposeful yet natural spontaneous character: these processes, going on under interaction with the environment, are more or less autonomic and relatively independent of the environment. (p.566)

Key words in this definition are: *complexity* - being interpreted as consciousness from philosophical point of view or at least its weak form as quasiconsciousness; *probability* - a description of randomness which can be considered (in a mathematical sense) as maximum complexity[12] - that is inability to describe a phenomenon by the final algorithm; *spontaneity*- as meta-randomness or inability to predict a change of the parameters of a distribution function.

Self-organization is most distinctly manifested first of all in human intellectual activity. This subject was analysed in detail in many works of mine. I recall here only some of the principal propositions:

a) Each serious text is a phenomenon of a maximum complexity. It cannot be written shorter (algorithmically) than it is. In my model (Nalimov, 1989, 1992) any text is given by a probability density $p(\mu)$, according to the initial premises that all potential meanings are primordially plotted on a numerical axis μ. Comprehension of any text is always very personal. It is described by the Bayesian syllogism:

$$p(\mu/y) = kp(\mu)p(y/\mu)$$

---

[12] According to Kolmogorov's definition

where the filter $p(y/\mu)$ appears spontaneously under the new situation y.
Spontaneous grasping of the text testifies to the process of self-organization within consciousness. Consciousness is our most familiar self-organization system. That system serves as a pattern which provides one with a self-organization image.

b) The Human *Ego* is a text $p(\mu)$, but a special one, alive - capable of incessant reinterpretation of itself under a new situation due to the spontaneous appearance of the filter $p(y/\mu)$. Making the model more complicated it is possible to describe *Ego* as a multidimensional structure. Here we face another degree of freedom - the change of a probabilistical structure of personality resulting from a turn of coordinate axes.

c) *Biosphere*. Biological systems are also given by the probability density $p(\mu)$, when the whole manifold of morpho-physiological attributes is originally packed on the numerical axis μ. The live world is then viewed as a probabilistically weighted unpacking of the continuum of biological meanings. Species are described by multidimensional distributions constructed on axes $\mu_1, \mu_2 \ldots$ corresponding to individuals. It is essential here that changeability has a spotty nature when the entire field of attributes is being changed simultaneously (Nalimov, 1985).

d) The spontaneous selection of *fundamental constants* in physics also can be described by the same model. But we need to assume here that each constant has the meaning of an internal rather than a point. But this is not available for our measuring systems. It is essential that the whole cluster of numbers is selected on the numerical axis. The selected amount of constants provides self-organization for the Universe.

Thus in my approach self-organization is given by the *selection* of the numerical values of the system: a process of spontaneous selection can be considered as a creative one since it appears to be a mechanism of universal creation at all levels of the existence.

It is natural to assume the existence of quasi-consciousness in the Universe which is the carrier of the creative principle. In other words the basis of the Universe is *mentality*. This idea is not a new one, but now it obtains a new background.

The main purpose of this paper is to show the ubiquity of consciousness, manifested at least in its weak forms. Hence we can speak of the ubiquity of self-organization. And if we agree to consider self-organization as a creative process, then we have to accept the ubiquity of creativity.

Such is the conception. To make it more clear it is enough to establish (ground) the idea of ubiquity of consciousness. That is the main subject of the paper.

2 The Regulating Role of Consciousness

"Truth did not come into the world naked, but it came in types and images. One will not receive truth in any other way (67, p. 140)."

(The Gospel of Philip, Robinson,1981)

Consciousness reveals itself in various ways. One of its functions is to regulate our sensual perception to such a form which allows to perceive the world through a system of images.

It follows from the Kantian Critique of Pure Reason that the image of the world contemplated by us is not a mechanistic reflection of external reality but its reconstruction. Man is not a passive observer but a great architect constructing the edifice of the Universe proceeding from the sensual experience of interacting with the external world processed by the filters of personal consciousness. We wish to attach to this subject, traditionally considered as philosophical, an unusual illumination, bringing it closer to the problems of comprehension of the nature of scientific knowledge, on one hand, and the conception of the Transpersonal Observer, on the other hand.

Now we shall discuss the main features of the filters:

Space. In our days the possibility opened up to deepen Kant's idea. When he spoke of space as an a priori given *form* of perception of sensual experience he naturally meant only two- and three-dimensional spaces given by Euclidean geometry which were known at that time. Now we can say that we are a priori given the possibility to interpret the perceived world via spaces of different geometries. Space can be multidimensional and even fraction-dimensional; its metrics may be homogeneous or heterogeneous; it may be non-positively defined; space curvature may be given in different ways; topology determining the neighbourhood of a point may be far from trivial; it may be fiber space; space elements may be represented not by points but by strings of Planck length.

Man extracted ideas of different geometries from the depth of his consciousness. For some reason it was prepared to reveal the manifold of geometries. The success of contemporary physics is to a significant degree determined by its *geometrization* (Kalinowski,1988).

But what exists in the physical reality of the world? Space without geometries? But such space is complete nothing, Void of any attributes, it cannot be discovered and, therefore, cannot be regarded as existing.

Thus, the world is as it is. But if the observer is within it, the consciousness carrier, the world proves to have a non-trivial spatial arrangement. But is the Observer transient or is he endowed with ontological status, is he a non-personified (universal) carrier of consciousness? Here the question naturally arises: what is non-personified consciousness? So far we do not know the answer.

Time. Time for Kant was also an a priori given form of sensual contemplation. Note that at the dawn of our era the Greek philosophers Antiphon and Cryptolaos regarded

> "...time as an idea or measure, not an objective substance (p. 39)"
>
> (Losev, 1964)

We can say now that time is a condensed experience of millenia of interaction between human consciousness and the World. It is an *image* allowing us to resort to a numerical measure when comparing different cyclic systems which are in different conditions. We know from physics that when the motion velocity of the system observed approaches the velocity of light, time slows down; the same happens in strong gravitation fields. Our personal time in some situations accelerates, in others, slows down; the limit of this slow-down is death (Nalimov, 1989) .

Probably the following can be said now: we know not time but the *idea* of time and can select a numerical measure corresponding to it. We feel it relevant to speak here again of universal (non-personified) consciousness supporting our expanding notion of time. Perhaps, the Universal Observer has a notion of extratemporality yet unrevealed to us? And if everything were otherwise, man would feel very uncomfortable in this World.

Number. Things exist without being calculable *per se*. A number, even a natural number, is a concept calculated by the Observer (the latter is here equivalent to consciousness, awareness) . In its simplest manifestation a natural number emerges as an answer to the question posed by the observer: how many objects of a specific type are here? The semantic nature of number is manifested even more explicitly when we are ready to expand this notion by introducing negative numbers, rational and irrational numbers, complex numbers, transcendental numbers (such as π and ε), cardinal numbers, transfinite numbers. The number π appeared as an answer to the question of squaring the circle. The Number ε is one of the most significant constants of mathematical analysis. How could it be recorded in the physical world: by figures, but of what length, or by the formula

$$\varepsilon = \lim_{n \to \infty} (1 + \frac{1}{n})^n$$

Any of those recordings testifies to its semantic nature.

The above can be illustrated by a joke:

The professor enters the classroom. It is empty. The professor starts to read the lecture he ought to. Two students come in. The professor is happy. Then suddenly five students leave. "Well" - the professor muses unhappily - "three more people will come and there will be nobody again".

This a fairly realistic but also a deeply metaphorical description of what goes on in the professor's mind. Perhaps many of our models proceeding from the numerical vision of the world look the same?

But what can be said about fundamental physical constants? They enter fundamental physical equations and thus determine the specific arrangment of our World. It appears that numerical values of the constants, being semantic entities, are directly built into the physical world.

The Greek philosophers, Plato and Plotinus who attached immense importance to number had amazing intuition.

Probability is a measure determined by a system of axioms of its calculus. Subject interpretation of probability may be very varied. In its broadest interpretation probability is a measure of fuzziness of our judgements or concepts of what happens. The word "probability" is related to such words as "predisposition", "inclination", "prevision", "ectation", "dispersion" (distribution along the numerical axis) and, certainly, "contingency". A random variable is considered to be given, if its probabilistic characteristic is given. If we have proved to be ready to acknowledge that the nature of number is *semantic*, it should be all the more so for a probabilistic measure whose calculability is set by a system of axioms.

Probabilistic thinking has essentially changed our vision of the world[13]. That was especially dramatically manifested in the physics of the microworld, a subject beyond the limits of everyday experience and common sense.

In quantum mechanics a particle is not localized in space and time. Its fuzziness is described by a probability density given by the square modulus of the ψ-function[14]. The familiar macroscopic concept of a particle emerges only after the observer's intervention, when some of the potentials are realised by the observer's choice. During the observation intervals we have only a description of the possibilities. In order to perceive what has been said as a reality of the Universe, one has to assume the existence of the *Universal Transpersonal Observer*. Otherwise the concept of a fuzzy particle is a fiction present only in the mind of a contemporary physicist. An attempt to directly include consciousness as a constituent of the physical world is found in the book *The Ghost in the Atom* (P. C. W. Davies and J. R. Brown, eds., 1989).

3 Summarizing the Idea of Consciousness Regulating Filters

All the aforesaid permits us to draw the following conclusions:

a) The classical Kantian conception of the Nature of Pure Reason can be completed now -
   - (a) *number* is also an a priori *form* for contemplation[15] of sensual experience (as well as time and space).
   - (b) the twelve Kantian categories of possible a priori synthetic judgements should be complemented by stochasticity or, more broadly, by *spontaneity*.

---

[13] Historical aspects of this subject are examined in the books by Nalimov (1981) and Gigerenzer et al. (1989)

[14] Note that the ψ-function entering the famous Schroedinger equation is called a probability amplitude. Its values are given by a complex number and therefore it is not directly interpreted within physical system of notions.

[15] Note that number is inherently connected with the notions of space and time. A variety of geometries with different spaces is given by number. Time is also manifested through number. That is why we cannot mention forms of consciousness without number.

b) As culture develops, our consciousness expands by mastering the new fundamental a priori structures.

c) The filters through which we perceive the World are mathematical by nature since they proceed from basic mathematical notions: space, time, number, probability and therefore, chance. This is how our Mind is arranged, but not everybody understands that.

d) But what is mathematical filtering? Perhaps, this is but a manifestation of an inborn mental derangement? Or the filtering is correlated to an independent reality which we are ready to call meta-consciousness regulating the World order? Is it possible to speak of the mathematical nature of the *Transpersonal*, non-personalized consciousness participating in what happens? We feel those questions are very important in revealing some new ways of developing the model of the Universe.

e) Spontaneity (or meta-randomness) manifests itself in the Bayes syllogism. It is a formal, i.e. mathematical, description of the self-organization process which is independent of cause-and-effect interpretation. Maybe it is possible to speak of meta-consciousness, but at present such talk is pointless. Here we are approaching the Ultimate Reality. We have to acknowledge the existence of *Mystery* in the world. Our consciousness, as was discussed above, is getting expanded (enlarged) through cultural development. But this process only moves aside the frontiers of Mystery, and does not destroy it.

4 Self-Organization in the Biosphere

Now a few words about the biosphere. Biological evolution seems similar to human creativity. The emergence of a sharply distinct principle would have been ruinous for an organism. The emergence of a new species can reasonably describe as the emergence of a new filter on the continuum of morpho-physiological attributes. It is no longer possible (after 30-years experience of cybernetics) to think of natural selection without a previously given system of optimality criteria. The question how these criteria are formed brings us again to the Universal observer-participant. A related subject is also the problem of the *aesthetic* in the biosphere and, even more, in the morology of our planet Earth.

"We must again come back to the ancient Greek notion of the Earth as a living organism, the Goddess named Gaia."

(Lovelock, 1988)

"Our approach to co-evolution is close to the well-known Bergson position, as well as to the one of Naess, a popular ecologist of nowdays."

(Naess, 1991)

Here are two relevant quotations from his book:

"Modern ecology has emphasised a high degree of *symbiosis* as a common feature in mature ecosystems, an interdependence for the benefit of all. (p. 169)

The door is open for the positive evaluation of an increase of the realisation of potentialities, that is, of the possibility that more potentialities will be realised. This is meant to imply *continued evolution at all levels,* including protozoans[16], landscapes, and cultures." (p. 201)

Such an enlarged viewpoint of evolution looks very appealing. It is quite natural to try to describe it by applying the concept of multiple self-consistent filters which select, by probabilistic weighting, from the totality of potentials that very one which is capable of creating a viable ecosystem.

5 Self-Organization on a Cosmic Scale

We can easily observe self-organization in the animate world. Though it is not so easy to see it in inanimate nature. But do we easily differentiate between the animate and the inanimate? Maybe we just define as the "inanimate" everything that we cannot recognize as "animate".

Will you imagine a landscape of a sea bay in rocky coast. All the processes within that environment are normally so slow that usually we overlook them. But changing the time-scale we can observe them from the position of a long life being - for example, one such as a sequoia. Within the new time-interval we shall observe some life-like changes, especially evident if we regard the bay from the position of integrity as an ecological system. And here arises a question - whether all the changes in the system keep going on according to the cause and effect relations, or whether they have instead a spontaneous character? Who can answer that question without a premeditated conviction?

But let us come back to the image of the Earth as an animate organism. I would like to appeal to the book by A. Mindell (1989). Referring to M.Eliade he reminds us of the mythological conception of the Earth as an animate body:

...In cosmological terms, we could say that the earth is a dreaming body, organized by divine patterns.

These myths picture the earth as a living body with a dreaming mind. It is not just an electrical nervous system, but a dreaming system, and the events and conflicts here on earth are manifestations of its dream. (p.38)

Earlier Hegel described the Earth as an animate body:

"The earth is a living whole or an individual organism because it is the totality of all of its own chemical processes, which it perennially keeps going on; it is not consumed by them as natural things are, but maintains itself through them as their organizing unity. The earth is both immediate subject of all its activities as well as its own object-product.

---

[16] Protozoans are the simplest single-cell animate organisms.

In its visible shape the earth is also the embodiment of its cosmic past and of cosmic powers beyond it." (261, p. 181)

(Hegel, 1959)

"Those ideas belong to the past. But now going beyond the frontiers of scientism, we seem to be ready to consider them."

(Lovelock, 1988)

If we accept the statement that the Earth has a weak form of consciousness, we can imagine its evolution being revealed according to the same scheme as the evolution of the animate. And here we face again Bayes syllogism, again variability (changeability), begotten by the spontaneous emergence of filters.

Now we'll climb up and have a look over the whole Universe. Let us refer to the attractive book by Barrow (1991), an English astronomer, and pay attention to the diagram in Fig. 5.1 where in coordinate system "masses/sizes"[17] there appeared to be almost linearly arranged principal objects of the Universe starting from proton and atom up to man, asteroids, planets, stars, galaxy and visible Universe. The author comments on this:

The fact that so many of Nature's most important creations owe their gross size and structure to the mysterious values of the constants of Nature places our own existence in a new and illuminating perspective. We can see how the conditions necessary for our own existence are contingent upon the values taken by the constants. At first one might imagine that a change in the value of a constant would simply shift the size of everything a little, but that there would still exist stars and atoms. However, this turns out to be too naive a view ... Were the fine-structure constant to differ by roughly one percent from its actual value, then the structure of stars would be dramatically different ... But, carbon, the crucial biological element which we believe to be essential for the spontaneous evolution of life, should really only exist as the minutest trace element in the Universe instead of in the healthy abundance that we find. (p. 94-95)

The mysterious character of fundamental constants was completed by the formulation of the *Anthropic* principle. I like the words written by J. A.Wheeler, a well-known American physicist, in the Preface to the book (Barrow and Tipler, 1986):

No! The philosopher of old was right. Meaning is important, is even central. It is not only that man is adapted to the universe. The universe is adapted to man. (p. vii)

---

[17] The masses (in grams) and average sizes (in centimetres) of a wide selection of all the principal objects we know to exist in the Universe. The composite structures are equilibrium states between different forces of Nature and their approximate sizes are determined by the fine-structure constant $\alpha = \frac{1}{137}$ and the gravitational-structure constant $\alpha_G = 5.9 \times 10^{-39}$ (which were introduced in the text); the dependence of the mass and size upon these two quantities is indicated for each object (in the book).

That is the highest most explicit manifestation of self-organization. It extends to the past, up to the very moment of value selection of the existing fundamental constants. But this topic is beyond the present discussion.

Self-organization theory if not yet able to explain more or less clearly such a magnificent phenomenon as spontaneous self-organization of the Universe as a whole (including the manifestation of consciousness), should be open enough to accept the possibility of such a phenomenon.

6 Conclusion: The Self-Conscious Universe

Now we have much more reason to ground the thesis of the ubiquity of consciousness, at least of its weak form. The Universe appears before us as a grandiose self-conscious structure.

Consciousness is spontaneous by nature. That is why I chosen as the title of my latest book *spontaneity of consciousness* (Nalimov, 1989).

Spontaneity brings forth self-organization. Self-organization is a selection or in other words - filtering from the initial potentialities. The process of filtering can be described in mathematical language, by reference to Bayes syllogism. The meaning of the model can be explicated especially well when we deal with high forms of consciousness such as the human psyche.

And that is all – the Frontier. Beyond it – we are facing the Ultimate Reality. Mystery.

References

1. Barrow, J. D.; *Theories of Everything*. The Quest for Ultimate Explanation.
   Oxford: Clarendon Press, 1991, 223 p.
2. Barrow , J. D. and Tipler, E. J.; *The Anthropical Cosmological Principle*.
   Oxford: Clarendon Press, 1986, 706 p.
3. Davies, P. S. W. and Brown, J. R. (Eds.); *The Ghost in the Atom*. A Discussion on the Mysteries of Quantum Physics.
   Cambridge: Cambridge University Press, 1989, 153 p.
4. Gigerenzer, G., Swijtnik, Z., Porter, T., Daston, L., Bentty, J., Kriger, L.;*The Empire of Chance*. How probability changed science and everyday life.
   Cambridge: Cambridge University Press, 1989, 340 p.
5. Hegel, G. W. F.; *Encyclopedia of Philosophy*.
   New York: Philosophical Library, 1959.
6. Kalinowski, M. W.; The Program of Geometrization of Physics: Some Philosophical Remarks. *Synthese*, An International Journal for Epistemology. Philosophy of Science, 1988, Nr. 9, v. 77, p. 129-138.
7. Losev, V. F.; Measure. *Philosophical Encyclopaedia*.
   M.: "Sovetskaya Entsyklopedia", 1964, v. 3, p. 389-394 (in Russian).
8. Lovelock, J.; *The Ages of Gaia*. A Biography of our Living Earth.
   New York, London: Norton and Company, 1988, 252 p.
9. Mindell, A.; *The year I. Global Process Work*.
   London: Arkana, 1989, 160 p.

10. Naess, A.; *Ecology. Community and Lifestyle.* Outline of an Ecosophy.
    Cambridge, Cambridge University Press, 1991, 223 p.
11. Nalimov, V. V.; *Faces of Science.*
    Philadelphia: ISI press, 1981, 297 p.
12. Nalimov, V.V.; *Space, Time and Life.* The Probabilistic Pathways of Evolotion.
    Philadelphia: ISI Press, 1985, 110 p.
13. Nalimov, V. V.; *Spontaneity of Consciousness. Probabilistic Theory of Meaning and Semantic Architectonic of Personality.*
    M.: "Prometei", 1989, 287 p. (in Russian).
14. Nalimov, V.V.; *Spontaneity of Consciousness.* An Attempt to Mathematical Interpretation of Certain Plato's Ideas, p. 313-324
    In: M. E. Crarvallo (Ed.); *Nature, Cognition and System II*, Current Systems-Scientific Research on Natural and Cognitive Systems, Volume 2: On Complimentary and Beyond, Dordrecht, Kluwer Academic Publishers,1992, 393 p.
15. *Philosophical Encyclopaedia*, 2nd edition; M. : "Sovetskaya Entsyklopedia", 1989, 815 p.
16. Robinson, J. M. (Ed.) *The Nag Hammadi Library.*
    San Francisco: Harper and Row, 1981, 493 p.
17. Wheeler, J. A.; World as System Self-Synthesized by Quantum Networking, p. 103-129
    In: E. Agazzi (Ed.); *Probability in the Sciences.* Dordrecht: Kluwer Academic Publishers,1988, 269 p.

Music and Mind – A Theory of Aesthetic Dynamics

Brian D. Josephson and Tethys Carpenter

Abstract

It is argued that purely perceptual or generative accounts of music are inadequate to account for its specificity, and that proper accounts of music must take into account also a more fundamental level of the mind (or of consciousness), a level we term the "aesthetic subsystem". The latter constitutes a domain of universality and of meaning that acts in conjunction with more peripheral aspects of cognition. Suggestive parallels with systems such as biosystems and lasers are used to account for a number of features of musical processes in terms of the model.

1 Introduction: The Problem of Music and Meaning

The question "what is music?" is a subtle one, which has been approached from a number of different points of view by different authors. Lerdahl and Jackendoff [1], for example, on the basis of examination of the structure of musical compositions, propose a number of "generative principles", which they hypothesise can serve to distinguish actual music from arbitrary sequences of musical elements; while Bregman [2] focuses his attention on the perceptual processes involved in musical perception, arguing that the latter can be expected to utilise the same cognitive mechanisms as those involved in ordinary auditory perception. The skill of the musician, according to this point of view, is in essence the skill of assembling musical elements in such a way that the mind will put them together to form a complex "auditory scene", analogous to an ordinary visual scene.

A third approach is that of Suzanne Langer [3], who regards a piece of music as something functioning as a symbol, and which can thereby create in our minds ideas related to feelings. This symbol aspect, implying a distinction between what we perceive as happening in the music (the statement and development of themes, for example), and the effects that these percepts have on us, is not explicitly addressed in either of the approaches previously mentioned. The concept that there is an "idea" and not just a structure or a scene can be illustrated by an example, such as that of a person listening to the second movement of Mozart's piano concerto no. 21 in C major. A few bars of this movement are sufficient to create a very specific mood, which is sustained and amplified through the whole of the movement. Listening to it being played, the dominant impression is of how well the musical choices made by the composer fit into the whole situation in this regard, the choices made having the appearance of being both appropriate and near optimal. These features apply equally well to music of a more intellectual character.

Such phenomena in music fit in with Suzanne Langer's ideas (particularly since one of the characteristic features of a symbolism is its specificity), but tend to suggest that there are more constraints on good music than can be accounted for in terms of approaches such as the two first mentioned. Again, music generated by applying such rules as have been proposed, while in some sense sounding like music lacks, to an experienced listener, the quality of real music.

We are faced, then, with the existence of a phenomenon that is apparently at once very specific and very arbitrary (the latter in the sense of there being no clear reason for the observed specificity). Langer's ideas, being essentially statements of listeners' intuitions, have little to say about underlying mechanisms. As a step towards the formulation of hypotheses concerning the latter, we take our cue from situations in physics that are analogous, such as the existence of highly specific and reproducible atomic spectra. The latter, while observable using very simple equipment operating on a macroscopic scale, can be comprehended only in terms of the physics of a different domain, the domain of the atom as understood in the light of the quantum theory. We seem virtually forced to view music in a similar fashion, that is to say as the outward expression of more fundamental phenomena occurring at deep levels of the mind, or of consciousness. This paper is devoted to the exploration of the feasibility and utility of such a concept.

The basis of the present formulation is the postulate of the existence of an aesthetic subsystem, which supports specific aesthetic processes, and is presumed to be distinct from both perceptual processes and processes of rational analysis (in which respect it should be noted that while we can analyse an art object with a view to understanding what factors are involved in our appreciation of that object, such rational analysis is not a precondition for such aesthetic appreciation to occur). Perception of music is not "mere perception' but perception allied to the presence of a different, more fundamental system.

In subsequent sections of this paper, this idea will be developed in some detail. But before doing this, we note the existence of a close parallel with natural language, in that comprehension of natural language involves more than mere perception of its structure: knowledge of meaning and knowledge of the corresponding object domain are involved also. We do not, in view of the subtleties of the constraints involved in both cases, consider that these additional factors can be adequately taken into account by means of a closed set of semantic "generative rules" as Lerdahl and Jackendoff [1] seem to believe, either in the case of language or in the case of music.

Some comment should be made finally on the work of Cooke [4], who attempted, in a somewhat simplistic fashion, to develop the idea that music is a specific language whose elements could be discovered by understanding their roles in particular pieces of music. This idea that music is a language has been challenged by Langer [3], although certain aspects of Cooke's ideas are consistent with the proposals developed here.

2 Details of the Aesthetic Process

2.1 As already indicated, the aesthetic process is conceived of as a special aspect of mind that deals with aesthetic matters (in an analogous way to the way that processes of the visual system deal with vision). We can be consciously aware of the outputs generated by the aesthetic processes, just as we can be consciously aware of the visual percepts that the processes of visual perception generate.

2.2 Activity of the aesthetic subsystem can be induced in various ways. One is the process of listening to music, and another the process of composition. These differ, obviously, in that in the former case the stimulus comes from outside, while in the latter case the aesthetic state is self-induced (while performance of music involves a combination of both features).

2.3 By virtue of the fact that music can be considered as information, it follows that information plays a significant role in the functioning of the aesthetic subsystem. In listening, information fed in determines the state of the aesthetic subsystem. Conversely, in composition the aesthetic subsystem generates information (see, however, Sec. 3.2). The importance of information is one of a number of aspects in which aesthetic processes parallel life, where information (e.g. DNA) plays a similarly important role (parallels with life have been noted from a different point of view by Langer [5] and Schoenberg [6].

2.4 We are concerned therefore with a two-way process, where information affects state and state generates information (for example, a particular rhythm in conjunction with other aspects of the music is a generator of a particular mental state (state of the aesthetic subsystem) and a mental state may tend to generate a particular rhythm. This leads one to hypothesise that self-sustaining loops may play an important role in aesthetics (see the subsequent discussion of composition in Sec. 3.2). A similar feature is found in life, where enzymes may catalyse particular subprocesses which in turn regenerate the same enzymes. According to this picture, the rhythms or other features (e.g. harmonies or melodic patterns) that tend to be adopted for music are the ones that tend to be self-sustaining in the manner indicated.

2.5 This picture can now be broadened to the more comprehensive one indicated in Fig. 1. This assumes that a variety of phenomena found in music are directly connected with unmanifest processes occurring at the deeper aesthetic level (in the same way that atomic spectra and the Raman effect are visible consequences of corresponding phenomena at the atomic level). We hypothesise, for example, that particular themes that are perceived in the music, and which are seen to develop in particular ways, are connected with particular subprocesses and their development at the aesthetic level. These subprocesses in the aesthetic system have specific effects on each other (parallelling analogous behaviour characteristic of biosystems), in a way that has visible consequences in the way themes or musical elements influence each other.

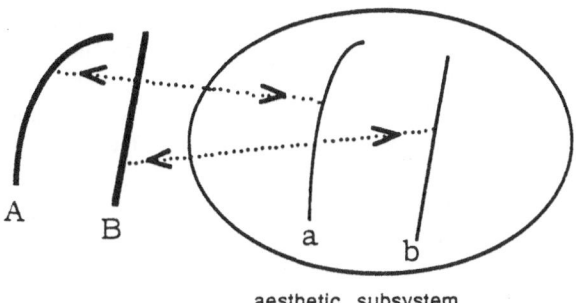

aesthetic subsystem

Fig. 1. Illustrating the way musical phenomena in perception (or in the imagination), such as A and B in the diagram are linked with deeper phenomena such as a and b in the postulated aesthetic subsystem. The interactions and dynamics in the aesthetic subsystem have a strong influence on the observed structure of music, in the same way that the dynamics of the atom influences the observed behaviour of atomic spectra.

2.6 A number of parallels between aesthetic processes and life processes have already been noted. An interesting further parallel is one between the process of tension creation and tension reduction, in the case of music, and the process of homeostasis in biology. Some aspects of biosystems can be best understood in terms of the existence of prescribed norms and the processes of homeostasis which seek to reestablish these norms when sufficient deviations from them occur. The balance-unbalance condition in biosystems has its musical parallel in tension and its absence, and the specificity of the processes by which biosystems restore balance has its parallel in the specificity of the ways in which musical tensions are resolved.

3 Global Level Features of Music

The previous section has been concerned mainly with local details of the structure of music. In this section we step back and examine more global features, rephrasing and extending discussions given elsewhere [7] [8] that were centred around the concept of musical idea. In the current picture music can be defined, roughly speaking, as activity in the aesthetic system which is, to a first approximation, self-sustaining under the given conditions. This condition applies both to a listener who is following the music and appreciating it, and to a composer whose compositional processes are flowing adequately. Thus the listener and the composer have related goals, connected with maintaining the activity of the aesthetic subsystem in an appropriate form.

3.1 The Listening Process

For the listener, the aesthetic system activity is the response to the structured perception of the sound, and he must adequately perceive its structural basis in order to create the full aesthetic response. Familiarity with the style of the music plays a significant role here in making it easier to discover the appropriate structure, while being habituated to a different musical culture on the other hand may have negative effects, in that it may mislead the listener into perceiving incorrect structures, that is to say into perceiving overall structure that does not activate the aesthetic subsystem in the way that it should. But in the long term, that is to say on repeated listening, feedback in terms of the degree of aesthetic response to the presumed structure may enable the listener to adjust his perceptions in the right direction, and thus in the end lead him to hear the music appropriately.

3.2 The Process of Composition

The composer's task is the reverse of the listener's. The principles involved here are considerably more subtle, involving a delicate two-way interplay between the aesthetic subsystem and peripheral systems that link with it. A physics analogy in the form of an analogy with a laser offers useful insights here (see Fig. 2). The characteristic patterns of music have been assumed to be reflections of the underlying dynamics of the aesthetic subsystem. Listening to music involves the response of the aesthetic subsystem to the music, a process parallel to the resonant response of an atom to electromagnetic radiation. If we feed a system of atoms with energy in an appropriate way ("pumping" the system), a process of stimulated emission occurs whereby the atoms emit more radiation than they take in by absorption, a state of affairs which tends to result in the system becoming unstable and generating radiation spontaneously at a frequency close to the frequency of the spectral line. But the precise details of the radiation, e.g. in terms of frequency and spatial distribution, may be strongly influenced by the macroscopic environment of the atomic systems as well; for example, a laser cavity consisting of two parallel mirrors may have a series of resonant modes, and if these are more precisely defined in frequency than is the relevant atomic transition then the laser will oscillate in one of these cavity modes. By analogy, we expect that in the musical case, under appropriate conditions, the aesthetic subsystem activity will take a form determined both by what is possible in terms of the aesthetic subsystem dynamics (corresponding to the atomic resonance) and the composer's learned musical repertoires (corresponding to the modes of the cavity). The composer's ability to create music depends on the appropriate musical environment existing in his peripheral systems, just as the physicist's ability to create laser radiation depends on his ability to provide an appropriate electromagnetic environment for the laser radiation; in both cases, adequate matching for the two subsystems must exist.

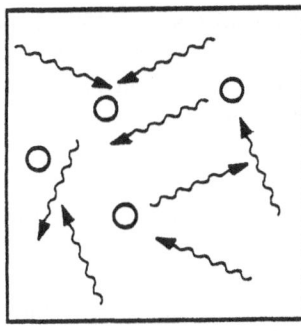 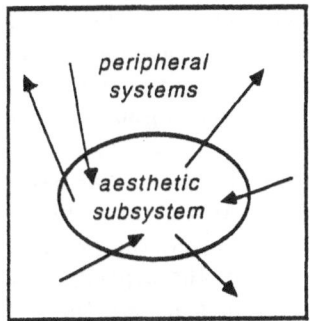

(i)   atoms and radiation   (ii)   interacting subsystems
                                        for music

Fig. 2. The composer's aesthetic system and the peripheral systems mutually influence each other, in the same way that in the laser the atoms and the radiation field mutually influence each other. In either case, the two subsystems in isolation have their own characteristic modes of behaviour, and when the subsystems are coupled together the observed behaviour has features deriving from the modes of behaviour of the isolated subsystems (see Sec. 3.2 for a fuller discussion).

3.3 Time Dilation

Some composers (e.g. Mozart) have claimed to have seen a whole new composition in a flash. This facility poses no particular problem for the present point of view, provided that the properties of the aesthetic subsystem and the linked peripheral systems are such that it is possible to accelerate their activities without any significant change in form.

4 Meaning in Music

The problem of the meaning of music is a very subtle one. Meyer [9] distinguishes between internal meaning or reference in music, and external reference. If music had only internal reference, listening to music would be a very introverted kind of activity. However, psychological experiments designed to discover specific external references, i.e. links between specific pieces of music and specific types of events, have been generally unsuccessful. There are various possible reasons for such a failure, one of them being that the abstract situation, where a subject is asked whether a piece of music A reminds him of situation B, may be significantly different operationally from the corresponding concrete situation of placing a subject actually in situation B, and asking him how he feels about music A then. Meyer's suggestion [9] that the reference of music may be archetypical in nature (e.g. the whole complex of feelings associated with death) rather than conceptual may also be relevant.

Let us now see what the picture presented here has to say about such matters. One way to describe meaning is to say that it is a way of talking about certain kinds of cause and effect in the context of information processing [10]. From this point of view, the initial meaning of a composition may be considered to be the corresponding activity in the aesthetic subsystem. Component parts of a composition equally have meaning in respect of the various specific effects that they induce in the aesthetic subsystem.

The "meanings" considered thus far have been in the category defined by Meyer as inner meaning. As already noted, the question of external meaning or reference is unclear. However, the problem can be approached via the fact that music appears to have the capacity to enhance other kinds of activities (for example, ballet and opera). Sometimes, for example, an opera may have an inadequate libretto or be performed or acted badly, but the performance remains effective because of the music. It might be said in interpretation of this fact that the music adds information that is missing from the performance (in the same way that a commentary or a caption adds missing information to a picture). The aesthetic subsystem can thus be considered to have an informative function, perhaps closely connected with intuition or understanding in general.

5 Parallels Between Music and Exposition

The use of phraseology such as "the statement of a theme" or "the development of a theme" points to the existence of parallels between music and the exposition of a collection of ideas. Such parallels are to be expected in a model where informational input is building up complex structures, and it may be anticipated that similar universal principles or processes might be involved in both situations. One such principle is the principle of reference, i.e. the connections that exist between simple surface forms (e.g. words in the case of language; particular patterns of notes in the case of music) and deeper entities. In the case of language, words are used to gain access to particular pieces of knowledge or activities based on that knowledge. In the case of music, patterns of notes provide access to musical processes (cf. Fig. 1), for example starting a process, developing it or bringing it to a close.

However, music and natural language are in many ways not at all equivalent. For example, with language, meanings (which are mechanisms whereby different processes may be linked) are learnt associations, whilst in the musical case the dynamical factors linking different processes in the aesthetic subsystem are assumed to be universal. This is not, however, to say that learning is unimportant or irrelevant in the musical case. But in the musical case learning, as has been explained above, is taken to be operative at more peripheral levels (e.g. the level of auditory perception).

6 Concluding Comments

We started off arguing that a conventional perceptual account of music was inadequate, and that something corresponding to the subtle influence of the products of the perceptual process on the mind ought to be included in cognitive models as well. A model incorporating an "aesthetic subsystem" was developed, and used to give an account of a number of processes associated with music. The essence of the model is that aesthetic processes have a certain universality, making them essentially independent of the ordinary biological domain but perhaps akin to mathematical intuition. While these are somewhat unusual hypotheses, it is hard to see how the specificity of music can be accounted for in any other way.

For the future development of the model, one possible avenue is the question of to what extent very specific regularities of music discussed by investigators such as Lerdahl and Jackendoff [1] and Narmour [11] can be accounted for in a transparent manner on the basis of the model. Such connections might enable one to characterise specific processes in the aesthetic subsystem, and to understand their cognitive function, perhaps producing ultimately accounts of the workings of the aesthetic (or intuitive) subsystem of a similar kind to those that we now have of life. Finally, attempts to account for the semantic aspects of music in detail, such as those mentioned in the introduction due to Cooke [4], can perhaps be usefully extended on the basis of concepts such as those developed here.

Acknowledgement

We are grateful to Dr. Ian Cross for helpful comments on drafts of the manuscript.

References

1. F. Lerdahl, R. Jackendoff: A Generative Theory of Tonal Music (MIT Press, London, Cambridge, Mass. 1983).
2. A. S. Bregman, Auditory Scene Analysis: the Perceptual Organization of Sound (MIT Press, 1990).
3. S. K. Langer, Philosophy in a New Key (Harvard University Press, London, Cambridge, Mass. 1951).
4. D. Cooke, The Language of Music (Oxford University Press, Oxford, New York 1989.
5. S. K. Langer, Mind: an essay on human feeling (Abridged edn.: Johns Hopkins University Press, Baltimore 1988, chap. 7).
6. A. Schoenberg, Folkloristic Symphonies, in Style and idea: Selected Writings of Arnold Schoenberg (ed. Leonard Stein, Faber and Faber, London 1975).
7. B. D. Josephson and T. L. Carpenter, New Scientist 129(1762), 2, Mar. 30th. 1991.
8. B. Josephson and T. Carpenter, New Scientist 131(1780), 51-2, Aug. 3rd. 1991.
9. L. B. Meyer, Emotion and Meaning in Music (Univ. of Chicago Press, Chicago 1959).
10. D. J. Bohm, Meaning and Information in - The Search for Meaning - (ed. P. Pylkkaenen, Crucible, Wellingborough, Northants., U.K. 1989).
11. E. Narmour, The Analysis and Cognition of Basic Melodic Structures: the Implication-Realization model (University of Chicago Press, Chicago 1990).

The Paradigm of Self-Organization and Its Philosophical Foundation

Eduard Zwierlein

1 Survey

This contribution falls into five small parts:

- it starts with a look to the architecture of science and the word "paradigm";
- in a second step I will introduce the so-called philosophical questions, beginning especially with ethics;
- in the third part I will briefly refer to anthropology and the so-called "Life-World Problem";
- in my fourth point I will draw the reader's attention to the epistemological question of how we understand;
- in my concluding part I will sum up my contribution and state two important remarks on my subject.

My larger goal is to clarify the background understanding against which the discourse about self-organization takes place, and to grasp its broader implications reflecting some aspects of the meaning, sense, and value of self-organization.

2 The Architecture of Science and the Term "Paradigm"

I would like to start with a look at the elementary architecture of sciences and the term "paradigm".

2.1 Usually all sciences consist of three elements: an object (or a class of objects) treated (I), methods, instruments, calculations etc. (II), and basic ideas, theories, principles, axioms (III). That is: (I) something that is (II) represented, interpreted or manipulated by something (III) in the light of something.

2.2 It looks like pointed out in figure 1

2.3 This elementary structure can be also illustrated in another way shown in figure 2 with the help of a light source (representing the context of basics), rays or beams (representing the context of statements), and objects illuminated (representing the context of phenomena):

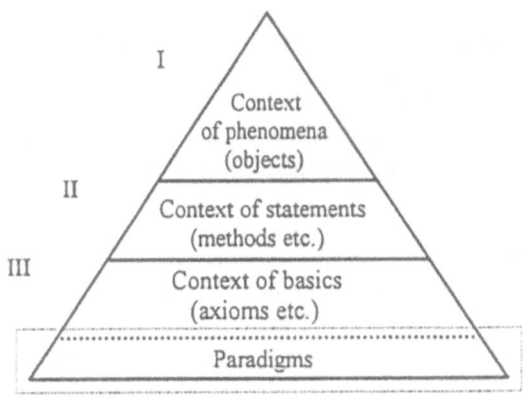

Fig. 1. The architecture of sciences

Fig. 2. The architecture of science, illustrated

2.4 This architecture only shows a rough framework. For example, layer II would include several contexts such as the contexts of invention, description, explanation, or application.

2.5 At the bottom of layer III you will find the fundamental belief that is called "paradigm".

    2.5.1 A paradigm is a meta-idea, a master-strukture or superstructure, that opens a window to reality.

    2.5.2 The usual characteristics of a paradigm are:

- It is comparable to the eye you look with and you are not looking at, i.e.: it normally is an unquestioned, doubtless, non-reflected basic idea or perspective. (Basic patterns, standards, and rules of the scientific game.)
- This perspective seems to be a cluster of experiences, convictions and expectations able to solve (many) problems and riddles: it constitutes phenomena as scientific objects within a notional and theoretical obligatory interpretation-framework.
- From an outward point of view you will call a paradigm a mighty historical idea that slowly changes within time.
- From a distance you can also see that there can exist (even within one science) a various number of paradigms, frequently believed to be autonomous and incommensurable (incomparable and incompatible).

- It might be possible to say that a paradigm (in the sense of T.S.Kuhn) when becoming conscious and chosen will hereby be transformed into a research program (in the sense of I. Lakatos) guiding the activity of scientists.
- It should be obvious that every external point of view is itself based on another paradigm not reflected at the time it is the eye you are looking with.

Perhaps the "paradigm" can be inserted into a picture having an "eye" symbolizing human mind, self-experience, and self-awareness, using "glasses" representing a (or several) paradigm(s), having "rays" symbolizing the stream of consciousness operating with methods, instruments, calculations etc., and finally using "objects" representing phenomena constituted or reflected (see figure 3).

Fig. 3. The term "paradigm", illustrated

With respect to self-organization as a paradigm in science you can hold two different viewpoints:

(1) an internal view looking at the material content or objects: self-organization as a unifying paradigm with reference to the explanation of objects;
(2) an external view looking at the formal process of scientific work: self-organization as a unifying paradigm in which the sciences start to cooperate as one big intellectual organism.

3 Philosophical Questions, Starting with Ethics

As a working hypothesis self-organization might be the phenomenon that some (i.e. nonlinear dynamic) systems may by irreversible processes spontaneously generate a structure of higher complexity than the original starting point was made of. This higher complexity might be called "order". "Spontaneously" here means that one cannot discover any kind of a plan, teleology, acting subject etc. "Self" means to be the result of internal mutual or reciprocal relations. The "paradigmatic" character of self-organization does not only mean that it is the basic idea of one science or a basic idea of several sciences but the underlying basis or unifying substructure of various sciences (concerned with systems).

As you know there are slight differences between scientific questions and philosophical questions.

Typical scientific questions are:

- What is an "x"?	(Stating and describing)
- How does "x" work?	(Analyzing functions)
- Why did "x" happen to become "x"?	(Explaining genesis)

As Kant says in his Lecture on Logic the typical philosophical questions are the following:

- What can we know?	(Epistemology)
- What shall we do?	(Ethics)
- What may we hope?	(Philosophy of religion)

According to Kant all these questions converge and cumulate in one question: What (Who) is man? That is: anthropology is the source and goal and key of all the other questions. For the purpose of this paper I touch on only some aspects of these philosophical questions. Let us start with ethics.

There are not only descriptive questions (like the scientific ones) but also normative and valuative questions: What is good and what is bad? What should we do and what should we abstain from doing? These questions have to do with our wishes and needs, our desires and goals, our intentions and aims in trying to answer the very fundamental question of a good and just life.

We therefore have to add on the top of the layers an ethical aspect (figure 4). Asking the ethical question with reference to self-organization amounts to asking: What is it good for? What is the use of self-organization, what are its benefits? Does self-organization serve a good life, i.e. does it support survival of mankind, or does it strengthen coexistence, the togetherness we share among all human people and with every being in the world? Does self-organization shed light on some of these ethical questions?

I would like to point out that the ethical aspect of self-organization includes several hidden problems, and I only want to mention two of them:

(1) The naturalistic fallacy or is-ought problem where you cannot conclude from what there is (facts), what there should be (norms); where you cannot derive prescription from description. So, the question is how to bridge this gap? (Is nature a model, a standard, a prototype for human action? And what would that mean?)

(2) Self-organization seems to be: Ethics seems to refer to:

a	a
description	prescription
of the	of the
(collective)	(individual)
rationality of systems	rationality of actions.

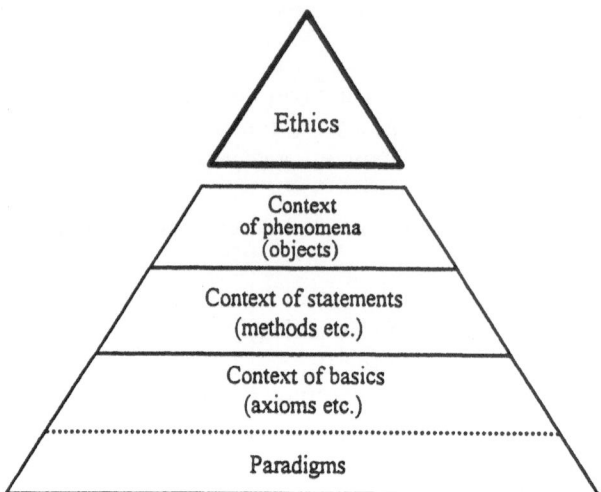

Fig. 4. Ethics within the architecture of sciences

Is it possible to combine the two "worlds"? Is self-organization a normative notion transferred to nature, or can we find a normative attitude towards a matter of fact, or can we integrate both aspects in a more complex context and look at the descriptive thing as a part or partial goal of ethics?

4 "Lebenswelt" and Anthropology

As Kant says: in the end it all comes down to the same for everyone and everything, in the end you have to consider carefully anthropology. Nature does not automatically or clearly speak to us. We have to translate nature by our language and into our language. We have to do the work of an interpreter/translator. Science is one of the ways to arrange this translation. To read in the book of nature demands that one understands the reader and the process of reading that is always an interpretation.

So, if you ask the decisive questions: How is it possible to understand systems or self-organization (or whatever) scientifically? What are the necessary preconditions of scientific interpretation with all its beauty and all its limits? On what is science based itself? What are its roots? it will finally take you to anthropology.

Most probably the reason for this is the following. There is one thing, and only one, in the whole universe which we know more about than we could learn from external observation. That one thing is Man. We do not merely observe man, we are man. In this case we have, so to speak, inside information; we are in the know. There is only one case in which we can really find a starting point of knowledge and understanding, namely in our own case.

To be a little more precise we can say that anthropology expresses and is situated or contextualized in what is called "Lebenswelt" (i.e. the Husserl-Heidegger idea of "Life-World", see figure 5). And we have to comprehend science as a function and aspect of the Life-World. How can this be understood?

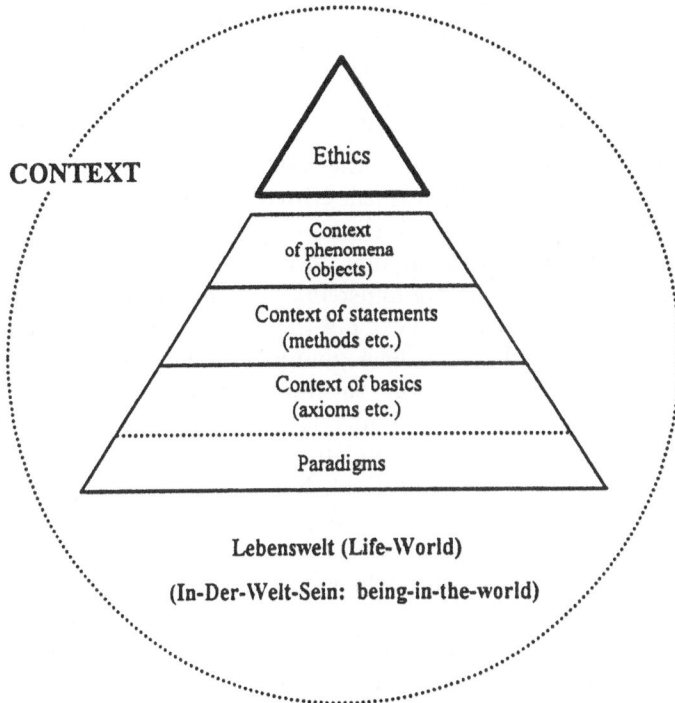

Fig. 5. The context of the scientific architecture

4.1. The answer may start with the statement that each science and each scientific paradigm stays within and results from (or originates in) a general context of introduction that is called "Lebenswelt".

4.2. "Lebenswelt" is the melting pot we are living in that includes history, language, culture, and prescientific theory and practice of ordinary life, the prescientific approach to the world and the mind. When you compare, as I did before, paradigms to windows and sciences to rooms you may then call the "Lebenswelt" a hall out of which doors open into several rooms with different windows. We are continously involved in this inevitable pre-understanding and everything falls within the scope of this commonly shared background. We are never standing "outside", but always stay "in" this pre-understanding.

    4.2.1. This "Lebenswelt" is the fundamental basis of all human events and actions. Science as a specific human action is based on this fundament, too, from which it is a specific (reduced and abstract and very successful) derivation. It is the final background that invisibly shapes – mediated through the paradigms and all the other contexts mentioned – scientific actions and perspectives. ("Invisibly" means that the fundamental beliefs and assumptions are implicit and that we are unable to achieve full explicit understanding of this concealed and hidden pre-understanding. We have to accept a particular blindness here.)

4.2.2. To realize the "Lebenswelt" as the real and inevitable starting point of every human action also means that science does only make sense within this context, that it has to acknowledge its genesis, its origin and dependence, that it has to serve the preceding and presupposed "Lebenswelt" and that it can be justified, criticized and controlled only as an element of this "Lebenswelt". One cannot approach any question at all from a neutral or objective standpoint. Every questioning grows out of a tradition and its underlying pre-understanding that opens the space of possible answers. To grow and to expand the horizons does not mean to surpass the condition of having a background of pre-understanding in principle. We will always operate within the framework the "Lebenswelt" provides for us. And it is definitely impossible that our understanding will ever be neutral or objective or complete.

4.2.3. The "Lebenswelt", though it is not rigid, but a living and changing thing, is the last horizon that human thinking can reach, and within it open all specific horizons. Therefore you can call this last horizon the metaparadigm of all paradigms. In the language of Heidegger you might think of the prereflective, prescientific "In-der-Welt-Sein" or "Being-in-the-World", a fundamental unity that goes beyond the separation of subject and object and describes our primary access to the world as being practically involved or "engaged" in the practice of the "Lebenswelt" building our last horizon possible.

5 Logical Anthropomorphism

Finally I would like to draw the reader's attention to the epistemological question of how we understand: What is the basic structure of understanding within the "Lebenswelt"?

As we have found, the starting point is man. I am explaining (according to a comparison used by C. S. Lewis) the unrevealed packets I am not allowed to open by the ones I am allowed to open. The only packet I am allowed to open is Man giving us a double (internal and external) insight. From here all analogy starts (and everything – however abstract it may be – is in the end a metaphor). In the view of Aristotle and Thomas Aquinas one might say "homo quodammodo omnia" (man is – in a certain way and to a certain degree – everything) being a real "microcosmos" including essentials of what is below and above mankind and therefore being a (limited) universal hermeneutic key within the "Lebenswelt" set out as a conceptual framework. Meaning (possessing relevance and having resonance) is grounded in understanding; understanding is grounded in anthropology; anthropology expresses and constitutes itself in the

"Lebenswelt" as its first and final background of interpretation. What is the epistemological structure of this metaparadigm? What are the basic elements and the structure of the "eye" we are looking with symbolizing human mind? (See figure 6)

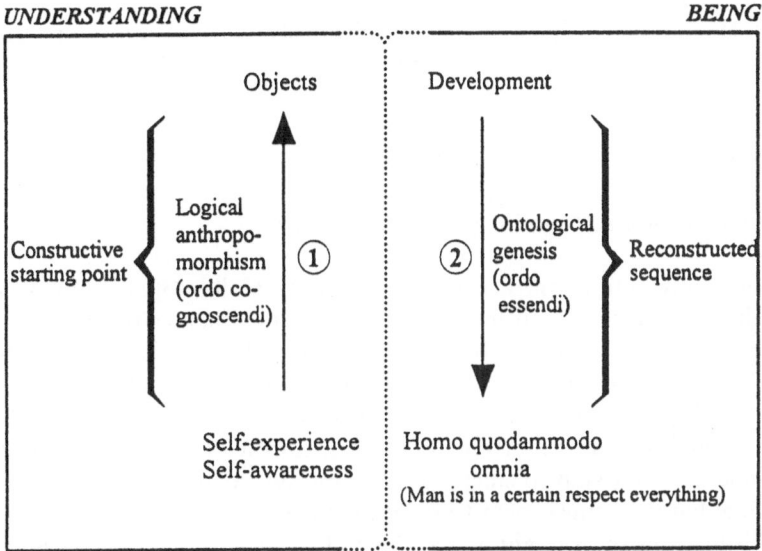

Fig. 6. The relationship between understanding and being

5.1. The structure, that is the working process or "logic", of this metaparadigm is what I would like to call "logical anthropomorphism":

  5.1.1. Every non-human thing and every system-datum has in the end to be translated and interpreted in a human language. (This is even true for mathematics according to what e.g. Pascal and Gödel have shown.)

  5.1.2. Above all concrete languages used (such as English, French, or German) the common and universal "language" shared by all human beings is the basic anthropological language reconstructed in philosophical anthropology.

  5.1.3. Only the self-experience and self-awareness of a living spatio-temporal teleological human being causes the resonance- and relevance-bottom for the possibility to understand and explain something else.

  5.1.4. Every human understanding finally works with an extrapolation (analogy transfer) of its own self-expierence and self-awareness. This fundamental hermeneutic process is what I call logical anthropomorphism. And it is also true for words like "self", "organization", "pattern", "order", "structure", and so on.

But as you can see in figure 6 there are two lines or arrows here (or you also could say: two movements), the stream of being and the stream of consciousness.

Nearly all philosophers from the very beginning in the early days until in our times have tried to connect these two lines, to shoot both arrows at the same time, to execute both movements simultaneously. They have tried to combine the two aspects in a perfect circle of self-organization. To enlighten this perfect circle completely is still a dream and a riddle unsolved. Nobody has so far found the key to the solution. Nobody has found the stone of wisdom for this old philosophical dream. The only thing we can obviously do is to jump into the circle and to engage ourselves in the movements of its components.

6 Conclusion

To apply these explications to self-organization means:

6.1. Epistemologically: self-organization is, according to a philosopher's point of view, in the end a term of self-interpretation of human beings trying to get acquainted with the world they live in and is originally based on their self-experience and self-awareness, i.e. based on anthropology within the ultimate context or metaparadigm of the so-called "Lebenswelt".

6.2. Ethically: to find out the importance of self-organization for a good and just life demands as a first step to elucidate what we can call a just survival and a good life of human beings coexisting with many other living beings. In the light of this first answer it might be possible to look at self-organization as an instrument, a way, or even a partial aim of the universal goal of a peaceful, friendly and loving togetherness based on our primary ethical insight.

6.3. These two points, the epistemological and ethical premises determining the sense and value of self-organization are to my opinion the foundation of all clear philosophical thinking about ourselves and the universe we live in. And I am strongly convinced that the old philosophical dreams will at least find new scientific wings and that the old philosophical riddles will gain at least new inspirations through the discussion of self-organization.

References

1. Bateson G., Ökologie des Geistes. Anthropologische, psychologische, biologische und epistemologische Perspektiven. Frankfurt am Main 1985. (Steps to an Ecology of Mind. Collected Essays in Anthropology, Psychiatry, Evolution and Epistermology, 1872.)
2. Bateson G., Geist und Natur. Eine notwendige Einheit. Frankfurt am Main 1987. (Mind and Nature. A Necessary Unity, 1979.)
3. Bohm D., Peat F.D., Science, Order, and Creativity. New York 1987.
4. Convey P., Highfield R., Anti-Chaos. Der Pfeil der Zeit in der Selbstorganisation des Lebens. Reinbek bei Hamburg 1992.
5. Cramer F., Chaos und Ordnung. Die komplexe Struktur des Lebendigen. Stuttgart 1988.
6. Davies P., Cosmic Blueprint. London 1988.

7. Ebeling W., Chaos - Ordnung - Information. Selbstorganisation in Natur und Technik. Leipzig 1989.
8. Foerster H.v., Zopf G. (Ed.), Principles of Self-organization. Oxford 1962.
9. Foerster H.v., Sicht und Einsicht. Versuche zu einer operativen Erkenntnistheorie. Braunschweig 1985.
10. Haferkamp H., Schmid M. (Hg.), Sinn, Kommunikation und soziale Differenzierung. Beiträge zu Luhmanns Theorie sozialer Systeme. Frankfurt am Main 1987.
11. Heidegger M., Sein und Zeit. Tübingen, 15. Aufl. 1979.
12. Husserl E., Die Krisis der europischen Wissenschaften und die transzendentale Phenomenologie. Eine Einleitung in die phnomenologische Philosophie. Hamburg, 2.Aufl. 1982.
13. Kuhn, Th.S., The Structure of Scientific Revolutions. Chicago, second printing 1970.
14. Lakatos I., Musgrave A. (Ed.), Criticism and the Growth of Knowledge. Cambridge 1970.
15. Luhmann N., Die Wissenschaft der Gesellschaft. Frankfurt am Main 1992.
16. Maturana H.R., Erkennen: Die Organisation und Verkörperung von Wirklichkeit. Ausgewählte Arbeiten zur biologischen Epistemologie. Braunschweig 1982.
17. Maturana H.R., Varela F.J., Der Baum der Erkenntnis. Bern-München-Wien 1987.
18. Prigogine I., Nicolis G., Self-Organization in Non-Equilibrium Systems. From Dissipative Structures to Order through Fluctuations. New York 1977.
19. Prigogine I., Stengers I., Dialog mit der Natur. München 1981.
20. Probst G., Selbst-Organisation. Ordnungsprozesse in sozialen Systemen aus ganzheitlicher Sicht. Berlin-Hamburg 1987.
21. Probst G., Gomez P. (Hg.), Vernetztes Denken. Ganzheitliches Führen in der Praxis. Wiesbaden, 2. Aufl. 1991.
22. Riegas V., Vetter Chr. (Ed.), Zur Biologie der Kognition. Ein Gespräch mit Humberto R. Maturana und Beiträge zur Diskussion seines Werkes. Frankfurt am Main 1990.
23. Roth G., Schwegler H. (Ed.), Self-organizing Systems. An interdisciplinary approach. Frankfurt am Main-New York 1981.
24. Searle J.R., Intentionalität. Eine Abhandlung zur Philosophie des Geistes. Frankfurt am Main 1987. (Intentionality. An essay in the philosophy of mind, 1989.)
25. Spaemann R., Löw R., Die Frage Wozu? Geschichte und Wiederentdeckung des teleologischen Denkens. München 1981.
26. Ulrich H., Probst G. (Ed.), Self-Organization and Management of Social Systems. Insights, Promises, Doubts, and Questions. Berlin, Heidelberg, New York, Tokyo 1984.
27. Varela F.J., Kognitionswissenschaft - Kognitionstechnik. Eine Skizze aktueller Perspektiven. Frankfurt am Main 1990.
28. Watzlawick P., Krieg P. (Ed.), Das Auge des Betrachters. Beiträge zum Konstruktivismus. München 1991.
29. Winograd T., Flores F., Understanding Computers and Cognition. A New Foundation for Design. New York, third printing 1988.
30. Wittgenstein L., Philosophische Untersuchungen. Frankfurt am Main 1977. (Philosophical Investigations, 1953.)
31. Zwierlein E., Das höchste Paradigma des Seienden. Anliegen und Probleme des Teleologiękonzepts Robert Spaemanns. In: Zeitschrift für philosophische Forschung, Bd. 41, H. 1, 117-130.

32. Zwierlein E., Die Idee einer philosophischen Anthropologie bei Paul Ludwig Lands-
    berg. Zur Frage nach dem Wesen des Menschen zwischen Selbstauffassung und
    Selbstgestaltung. Würzburg 1989.
33. Zwierlein E., Künstliche Intelligenz und Philosophie. Zur Debatte um J.R. Searle's
    Einwände gegen harte KI-Versionen. In: Journal for General Philosophy of Science
    21, 1990, 347- 358.
34. Zwierlein E., Chaos und Kosmos. Nachwort zu J. Briggs und F.D. Peat: Die Ent-
    deckung des Chaos. Klassiker des Modernen Denkens. Gütersloh 1992, 361-381.